**Project Management in the
Construction Industry**

Project Management in the Construction Industry

From Concept to Completion

Saleh Mubarak

Library of Congress Cataloging-in-Publication Data:

Names: Mubarak, Saleh A. (Saleh Altayeb), author. | John Wiley & Sons, publisher.
Title: Next generation construction project management : from concept to completion / Saleh Mubarak.
Description: Hoboken, New Jersey : Wiley, [2024] | Includes index.
Identifiers: LCCN 2023048972 (print) | LCCN 2023048973 (ebook) | ISBN 9781394221127 (hardback) | ISBN 9781394225200 (adobe pdf) | ISBN 9781394225217 (epub)
Subjects: LCSH: Construction projects—Management. | Construction projects—Risk management. | Construction projects—Cost control.
Classification: LCC TH438 .M75 2024 (print) | LCC TH438 (ebook) | DDC 624.068—dc23/eng/20231122
LC record available at https://lccn.loc.gov/2023048972
LC ebook record available at https://lccn.loc.gov/2023048973

Cover Design: Wiley
Cover Image: © Lya_Cattel/Getty Images

Set in 9.5/12.5pt STIXTwoText by Straive, Chennai, India.

SKY10069779_031524

To the soul of my father, my role model, who taught me the importance of both knowledge and ethics

To the soul of my mother, who taught me the ethics of discipline and hard work

To my wonderful wife, who persevered with me through my path since graduate school and provided me with the atmosphere to work and produce

To my children, grandchildren, extended family, friends, colleagues, and students, who surrounded me with love and support

Contents

Special Request

To all colleagues and users of my book:

I did my best in writing the book and its supplemental materials. However, we are human: We strive for perfection, but we can't reach it.

I hope you can help me and help other users by sending me any correction or suggestion you may have that may improve the book and/or the supplemental materials to my email:
cpmxpert@gmail.com.

This request extends to my two other books:

How to Estimate with RSMeans Data: Basic Skills for Building Construction, 5th Edition
Construction Project Scheduling and Control, 4th Edition

Introduction

My journey with construction probably started in my childhood when I was building "houses" of pillows, blankets, chairs, and any object I could get. After finishing my bachelor's degree in civil engineering, I did my master's degree in structural engineering, which gave me a good foundation but did not attract me to a career path. When I started pursuing my PhD in construction project management at Clemson University, I felt a match with my interest. It was like "This is what I like to do for life!"

Over three decades later, I still have the same passion and attraction to the field, but from a different perspective. With over 33 years of diversified experience, I am grateful to be able to give to others. I worked in different positions and roles, in private and public organizations, in the industry and academia, and inside and outside the United States. My earlier two books experienced tremendous success: *How to Estimate with RSMeans Data: Basic Skills for Building Construction* is in its 5th edition, and the 5th edition of *Construction Project Scheduling and Control* is coming soon. I am blessed with the art of simplifying information, which is a skill we badly need in both colleges and construction sites. Effective educators possess two major qualifications: knowledge in the subject matter and the skill in conveying this knowledge to the recipients. Educators must put themselves in the shoes of their students and start with the basics and then build knowledge, one simple step at a time, with no shortcuts and no "you know what I mean!"

Now, the question is this: Why did I write this book with so many other books in the market? Many of these books are very good, and I benefited from them, but I had different issues with each of them. Here is a summary of the features in my book:

- Contains all the traditional chapters that are expected in a book on this subject.
- Starts with the discussion of design, architecture, and engineering and then ties it with the field execution (construction).
- Special focus on cost and time management as the two most important areas of construction management.
- A full chapter dedicated to soft skills needed by the project manager and the team.
- A chapter discusses project management from the owner's perspective.
  A chapter briefly discusses construction in the international and multi-cultural environment.
- A chapter briefly discusses the management of remodeling, renovation, restoration, expansion, and demolition projects.
- A chapter on construction and evolving technology.
- A chapter dedicated to real estate development.

The book also includes many sections that I will leave to the reader to explore. Each chapter has a number of exercises, many of which need research by the student. I am also providing instructors who adopt the book with many support items.

I used many references from major professional organizations such as the PMI, AACE International, CII, and CMAA. I also made many references to computer software programs. However, I did not confine my book to a specific organization or software. I felt I needed to pick the best and most suitable materials, so I picked from all sources available, and sometimes I created my own definitions and principles.

The world is changing rapidly, both in its own conditions and in technology, which directly affects the supply and demand for almost everything. Those who do not keep up with changes will be left behind and get pushed out of the race. The construction industry is no different, with several factors influencing and dictating changes. Global warming is a reality that is hard to deny, which resulted, according to several studies, in harsher weather conditions and stronger and more frequent natural disasters. This, in turn, requires stronger and more resistant structures. Scarcity of resources and competition are also pushing the construction industry to be more efficient and sustainable. On the other hand, technology is advancing in all directions and allowing us to do what was science fiction until recently.

As mentioned in Chapter 18, the relationship between the construction industry and technology innovation is reciprocal: pull and push. These innovations allow designers and contractors to improve their work in so many ways and do what was yesterday impossible. At the same time, new challenges are emerging in construction that demand solutions. As long as we stay in this circle, taking advantage of every available tool and challenging the traditional methods, we should be able to stay in the race and even lead.

I am looking forward to seeing this book not only teach students and professionals but also provoke their minds to challenge and explore.

About the Companion Website

This book is accompanied by a companion website:

www.wiley.com/go/nextgencpm

It gives me pleasure to have you use my book. I am providing supplemental materials for college professors and instructors who adopt the book:

1. PowerPoint slides for all chapters of the book.
2. Solution manual to all exercises in the book. Some solutions were done using Excel sheets, that are also provided.
3. Sample questions, mostly multiple-choice, for use in exams and quizzes.

1

Introduction to Construction Project Management

Introduction

Management has been defined as the process of dealing with or controlling things or people[1]; however, the term management itself is widely understood and used in almost every facet of our personal and professional lives, for example:

- Managing one's work duties.
- Managing personal and family tasks.
- Managing an organization for those who own or manage their businesses.
- Managing personal finances.
- For students, managing school duties.
- For an administrator of an organization, managing the department under their responsibility.

This means that the person in charge (the manager) has to deal with all the human, financial, and other resources under his/her jurisdiction to achieve the set objectives.

What we manage ranges from simple tasks such as preparing a sandwich or a meal, to a complicated mega-project or program such as building a skyscraper or manufacturing a commercial jet. This takes us to the classification of these "pieces of work." Depending on size and complexity of this piece of work, we give each a different name, although this matter is somewhat subjective. In project management, we may call the small components an activity. In general, an activity (also called task) can be independent, such as making coffee or brushing one's teeth, or it can be part of a bigger and more complex scope of work, such as baking a cake for a birthday or wedding party or trimming a tree as part of garden maintenance.

In the management hierarchy, the higher the manager is, the greater his/her responsibility, authority, and jurisdiction get. In addition, especially in larger organizations, management is divided horizontally by specialty: Finance, IT, human resources, and so on.

---

1 Oxford Dictionary, https://www.lexico.com/en/definition/management.

Project Management in the Construction Industry: From Concept to Completion, First Edition. Saleh Mubarak.
© 2024 John Wiley & Sons, Inc. Published 2024 by John Wiley & Sons, Inc.
Companion Website: www.wiley.com/go/nextgencpm

Basic Definitions

In project management, a *project* is defined as a temporary endeavor undertaken to create a unique product, service, or result[2]. There are two keywords in the definition: temporary and unique:

- Temporary because it has a start point and a finish point, and terminates in a deliverable: product (building, road, or refinery), service (repair, remodel, or demolish), or result (improve efficiency or reduce cost).
- Unique because there are no two projects that are identical. Many people think of the project only from a design perspective (how it looks after completion), but the *project* in this context is the process that ends with the finished product. There are similar projects but not identical. They may differ in soil conditions, weather (at time of construction), labor productivity, cost, type and quality of materials, equipment used, site conditions, and circumstances (e.g., accidents).

> Every project must have a starting point, a finishing point, and a deliverable (a product, service, or result).

> Many of what we do in life are projects, but we never think of them this way. Think about it and give a few examples.

A typical project has components of different sizes and complexity, including activities, work packages (assemblies), and subprojects. An *activity* is a component of the project, which serves as a basic unit of work as part of the total project that is easily measured and controlled. It is time and resource consuming. This activity can be small such as stripping a column formwork or plumbing a vertical member, or as large as excavating 50,000 cubic yards of soil. However, for practical purposes, we like to limit the activity size to a reasonable range. This will be discussed later in subsequent chapters, mainly under the scheduling topic. Activities within a project are usually defined by breaking down the project into components we define as activities. This process depends on several factors that will also be discussed later.

Even though we look at the activity as the smallest component in construction project management, sometimes, we can even divide the activity into smaller components called steps. This may be necessary to distribute the cost or resources (called resource loading) over the duration of the activity.

A *work package* or *assembly* is simply a collection of related items or activities within a bigger scope of work (usually the project). The work package includes items assembled to form more comprehensive items to facilitate and speed up the construction process. It can be as small as a simple partition assembly or as large and complex as a 3D building component. It can be assembled onsite or delivered pre-assembled to the construction site. Compared to the auto industry, we can look at the finished automobile as the project, and the components such as upholstery, suspension systems, and the engine as assemblies.

A *subproject* is a smaller portion of the overall project created when a project is subdivided into more manageable components or pieces. It is *not* a stand-alone piece of work regardless of its size. A project

---

2 PMI, PMBOK, 7th edition, 2021.

may be divided into subprojects based on specialty, location, phase, or size. The subproject typically is composed of activities and work packages, and sometimes sub-subprojects.

A subproject is usually managed independently; however, it may affect and/or be affected by other subprojects in the overall project. It may even have smaller components, sub-subprojects. In fact, a large/complex project may be divided into subprojects, or it may be considered a program with its components as projects.

The idea of breaking up the project (or large/complex work) into smaller and more manageable and controllable components, is well-known to simplify and facilitate the management of the project.

We are now defining two terms representing a piece of work larger than a project: a program and a portfolio. There are similarities between them but there are traits that make them different.

> A subproject cannot be considered a stand-alone project. It has to be a part of a project.

A *program* is a group of related projects, subsidiary programs, and program activities that are managed in a coordinated manner to obtain benefits not available from managing them individually[3]. Programs may include elements of related work outside of the scope of the discrete projects in the program.

Programs, such as projects, are temporary (start point and finish point), although many organizations have departments called programs that are continuous/perpetual. For example, a municipal government may have a stormwater program and a road maintenance program. These "programs" usually have an annual budget as well as an annual list of projects that can be performed within that budget. If we imagine a conventional program such as a box with many components inside (projects, activities, and subprograms), the "perpetual program" will then look like a conveyor belt: loading components and budget at its start and unloading finished projects at its end. A one-year snapshot of this conveyor belt will be a conventional program (Figures 1.1 and 1.2).

A *portfolio* is a collection of projects or programs and other work that are grouped together to facilitate effective management of that work to meet strategic business objectives. The projects or programs in the portfolio may not necessarily be dependent or directly related. An example of a portfolio is a group of projects that have one element in common, serving a strategic goal for the organization. It is like the retirement portfolio for a person: it may contain investment in different types of investments, for the purpose of providing the best and most secure retirement for the person.

Example on a program through a step:

- Program: Summer Olympic 2028 in Los Angeles
  - Project: Constructing a new building to house athletes
    - Subproject: Electrical work in the building project
      - Work package: Foundation or roof assembly
        - Activity: One spread footing
          - Step: Setting the formwork for the spread footing

---

3 PMI, PMBOK, 7th edition, 2021.

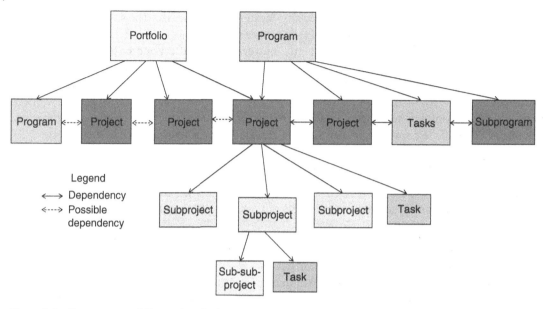

Figure 1.1 Programs, portfolios, and projects.

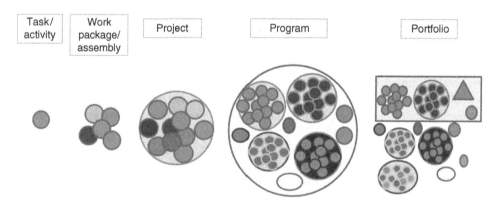

Figure 1.2 Activities, programs, portfolios, and projects.

Relating to the same example, think of a construction company that is contracted to perform part of the infrastructure work for the Summer Olympics in 2028, and the company may have a "City Works portfolio" that includes its portion of the Summer Olympics 2028 program in addition to other projects/works with the city outside the scope of the Summer Olympics. The entire portfolio serves a strategic goal for the company in its relationship with the city government. The city, on the other hand, may also have a portfolio that includes part of the Summer Olympics program as well as other projects/works that serve a strategic goal of "improvement of city infrastructure." In both examples, the portfolio overlaps with the program, but each is defined differently with different components and objectives.

Another example is obtaining a bachelor's degree for a high school senior. Let us assume the student is "good average," meaning he will pass all classes and graduate in 4 years:

- Program: Graduating with a bachelor's degree in civil engineering[4]
  - Project: One semester
    - Subproject: One course
      - Work package: Homework (for the entire course)
        - Activity: One set of homework

There is one reservation in this example: All projects, subprojects, and to some extent, work packages, have the same predetermined duration. This is not the case in construction and other industries.

Where to draw the line? In many cases, the line between an activity and an assembly, an assembly and a project, or even between an activity and a project, is subjective and not so clear. Many chunks of work can be classified as an activity, assembly, or a project, depending on the overall context and the user's preference. For example, baking a birthday cake can be considered a project, or an activity within the "the party project."

Other examples:

1. Building concrete block wall around a building: An activity, subproject, or a project?
2. Converting my garage into an office: An activity or a project?
3. Replacing the carpet in my house: An activity or a project?
4. Changing a flat tire: An activity or a project?

In almost everything we do, we can break down the work based on the nature (specialty) of work, location, phase, or size. The "independence" of the work item can make a difference in labeling it. For example, in example 1 above "Building concrete block wall," if it is part of a bigger work assignment (project), then it can be an activity (task) or work package. If it is an independent work assignment, it can be considered a project.

The components may also differ in resource requirements, importance, urgency, and other criteria, but they still have one thing in common: their temporary nature (start + finish). However, what if this activity/project has repetitive, perhaps routine, nature, such as mowing the lawn, cooking dinner, changing air conditioning filter, or cleaning the house? In this case, we consider each occurrence as a "stand-alone" activity or project with its start and finish points, and the deliverable (what we have achieved). We deal with the multiple/frequent occurrences within the context of "time management," which comes within the bigger context of "life management." Such activities may be done routinely (daily, weekly, and monthly) or sporadically/occasionally. In business, these repetitive/routine/sporadic activities become part of *operations management*.

Unlike project management, operations management does not have a starting point and finishing point because it continues with continuous input and output. There are many examples of operations management such as:

- Managing a department store or a supermarket, or a specific department in the store or the supermarket.
- Managing a company, restaurant, bank, clinic, or any other business.

---

4 S/he can add constraints such as: Time = 4 years, tuition cost $\leq$ $12,000, and GPA $\geq$ 3.0.

- Managing a department (such as IT, finance, and human resources) in a public or private entity.
- Managing a professional society or charity organization.

> No two projects are identical, even if they look so. They may differ in the type of soil, labor, equipment, methods, management style, location, regulation, weather (when erected), productivities, or even accidents or incidents during construction.

Operations management has its own methods and rules for running and evaluation/measuring success. Project and operations management have similarities as well as differences. This book focuses on the management of construction projects. After the completion of the construction project, its operation and maintenance (building, road, or refinery) through its lifecycle is another topic that is outside the scope of this book.

Why and How Projects Are Initiated

Construction projects get initiated for a variety of reasons. There could be a need for the project itself or it could be an investment project. In the latter case, its expected rate of return, ROR, has to be greater than or equal a minimum threshold called the minimum attractive rate of return, MARR.

Some investors and owners like to do a pro forma for the project, which is a projection or estimate of the cost/benefit ratio or ROR on an investment. Typically, a pro forma includes all expected costs (initial, recurring, and occasional) and expected revenues throughout the life cycle of the project, and then calculates the ROR. It helps project owners and investors decide on whether or not to carry out the project.

The process of initiating construction projects also differs. A private owner has the authority to initiate a project regardless of the formalities, as long as the project satisfies legal requirements and falls within the owner's financial capacity. Corporations, privately or publicly owned, have their own rules and regulations for initiating projects, with authority usually balanced between the executives and the supervising boards. Public agencies are even stricter with the process because they deal with public money. There are many restrictions in public agencies regarding the approval of a project, the funding, the selection of a contractor, and more. In addition, public agencies must be transparent in all their dealings.

In most cases, the project starts with a proposal, and then the proposal gets vetted in what is known as feasibility study, which includes all legal (such as zoning, building code, and environmental), financial, and other aspects. For example, an owner wants to build a shopping mall on land that is currently zoned as agricultural. The feasibility study must address issues such as "Is it possible to rezone the land for commercial use?" Considering that part of the land is classified as "wetland," will there be enough land for the buildings, parking, roads, and green areas? Can this wetland be "mitigated"? Or what is the maximum building area possible there? Will the project generate net ROR greater than the set MARR? There may be more questions about the details, but the process must help the owner make the final say on the "Go/No Go" decision. Keep in mind that some issues in the feasibility study may require conducting geotechnical and other testing/studies. This is part of the planning process, which costs money, only to provide better guidance on the decision to go or not to go ahead with the project.

The feasibility study may differ in content and effort based on the type of project: residential, commercial, industrial, or other. Public projects also must go through this process, but if the project is identified as "eminent domain," it will have different and "more forceful" approach. Eminent domain refers to the power of the government to take private property and convert it into public use when the project is deemed a must for the interest of the citizens in the public agency's jurisdiction. Of course, the government must compensate the owners of the land taken by the government for the project, according to the local law[5].

If the project is approved, the next step is to do the project charter. The project charter (also called project definition or project statement) is a statement of the scope, objectives, and participants in a project. It provides a preliminary delineation of roles and responsibilities, outlines the project objectives, identifies the main stakeholders, and defines the authority of the project manager (PM). It serves as a reference of authority for the future of the project[6]. The Project Management Institute (PMI) defines project charter as a document issued by the project initiator or sponsor that formally authorizes the existence of a project and provides the PM with the authority to apply organizational resources to project activities[7].

The formalities of the project charter may be more emphasized in public and private corporations. Private owners may or may not do the project charter, but they are strongly encouraged to do proper planning, including defining the project scope along with major constraints and conditions (Figure 1.3).

Defining the scope and constraints of the project, as in the charter, will be the foundation of several project disciplines to be managed during the execution.

Project Lifecycle

The project usually starts with a proposal followed by feasibility studies. After that, the decision is made to go ahead with the project or not. If not, the proposal is either abandoned or returned for modification that hopefully makes it acceptable in the resubmission. Once it is approved, the project starts for the owner in the planning phase. This is an important phase that is underestimated by some owners. Planning may get combined or coupled with the design phase in a way that allows the owner to modify the scope and constraints of the project based on the design development. After the design is completed, the owner will acquire a contractor to build the project. In Figure 1.4, the owner is using the traditional (Design-Bid-Build) delivery method[8], where design and construction are two separate contracts, performed sequentially. The contractor's project starts with signing the contract, but the contractor's planning may start when the bidding or negotiating process starts. After the final completion of the project, it gets handed from the contractor to the owner. This is an important legal

---

5 The term "eminent domain" is used in the United States of America. In other countries, there may be similar laws under different names.

6 https://www.sciencedirect.com/topics/computer-science/project-charter.

7 PMI, PMBOK, 7th edition, 2021.

8 To be explained in Chapter 3, Contracts and Contracting.

Project Charter			
Project Name			
Project Description			
Project Manager		Date Approved	
Project Sponsor		Signature	

Business Case		Expected Goals/Deliverables	

Team Members			
Name	Role		

Risks and Constraints		Milestones	

Figure 1.3 Project charter sample.

step where the owner takes possession and responsibility for the project. This step ends the "project" for both.

In Figure 1.5, the owner is using the Design–Build delivery method, where design and construction come in one contract, and may overlap. The contractor's project starts also with signing the contract, which includes design and construction.

Project Management

The Project Management Institute (PMI) is the leading organization in the field of project management, and it defines it as application of knowledge, skills, tools, and techniques to project activities to meet the project requirements. Accomplished through the appropriate application and integration of the project management

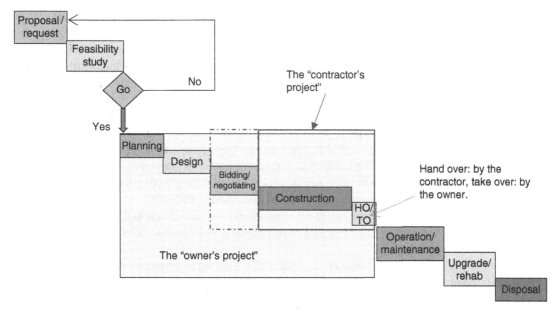

Figure 1.4 The lifecycle for a traditional (Design-Bid-Build) project.

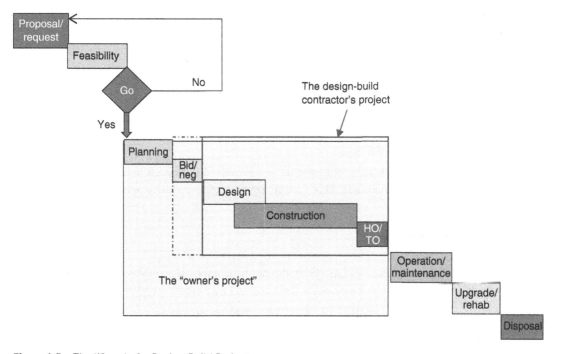

Figure 1.5 The lifecycle for Design-Build Project.

processes identified for the project (initiation, planning, execution, monitoring and controlling, and closing). Project management enables organizations to execute projects effectively and efficiently[9].

Project management, in general, is a combination of experimental/empirical science, trial and error, rules of thumb, experts' opinions, and even art. It also contains certain scientific components. In many issues, there is no unique right way to doing things. However, there are always guidelines and standards set by professional organizations and successful individuals.

When project management is broken into areas of specialty, and each area is broken into specific topics, we can then tie each topic to the principles and rules that are most applicable.

Keep in mind that in almost every facet of project management, there are many good rules but there is no such thing as "one size fits all." The industry is so dynamic and versatile, and each project has its own characteristics and conditions that make a specific concept or tool more effective than others for the situation. Even this best concept, tool, or practice, may and does change with time.

The PMI also defines *Program Management* as the application of knowledge, skills, and principles to a program to achieve the program objectives and obtain benefits and control not available by managing program components individually[10].

Project management focuses on interdependencies *within a project* to determine the optimal approach for managing the project. Program management focuses on the interdependencies *between projects, and between projects and the program level* to determine the optimal approach for managing them.

According to the PMI, the 10 Project Management Knowledge Areas are:

1. Project Integration Management
2. Project Scope Management
3. Project Time Management
4. Project Cost Management
5. Project Quality Management
6. Project Human Resource Management
7. Project Communications Management
8. Project Risk Management
9. Project Procurement Management
10. Project Stakeholder Management

These 10 areas apply to all industries but need to be customized to fit each industry's characteristics and traits. We will cover these areas in this book, from a construction industry perspective.

The Construction Industry

The construction industry represents a large component of the national economy. In the United States, as of January 2021[11]:

- Value of Construction = $1.516 trillion, up 15.6% from 2020
- Value of gross domestic product (GDP) = $20.991 trillion in 2021

---

9 Project Management Institute, PMI, PMBOK, 7th edition, 2021.
10 Project Management Institute, PMI, PMBOK, 7th edition, 2021.
11 https://www.abc.org/News-Media/News-Releases/entryid/9801/constructions-contribution-to-u-s-economy-highest-in-seven-years.

- Construction = 7.2% of GDP (was 6.6% in 2019)
- From 1999 through 2015, real (inflation-adjusted) construction investment varied from 5.1% of real GDP in 2010 and 2011 to 9.4% of GDP in 1999. In 2015 and 2016, construction investment was 6.2% of GDP.

In addition to the economic impact, the construction industry touches the lives of every person and many businesses. It is comprised primarily of small companies, though there are some very large companies (Figure 1.6).

What Makes Construction Projects Unique?

1. Projects are intricate and heterogeneous, involving many different specialties (crews/subcontractors). New specialties are added as technology advances in materials, equipment, and methods.
2. Projects take from a few weeks to several years to complete and cost up to billions of dollars.
3. Projects live from years to centuries. Many of them become eminent or historic and attract visitors.
4. Most projects are visible and usable by the public. For the project team leaders, there is excitement in watching the project get built! They are also likely to take pride in being part of the project creation team, especially for eminent projects.
5. Projects are characterized by their uniqueness, non-standard production, and infinite variability.
6. Projects (while construction and after completion) are subject to weather and climatic conditions. Site safety is a major concern: unlike factories, it is hard to see all actions.
7. The "as-built" rarely, if ever, matches the "as-planned." This is because the design and construction plan are predictions of a future event, which involves a lot of uncertainty.
8. Less "automation" and more "human touch" than most other industries, although this is changing with more modularization, 3D printing, and other technological applications.

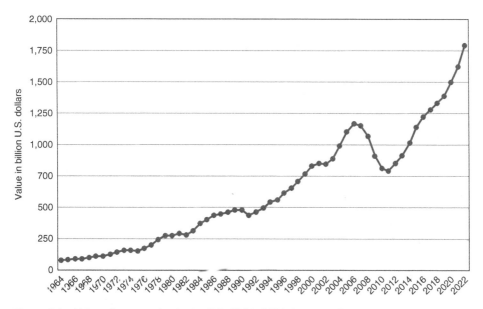

Figure 1.6 Value of construction in the U.S., 1964–2022. Source: U.S. new construction value 1964–2022 | Statista.

9. There are many project risks that are usually distributed among project stakeholders according to the contract and local laws.
10. As technology advances, the race to build structures that are bigger, taller, longer, stronger, or challenging in any way, continues. The challenge also is in the more efficient utilization of the resources and the preservation of the environment.

The definition of project management mentioned earlier applies to all types of projects, including construction and others. Managing construction projects has distinctive characteristics and guidelines. It has been described as one of the most exciting, challenging, and rewarding professions; construction management (CM) is a professional service that uses specialized, project management techniques to manage the planning, design, and construction of a project, from beginning (pre-design) to end (closeout). For all types of projects – large, small, vertical, horizontal, domestic, or international – a construction manager ensures the scope of work is skillfully adhered to, and the project is successfully delivered[12].

Managing construction projects from a general contractor's (GC) perspective is the overall management of the construction operations; technical; administrative; and others, in order to deliver the project to the client on time, within budget, and in conformance with the specifications and other contract terms, and within the boundaries of the law. This undertaking must be done while taking care of the welfare of the GC's own organization in terms of profitability, reputation, competitiveness, and long-term success, as shown in Figure 1.7.

Construction projects include a wide variety and can be divided according to several classifications. One classification includes these four categories:

1. Residential Building
2. Institutional and Commercial Building

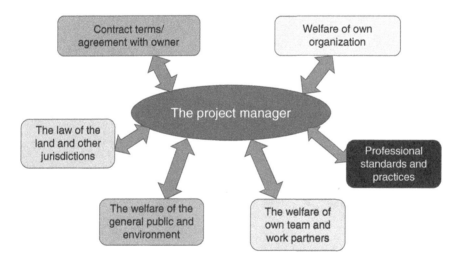

Figure 1.7 Responsibilities of the GC's project manager.

---

12 Construction Management Association of American, CMAA, https://www.cmaanet.org/about-us/what-construction-management.

3. Specialized Industrial Construction
4. Infrastructure and Heavy Construction

Another classification breaks projects into two groups: horizontal and vertical construction. Horizontal projects include highways/roads, bridges, tunnels, mass transit, railways, and airports. These are usually led by a civil engineer. Vertical projects include all types of buildings, with architect as the lead designer.

Another classification is infrastructure or heavy construction projects, which include:

- Highways/roads
- Bridges and tunnels
- Mass transit and railways, airports,
- Airports
- Water supply/resources, hydro-electric structures (dams)
- Waste management (solid and wastewater management)
- Dredging/flood control projects
- Power generation and transmission
- Telecommunications
- Process industry (oil and gas)
- Hazardous waste removal and storage

Another group is commercial and institutional projects, which include clinics, sports facilities, large shopping centers, hospitals, jails, universities, banks, libraries, warehouses, retail chain stores, skyscrapers, schools, and other projects of various sizes and types.

Residential construction seems to be a category by itself. Projects include single-unit homes, subdivisions, cottages, apartments, townhouses, and condominiums. Most home builders are specialized: tract housing (also known as cookie-cutter) or custom homes.

Pre-engineered/premanufactured/prefabricated homes and trailers belong to manufacturing industry and not the construction industry. We are seeing more and more overlap between the two industries in construction projects. In many cases, preassembled 2D or 3D units are used in construction. We can still count these projects as "construction projects" because the final process takes place at the permanent location of the project. The main difference between manufacturing and construction is not in what we do but rather in where we do it.

Professional Organizations Related to the Construction Industry

In the United States, there are many professional organizations and societies related to the construction industry. For example, these two organizations are very important for official certification of architects and engineers:

American Institute of Architects (AIA)
National Society of Professional Engineers (NSPE)

They also produce professional documents to standardize their profession, and work between their constituency (architects and engineers) and legislators.

Other professional organizations, general or specialized, work as a hub for professionals in the same specialty. They also advance the profession in a variety of ways. For example:

The American Society of Civil Engineers (ASCE)
The American Society of Mechanical Engineers (ASME)
Institute of Electrical and Electronics Engineers (IEEE)
National Institute of Building Sciences (NIBS)

There are professional organizations that work to advance the industry and professions under the industry by doing research and producing technical and other standards, such as

The Project Management Institute (PMI)
Construction Management Association of America (CMAA)
AACE International

The following are professional organizations for professionals in the construction industry, specialized or general:

Associated General Contractors of America (AGC)
Associated Builders and Contractors (ABC)
National Association of Home Builders (NAHB)
American Subcontractors Association, Inc. (ASA)
The Mechanical Contractors Association of America, Inc. (MCAA)
American Institute of Constructors (AIC)
Design-Build Institute of America (DBIA)
The Engineering and Construction Contracting (ECC) Association
National Association of Women in Construction (NAWIC)
National Contract Management Association (NMCA)
National Electrical Contractors Association (NECA)
National Utility Contractors Association, Inc. (NUCA)
American Water Works Association (AWWA)
Institute of Transportation Engineers (ITE)
The American Academy of Environmental Engineers and Scientists (AAEES)
American Society of Professional Estimators (ASPE)
Construction Financial Management Association (CFMA)
The American Arbitration Association (AAA)
Modular Building Institute (MBI)
American Wood Council (AWC)
The Society of Construction Law North America (and The Society of Construction Law, International)
The U.S. Green Building Council
SAVE International (formerly The Society of American Value Engineers)
College of Performance Management (CPM)
The North American Society for Trenchless Technology (NASTT)
National Contract Management Association (NCMA)

These organizations focus on construction education and/or research:

American Council for Construction Education (ACCE)
Associated Schools of Construction (ASC)
American Society for Testing and Materials (ASTM)
American National Standards Institute (ANSI)
Construction Specifications Institute (CSI)
The Construction Industry Institute (CII)
American Concrete Institute (ACI)
The National Fire Protection Association (NFPA)

Although our list includes only organizations either practicing in the USA or based in the USA, we must mention the International Organization for Standardization (ISO), based in Geneva, Switzerland, but its influence covers most of the world. It is focused, as its name implies, on standardizing products and specifications.

We may also mention the Royal Institution of Chartered Surveyors, RICS, based in the UK but influential in many parts of the world. It covers many construction-related specialties and issues certification for them. RICS is now expanding its activities to the USA and the Americas.

This book deals with all these areas, at varying depths, from the construction industry's perspective. The author quotes from and refers to the PMI's Project Management Body of Knowledge (PMBOK) with the reservation that PMI and its main reference, the PMBOK, serve the entire project management universe, although the PMI has published a Construction Extension to the PMBOK Guide that targets the construction industry. The construction industry, as discussed earlier, has its own characteristics, and its areas of management need to be "customized" accordingly. The author will use the PMI standards and guidelines as well as sources from other professional organizations.

Project Manager's Qualifications

Do project managers acquire their qualifications through education or experience? How do education and experience relate to each other? In fact, education and experience complement each other: Experience tells you *how,* but education tells you *why.* Experience tells you to put the reinforcement rebar at the bottom of the beam or suspended slab in the midspan, and at the top near the support. But education justifies this by teaching you that rebar must be placed where tensile stresses are high because concrete is weak in resisting tensile stresses.

Experience follows the work demand while education follows the learning process. So, in real life, you may have to jump into a situation without preparation but in education, there are curricula and prerequisites. It is wrong to compare experience to education when it comes to importance, as they complement each other. This is especially important in applied sciences. A fresh college graduate with no or little experience may get intimidated by someone with no college or even high school degree but has 30 years of experience! Of course, young college graduates need to work with others on the basis of respect, especially those with long experience, but they must also believe that their education gives them an edge that should help them reach higher professional levels. In the author's opinion, a college graduate with a few years of experience may surpass someone with 25 years of experience and no college education. Of course, there are exceptions to every rule, but we make deductions based on the population or the general case, not the exceptions (Table 1.1).

Besides, not all types of experiences are meaningful. Experience without education may become just repetition: 25 years of experience may be indeed one year repeated 25 times. In fact, it may become a barrier to advancement if you stick to the mentality of "but we've been doing it this way for 20 years!"

Table 1.1 Comparing education and experience.

Category	Academic education	Vocational education	Work experience
Type of knowledge	Mostly theoretical	Practical	Practical, hands-on, no theory
What does it teach?	Why we do things this way	How to do things	How to do things
Instructor/boss	Qualified faculty who must meet certain standards	Qualified professionals	May or may not be adequately qualified
Formal?	Yes	Yes	Not necessarily
Your choice in what you take	You choose major and courses within restrictions	You choose the specialty	Not much of a choice
Knowledge repetitive?	No	No	Yes
Sequence of knowledge	Systematic, logical	Systematic, logical	Random, as the job requires
Certificate of successful completion?	Yes	Yes	Maybe
Keeps up with changes?	A must	A must	Mostly yes but not always
Your "colleagues"	Your classmates are close in age, knowledge, and level	Your classmates are close in knowledge	Great deal of variance: Age, qualifications, and rank
Needed to get a job	Most likely	Sometimes required but always helps for related jobs	Most likely
Continuity of learning	A "project" with a target, plan, and end result	A "project" with a target, plan, and end result	May become just repetition
Longevity	Temporary; ends with graduation or dropout	Temporary; ends with graduation or dropout	Permanent, with possible horizontal and vertical moves
Knowledge advancement?	Possible via graduate study and research	No (terminal degree) but can learn new skills	Yes informally, depending on the person job, and organization
$$$	You have to pay unless exempted	You have to pay unless exempted	You get paid
Objective	To educate YOU	To qualify you for a technical job	To get the job done

Some people pick exceptional cases when someone achieved high professional success even though did not complete his/her college degree. Scientific observations and rules are based on populations and large samples, not exceptional cases!

Project Manager's Team

The team under the construction PM typically includes technical and nontechnical personnel. Team members may be dedicated to the project and report to the PM or borrowed from the main office and still report to their functional manager in the main office. The team also gets larger or smaller depending on the project needs, which – in part – depends on the stage of the project.

The formation of the project team follows the project management organization structure that the contractor's organization follows. Basically, there are these structures:

A. Functional project organization where all or most of the PM's team members belong and report to their functional managers in the main office.
B. Pure project (projectized) organization where the PM's team is almost autonomous. The PM acts like the CEO of his/her own organization.
C. The Matrix project organization is a hybrid between the above two opposite structures, with varying degrees of leaning toward one of them. The matrix structure closer to the functional organization is considered "weak matrix." Conversely, the matrix structure closer to the projectized organization is considered "strong matrix." (Figures 1.8 and 1.9)

Each structure has its own advantages as well as disadvantages. Also, each construction project lends itself to choosing the best structure for the situation. For example, the PM can have those needed on a full-time basis for the entire duration of the project as own team members. Others that are needed temporarily or on a part-time basis, can be borrowed from the main office or even a third-party source.

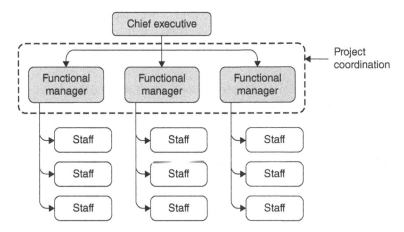

Figure 1.8 The functional project organization.

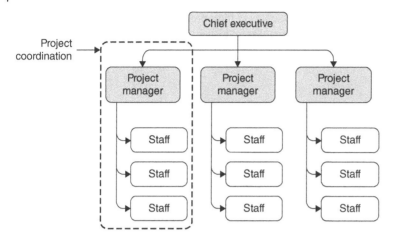

Figure 1.9 The pure project organization.

The Project Management Office, PMO

PMO is a management structure that standardizes the project-related governance processes and facilitates the sharing of resources, methodologies, tools, and techniques[13]. PMO is usually a permanent part of the organization that can go up and down in size, depending on the organization's needs.

A PMO may exist in an owner or contractor's organization, but there is a major difference between the two cases. An owner's organization that frequently executes construction projects may have its own PMO to provide services in both the planning and execution phases, such as preliminary design which helps define the approximate budget and timeframe. Also, it helps in the acquisition of a designer and contractor and in project management, from the owner's side. The contractor's PMO is a major part of its organization. It is like the central command or brain for standardizing the execution of the projects.

No two PMOs operate in exactly the same way, but they can be generally divided into three PMO types[14]:

1. Supportive: A supportive PMO collects all projects in an organization, and supplies best practices, templates, and training, but with a low degree of control.
2. Controlling: A controlling PMO checks if the project management tools, processes, and standards are being applied in the projects with some degree of control.
3. Directive: A directive PMO maintains a high degree of control in the project management process within the organization (Figure 1.10 gives a general idea of a PMO).

The size and structure of the PMO are directly related to the organization's culture and structure, but the PMO – by nature – lends itself more toward the functional project organization. Figure 1.10 gives a general idea of a PMO.

---

13 Project Management Institute, PMI, PMBOK, 7th edition, 2021.
14 https://www.projectmanager.com/guides/pmo.

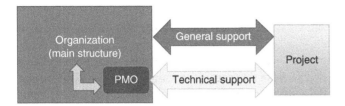

Figure 1.10   The relationship between the organization, the PMO, and the project.

Training, Continuing Education, and Certification

Training for college students is a great idea as it puts classroom education into application. They would know what setting the concrete formwork, framing a window or door, or installing a roof truss means. Internship is a wonderful idea, especially when the student is close to graduation. In addition to the experience it provides, it gives the company that hosts the internship an opportunity to closely "check" this potential job applicant, without a commitment. Many college students get job offers from those companies that hosted them previously during their internship. This is an indirect benefit of internship, provided that the student shows honesty, dedication, and a positive attitude.

Certification is another important milestone in the credentials of any professional. Certification simply means that the person who holds this certificate meets the minimum required level of knowledge in the field. It also helps standardize concepts and practices within the profession. It does not, however, guarantee good performance. In the authors' opinion, certification is more important and meaningful for a young person with little experience. It assures the potential employer that the candidate knows the basics of his/her field. For a senior professional with lots of proven experience, the certification is less meaningful, although it is still a "feather in his/her cap."

The most common certifications in construction project management are the PMI's project management professional (PMP) and the CMAA's Certified Construction Manager (CCM). In the UK, MRICS has several certifications in the field of construction.

For architects, the National Council of Architecture Registration Board (NCARB) is the main certification in the US. Engineers get certified as professional engineers (PE) by NSPE. Both certifications, NCARB and PE, are also a license to practice the profession and stamp (certify) designs. They both require an academic degree (bachelor's degree, as a minimum) and minimum number of years of experience in the field, supervised by a certified/licensed professional. Also, these certifications may have "step exams" toward the final certification.

In addition to the PMP certification, the PMI, has also a number of specialized certifications:

1. Program Management Professional (PgMP)
2. Portfolio Management Professional (PfMP)
3. Certified Associate in Project Management (CAPM)
4. PMI Professional in Business Analysis (PMI-PBA)
5. PMI Agile Certified Practitioner (PMI-ACP)
6. PMI Risk Management Professional (PMI-RMP)
7. PMI Scheduling Professional (PMI-SP)

The author was a member of the committee that put the first PMI-SP exam in 2007 (was offered in 2008 to the public for the first time.) Other organizations such as AACE International and CMAA, have their own certifications.

The contractor also needs a "General Contractor" license, which is more of a business license than a professional certification. It still requires passing an exam that covers the main areas of construction and contracting. This license usually falls under the jurisdiction of the local government. The requirements differ by local law. In the state of Florida, USA, it is under the Florida Department of Business & Professional Regulation and has several requirements that guarantee the fulfillment of both the legal, technical, and business needs.

Keep in mind that all types of professional certifications are good within a certain state or country. Sometimes, states or countries have reciprocity rules, where a certified professional in one state/country can practice in another state/country with the same certification. Even with such a rule, the professional has to be knowledgeable of the law, codes, and standard practices in the new location.

Continuing education is another important part of any professional career and certification. It can be both optional for adding knowledge or mandatory to renew a certificate. We all need continuing education for many reasons:

1. Because knowledge keeps advancing and changing, and we need to keep up with it from experts in their field
2. Because people may forget and need a refresher
3. Because renewal of certification likely to require a minimum number of continuing education hours (CEUs)
4. You can also pick up new skills or "support topics" (software, BIM, soft skill, etc.)

In most cases, we have options for continuing education seminars: online or physical, live or recorded. Many of them run concurrently during conferences, exhibits, and other events.

Construction Management College Education: For a long time in the past, construction and construction management were looked at as empirical/experimental science, or something you can pick up by practicing in the field. However, many realized the need for formal education in construction and construction management to standardize and formalize the discipline. In the United States, most construction management (CM) programs started under other older and more established programs such as architectural or civil engineering. Then, gradually they started becoming independent under different titles such as building construction, building science, construction management, construction science, construction engineering, construction technology, and others. These majors may have subtle differences; however, they all aim to pick up from the point architects and engineers left: the execution of the design.

Construction management college education also differs in width and depth. Some schools offer a CM class to their architectural or civil engineering students. Such course is "mile wide, inch deep." Some schools go further by offering a minor in CM, and others offer it at the master's or doctorate degrees. Graduate programs in CM opened the doors for research, which added strength to the discipline, especially when academics collaborate with the industry. In this regard, we need to mention the Construction Industry Institute, CII, which was created to bridge the gap between academia and industry.

In addition to the bachelor's degrees in construction management (with different titles), there are many vocational and training programs in construction trades in schools, union halls, or by professional organizations.

Project Participants and Stakeholders

A few terms are used, sometimes interchangeably to indicate the people or organizations involved in the project. Some are involved in designing or executing the project, some are involved in using/utilizing it after its completion, and others may be involved in both stages. During the construction stage, the term "project team" means the contracting parties: owner, designer, GC, subcontractors, and project management consultant. We may use the term "project participants" to indicate all those who have a role in the project: providing services, goods, finance, and/or permission. This includes the project team in addition to others such as vendors, equipment suppliers, test labs, government agencies (with jurisdiction over the construction process), investors/financiers, and those who may be consulted on specific issues (technical, legal, or other).

A project stakeholder is an individual, group, or organization, who may affect, be affected by, or perceive itself to be affected by a decision, activity, or outcome of a project, program, or portfolio[15]. Stakeholders are typically the members of a project team, PMs, executives, project sponsors, customers, and users. In other words, project stakeholders are the people, groups, or organizations who may have something to gain or lose from your project's outcome. These stakeholders include:

- The Owner/Client
- The General Contractor
- Designers (Architects and Engineers)
- Subcontractors
- Suppliers
- Government Authorities
- Local Residents and Business Owners
- Activists and Lobby Groups
- Labor Unions (if applicable)
- Professional bodies with jurisdiction over the design/construction process
- End Users

If we mention "key project stakeholders," we are likely talking about the main contracting parties. The number of stakeholders increases with large and high-profile projects, especially public projects and those owned by shareholders. For example, a road project may include all those residents and businesses along the sides of the road who will be affected by the road. Stakeholders for a factory project will include all residents who may be impacted positively or negatively. It is typical to see two opposing groups of stakeholders for the same project. It is unfortunate that stakeholders do not include the animals whose habitat and livelihood will be impacted by the project!

Types of Owners

1. Public (government): This includes all levels and types of government agencies.
2. Private corporations owned by shareholders, as well as nonprofit organizations. These corporations usually have a board of directors (or trustees) that oversees the running of operations as well as major decisions.

---

15 Project Management Institute, PMI, PMBOK, 7th edition, 2021.

3. Private individuals or corporations/businesses owned entirely by an individual or group of individuals.
4. Public Private Partnership (PPP): This is a common type in some states/countries where a private entity funds a public project in exchange for a percentage of the revenue for a period.
5. International or multi-national organizations such as the United Nations and its suborganizations. In this case, the project will be subject to rules and regulations of both: the owner (international organization) and the host country.

Impact of type of owner on many aspects of the project, particularly:

1. Project initiation process and justification. For public agencies, there are strict guidelines for why and how a project can be initiated. These guidelines include cost threshold for any specific process. For example, if the cost exceeds a certain amount, usually $100,000–200,000, a minimum of three qualified bidders is required. Even when the project is under that threshold, there is still a strict policy for awarding. Any changes to the project later after its initiation, including the budget, must also follow the official guidelines.
2. Authority and the decision-making process: While private owners have the liberty in making decisions on the project, designer and contractor acquisition, or others; public agencies are restricted to protect public money. The authority in making decisions for public agencies is usually well defined and varies at every level of the agency's hierarchy. This applies also to delegation of authority, which has its own guidelines in public agencies.
3. Stakeholders: One main difference between public and private organizations is that the owner and the client can be the same entity in the private sector. In the public sector, the owners are the citizens that the public agency serves. The officials of the public agency are the "clients" who deal with designers and contractors on behalf of the "owner." Stakeholders for public agencies may include a wide group of parties. In fact, every taxpayer among the constituents of the agency can be considered a stakeholder.
4. Source(s) of funding. Public agencies get their funds usually from a variety of sources including taxes, fees, grants, loans, and others. Many types of funds, particularly grants, have strict guidelines that can impact the project scope and constraints. Market trends may impact future projects for public agencies in more than one way: The estimated cost of the project may change before signing the contract. Also, these trends may impact the sources of funding for the project.
5. Transparency: For public agencies, all actions including contracting and spending must be transparent to the public. This is not necessarily required in private projects, especially when the owner and client are the same.

Capital Improvement Projects

A capital improvement project is any major improvement to facilities and infrastructure. Projects may include construction and renovation of roads and streets, bridges, intersections, traffic systems, recreation centers, libraries, parks, automobile parking garages, water and wastewater treatment facilities, solid waste and recycling plants, and power stations. It may also include the purchase of new fleet vehicles and IT networks. A capital improvement project is usually proposed and approved by the government agency that has jurisdiction when it can prove that the project fulfills a need or improves the quality of life for the citizens in that area. These projects are usually included under the capital improvement programs. Such programs get an annual budget that may not be sufficient for all proposed projects, so these projects are ranked in terms of priority. Some projects get priority in approval and execution, others are postponed or scaled down, while others are canceled.

Most governments put guidelines for projects in order to be called capital improvement projects and thus financed by the capital improvements program, such as a minimum budget and/or minimum lifespan.

Although capital improvement projects usually are created and financed by governments (city, county, state, and federal), it is possible for private organizations to create their own capital improvement projects. Some capital improvements are given favorable tax treatment and may be exempted from sales tax in certain jurisdictions.

The Construction Industry: Past, Present, and Future

Construction is one of the oldest professions and will continue as long as there are humans on the planet. People always need homes and places for work, recreation, worship, and other activities. They need transportation systems, water plants, power generation plants, and infrastructures for other utilities. Perhaps with time and advancement of science and technology, materials, equipment, methods, and concepts change, but there will always be a need for construction projects. Those who want to not only survive in the market but also advance and lead, need to be futurists, creative, and one step ahead. Several trends have already started to advance in the construction industry such as modularization and IT tools and computer software that were not even imagined a few decades ago. At the same time, challenges grow bigger such as global warming (that increased the intensity and frequency of certain natural disasters), shortage of fuel and certain materials, and environmental restrictions. Tomorrow's construction professionals will need to meet these challenges with innovative and efficient solutions to make sure products have quality, durability, and cost-efficiency.

Exercises

1.1. Define a project and give an example.

1.2. Give an example of a program that contains projects.

1.3. Can taking care of your garden be considered a project? Mention two cases where one answer is yes and the other is no.

1.4. You are thinking of converting part of your house attic to an office. How can you make this idea into a project?

1.5. Do you consider the following as projects (with the description as is)? If not, modify it to make it qualify as a project.
   a. Improving your education
   b. Changing a flat tire
   c. Losing weight

     d. Building a 20 feet by 25 feet expansion for your house

     e. Cleaning the gutters of your house periodically

     f. Maintaining your house garden in good shape

     g. Organizing your garage

     h. Going on a fun trip for a week

     i. Doing a birthday party for your younger sister

1.6. How does the owner's "project" differ from the contractor's "project"?

1.7. Why must an owner do feasibility study before approving or making a final decision on the project? What are the main objectives of this feasibility study?

1.8. Mention the main stakeholders in a private project.

1.9. Mention the main stakeholders in a public project.

1.10. In what aspects do public and private projects differ?

1.11. What is a capital improvement project?

1.12. What are the types of owners of a project?

1.13. What benefits does certification in construction management, or one of its specialties, bring to the bearer of the certificate, to the employer, and to the client?

1.14. What is a PMO? Why would an organization have a PMO as a part of its structure?

1.15. For the same construction project, the contractor has a PM, and the owner may have a project manager. What are the major differences in their roles?

1.16. Mention two types of each category:

     a. Vertical construction

     b. Horizontal construction

1.17. What are the two main types of project organization? How do they differ? What is the project organization that is in between the two types?

2

Architecture, Engineering, and Construction

Introduction

In the beginning of history, people used to build their own homes themselves, using local materials such as stones, mud, clay, tree branches, and even animals' ribs and skin. They later developed more sophisticated materials, such as blocks made of burnt mud and more geometric-sized stones, with the development of better tools. These building materials also included bonding, sealing, and other agents. They also introduced doors and windows for protection and security, as well as other methods for lighting and ventilation. Humans transitioned through time from the cave to more comfortable homes.

Usually, there was one person called "the builder" who was in charge of the entire design and building process. This person (the builder) would act alone or with helpers (team), depending on the size of the project. Talented builders were utilized by rulers and rich people to build homes and monuments for them.

Many of us visit, see, and admire old monuments that are still standing through hundreds or even thousands of years, resisting not only weather erosion factors but also natural disasters such as earthquakes. Some people naively use them as evidence that construction in the past was better than today. It is true that few structures, such as Egypt's pyramids, Greece's Acropolis, China's Great Wall, and India's Taj Mahal, are indeed great structures, but what is their percentage out of all structures built in the past? Those structures that survived represent only a tiny percentage of structures built in the past. This is simply because these monuments had unlimited resources and the best builders. They either belonged to the king of the land or were high-profile projects with mostly huge "carrots and sticks" for the builder. Many of these took years to build, consumed huge amounts of resources, and did not care about the safety of workers or the welfare of the environment. If you measure the factor of safety, used by structural engineers, in these structures, you will likely find it extremely high, which means – in todays' terms – economically inefficiency. The main objective for the builder then was to please the client, who likely was the ruler of the country.

In modern times, we require any structure to be designed and certified by a licensed professional. In fact, in most cases, it has to be designed by a group of professionals, each with their own specialty. For buildings, the design starts with the architect who designs both the external shape and the internal plans. We often use the A/E to designate the designer, Architect/Engineer.

The architect balances the project's functionality and aesthetics within the owner's constraints and preferences. The structural engineer designs for safety at minimum cost. Altogether, designers may

Project Management in the Construction Industry: From Concept to Completion, First Edition. Saleh Mubarak.
© 2024 John Wiley & Sons, Inc. Published 2024 by John Wiley & Sons, Inc.
Companion Website: www.wiley.com/go/nextgencpm

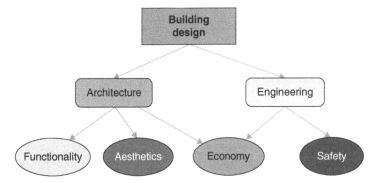

Figure 2.1 Building design: balancing functionality, aesthetics, safety, and economy.

juggle functionality and aesthetics, which will have an impact on cost, but cannot juggle or compromise safety, as shown in Figure 2.1.

> Owners and designers can joggle aesthetics, functionality, and cost… but they can't joggle safety or compromise it!

As an example of this joggling, imagine the four-column sections shown in Figure 2.2. Let us assume that they have the same loading capacity. Section A, the circular, is the optimum one from a structural (and specifically buckling) point of view. Section B, a hexagon, is second best then section C, the square, and finally section D, the rectangle. From a construction perspective (formwork), it may be the other way around, with sections C and D tied at "best." However, functionality and aesthetics are a different matter: sections A and B may be more appealing, but section D can be hidden in the wall/partition. All of these arguments, including the cost, must be considered, and the owner will make the choice based on all of these issues.

Design team members often compete for "discipline dominance," mainly the architect and structural engineer. Architects usually like clear spaces, wide spans, and attractive aesthetics, while structural engineers concern is safety. While structural safety is always a given, and is not a matter of negotiation or compromise, this competition can bring the best of both parties together for a design that satisfies all.

An example is shown in Figure 2.3, representing four different sections of a concrete beam used to carry a balcony. From a structural point of view, section A, a simply supported beam, is the best, but the architect strongly objects because she/he likes the clear space under it. This narrows the choice to a cantilever, but there are three possible sections: B, C, or D. Again, the choice must consider functionality, aesthetics, and cost.

A   B   C   D

Figure 2.2 Four different column sections.

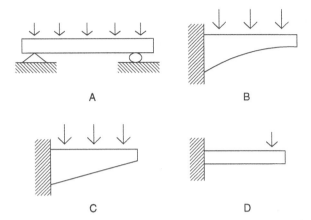

Figure 2.3 Four different beam sections.

The building "chief designer" is usually the architect who leads a team that includes several specialties, such as all types of engineers (structural, civil, mechanical, and electrical), interior designers, hardscape/landscape designers, and other specialties as the project needs.

While there are several arrangements for how an owner retains a designer and builder (contractor), the traditional practice for an owner is to hire a designer (A/E) and a general contractor (GC) in separate contracts. In addition, to monitor the construction process, the owner can do self-monitoring (if qualified or has a qualified team), hire the same A/E, or retain the third party for construction management services. There are other types of arrangements between the owner, on one side, and the designer (A/E), builder (GC), and construction manager (CM), on the other side[1].

Before we construct a project, it has to be designed. The term "engineer," as a verb, is often used to mean design, but the author prefers the term design because it is more generic and includes architectural and other disciplines which is the main step in building projects. The sequence we see in Figure 2.4a is the typical process in any construction project. The middle phase in Figure 2.4a, Procure, indicates the process of choosing and hiring the construction contractor[2]. This process will be explained in detail in Chapter 3. Figure 2.4b shows the same process in a different contractual type, Design-Build (DB), when the owner signs a contract with one entity to do both design and construction. Figure 2.4c shows a DB contract with an accelerated process called fast-tracking, where construction starts before the design phase is completed.

Depending on the type of project, the design team may be led by an architect or engineer but is likely to include several specialties such as structural, mechanical, and electrical.

Design and construction are not necessarily autonomous and independent processes. In fact, they must be integrated. The designer must take the construction process into consideration while doing the design. The contractor needs to understand the design well in order to build it. But, there is also a great benefit if the construction team is involved during the design phase. It can provide practical input that

---

1 This will be explained in subsequent chapters.
2 The term EPC in the construction industry is commonly used to denote "Engineering, Procurement, Construction."

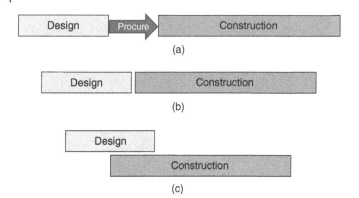

Figure 2.4 (a) Typical process in traditional construction projects, (b) typical process in Design-Build projects, and (c) typical process in Design-Build projects with fast-tracking.

can facilitate and optimize the design. Conversely, the designer may be involved in monitoring the construction process on behalf of the owner, to make sure it is performed according to the design standards.

This book is not about project design; however, we need to cover the transition, overlap, and interface between design and construction. The construction process is simply the execution of the design, or transforming it from paper (or digital) to reality. During this transition, many changes and deviations may happen, and the construction team has to manage them.

Clarification of Terminology: The following are the main players in any construction project. Titles may differ depending on several factors such as the type of contractual relationship and the local jargon. Let us try to clarify the identity and roles of each party:

- The client or owner of the project is the party (person, persons, or organization) who initiates the project, makes the decisions regarding design specifications and construction process, within legal boundaries and industry standards, pays all project costs, and takes ownership of the project after its completion.
- The term client is a generic term, referring to the party who hires other party(s) to provide him/her goods and/or services. The client can be doing so on their own behalf or behalf of others. When doing so on their own behalf, the client and owner become the same entity. As an example, if a person wants to build a house and hires the designer and contractor directly, he/she will be the client and owner at the same time. In public agencies, the "owner" is the taxpayers or citizens living in the jurisdiction of this agency. The person or department that will act on behalf of the owner in acquiring designers and contractors is the client. This applies also to private entities that are owned by shareholders.
- The designer is the design professional hired to do the project design. In most likelihood, this includes a group of designers based on the specialties required for the project. The architect leads the design of almost all building projects. That is why we often use the A/E designation to refer to the designer, meaning Architect/Engineer. In heavy construction projects, including roads and bridges, the architect is rarely used, so the lead designer is usually the civil engineer. In the process (oil and gas) industry projects, the chemical engineer is likely to be the lead designer. For any design

to be officially accepted by local authorities, it has to be signed and sealed by a licensed professional, architect, and/or engineer, depending on the type of design.

- The contractor's more accurate title is the constructor; however, it is common practice to say the contractor and sometimes the builder. For the general contractor, we use the designation GC. We also sometimes abbreviate the title subcontractor as sub.
- In project management, there are several designations for the roles of both individuals and organizations. The GC's team is usually led by a project manager, PM, but the owner likely has someone monitoring the construction process on their behalf. This person is usually called "owner's representative, or rep" but may also get the title "PM." If this role is delegated to an organization, it will be playing the role of "construction project management[3]," construction management, CM, or project management consultant, PMC. In some parts of the world, the generic term "consultant" is used to denote the person/entity monitoring the contractor's work onsite and looking after the owner's interests. Roles and titles differ, so it is important to clarify the title by defining the roles, responsibilities, and what party it represents.
- Traditionally, the owner has three choices in monitoring the construction process and the contractor's adherence to the contract terms:
  1. Do it himself/herself, if qualified and dedicated,
  2. Hire the same firm that designed the project for additional services: construction management, or
  3. Hire a third-party construction management firm.

Commitment to the Owner/Client

With a variety of different approaches, the entire project creation can be summarized as shown in Figure 2.5.

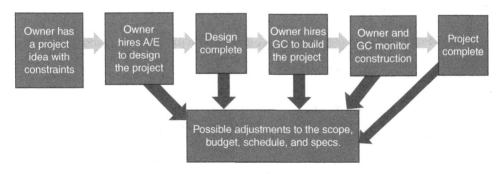

Figure 2.5 The construction project process.

---

3 Do not abbreviate "construction project management" as CPM, since this acronym is reserved for the Critical Path Method in the construction industry.

The A/E (designer) offers professional services to the owner in exchange for an agreed fee. The A/E, thus, has the professional commitment to provide these services (design) within both the owner's requirements and the legal and professional standards. In many situations, the owner's demands may be infeasible for whatever reason. The A/E has the obligation to explain the feasible options to the owner, along with their pros and cons and costs. There is also a responsibility and limited liability on the A/E to help the owner find a contractor to construct the project within the cost and timeframe the A/E estimated, with a small margin of variation (usually $\approx 10\%$) from the A/E's estimate. There are exceptions, of course, when unforeseen events happen, such as the sharp and unexpected increase in real estate prices and construction costs in 2020–2022.

On the other hand, the GC (contractor) has the legal and professional obligation to construct the project according to the terms of the contract signed with the owner, and within the legal boundaries and professional standards. However, variances from the original scope can and do happen for a variety of reasons, some of which are deliberate while others are not. In the end, the objective is to complete the project to the terms of the contract agreement, and the satisfaction of the owner/client.

Designers and Field Operations

Although the designer's job does not directly or explicitly require visiting the site, it is strongly recommended for every designer to do so. Most likely, the design effort takes place in an office behind a computer screen and away from the construction site. In fact, in most cases, the design occurs and finishes before even construction starts. However, it is important for designers to engage in the construction process and visually see the development of the project from design to construction. It is important to see how the design is being carried out, not just how it later became a reality. Field experience is a great asset for a designer. There are several ways for a designer to stay engaged in the construction effort such as:

A. Take courses in construction and construction management while in college.
B. Do internships also while in college.
C. Get a field job after graduation.
D. If you get a job with a large company, do your best to have a rotation during the first year or two, where you can spend a few months in different departments.
E. While working as a designer in the office, visit the site occasionally and learn how things work out in reality.
F. Watch educational videos on construction.
G. After acquiring some field experience, work on construction management, especially for projects your company has designed.
H. Keep up with new developments and inventions in the construction industry.
I. Attend continuing education seminars and professional conferences relating to construction matters.

The same argument can be said, to some extent, about construction professionals: they need to educate themselves in the design process. There is an overlap between studying design (architecture or engineering) and construction, with the latter focusing on executing what the design produced. Sometimes, the courses' titles look similar, but they are different. Take, for example, reinforced concrete courses: The civil/structural engineering students study the design of the concrete member for resisting service loads, such as cross-sections and reinforcement. The construction/construction management students study how to build that member, mainly the design of the formwork system to carry the load of the freshly placed concrete.

The author believes that the best professionals in both worlds, design and construction, are those who have deep knowledge in both fields, or at least, a deep knowledge of the main specialty and good understanding of the other area.

Exercises

2.1 Who is the "chief designer" for a typical building project?

2.2 Who is the "chief designer" for a typical road and bridge project?

2.3 The architect designs and balances two important criteria, while the structural engineer designs two other criteria. Discuss the balance of these four criteria. Which one of them is not subject to compromise?

2.4 Design and construction go in sequence: design first and then construction. Can they overlap? Explain.

2.5 In the design of building projects, the architect and structural engineer complement each other's work, explain.

2.6 The designer will attempt to optimize the design from the client's perspective but can also help optimize the design from the contractor's perspective. How can the designer do this?

2.7 What are the owner's choices in monitoring the construction process and the contractor's adherence to the contract terms?

3

Contracts and Contracting

Introduction

After the project design is completed, the owner will try to find the contractor who can build the project in a manner that is in the owner's best interest[1]. There are several approaches to how an owner selects the appropriate contractor. There are also several types of contracting arrangements between the owner and the contractor. In this chapter, we will cover all these approaches and contract types, in addition to other related topics.

What is a Contract?

A contract, in general meaning, is an agreement between two or more parties, where each party has responsibilities toward other parties and is entitled to rights from them, according to the terms of the contract agreement. It is a legally enforceable agreement that creates, defines, and governs mutual rights and obligations among its parties. A contract typically involves the transfer of goods, services, money, or a promise to transfer any of these at a future date[2]. In business, including design and construction, it is an agreement between an owner and a service provider (designer, contractor, and construction management consultant) for services in exchange for the agreed monetary amount (lump sum or percentage).

Following are the requirements for any contract to be legally enforceable:

1. Mutual assent, expressed in an offer and acceptance, done willingly.
2. Exchange of value (consideration). In the general case, one party pays money while the other party offers services and/or goods.

---

1 With the exception of Design–Build contracts, where the owner signs a contract with one entity to design and build the project.
2 https://en.wikipedia.org/wiki/Contract.

Project Management in the Construction Industry: From Concept to Completion, First Edition. Saleh Mubarak.
© 2024 John Wiley & Sons, Inc. Published 2024 by John Wiley & Sons, Inc.
Companion Website: www.wiley.com/go/nextgencpm

3. Capacity: Competence of parties involved (including sanity, legal age, and authority in case one party is acting on behalf of someone else.)
4. Legality: All actions taken by all parties must be legal.

Even though the law considers verbal agreements to be also legal and biding, having this agreement written in a formal contract, and signed by both parties, is extremely important.

In addition to these general rules, federal and state laws may impose more requirements on particular types of contracts.

Types of Contracting Approaches (Procuring Contractors)

There are several ways an owner can procure the services of a contractor, some of which may not be available to public agencies, except under certain circumstances:

1. Competitive (Public) Bidding
2. Selective Bidding
3. Negotiation/Direct Award

Competitive (General /Public) Bidding: Under this arrangement, the owner makes a public announcement for the project, with a chance for any qualified contractor to bid. This can be done in one or more of multiple ways for publishing: a public platform (website or newspaper), the organization's own platform (website or media outlet), or through a specialized publication such as the Dodge Report[3].

All public projects above a set monetary threshold must go through public bidding for the protection of public money, except in justifiable cases such as emergencies and national security projects. Even projects under that threshold still have a competitive process, most likely a presigned unit price contract, but no requirement of public bidding.

Furthermore, public agencies and many private owners require a minimum of three qualified bidders for the process to continue. Otherwise, the bidding process will be canceled and perhaps re-announced later with the hope of attracting more bidders. When there are only two bidders, some contractors may resort to soliciting a "courtesy bidder," who would submit a qualified but high bid, in order to legitimize the process and allow another bidder to get the project. This is unethical and illegal conduct. Bidders have to be serious in their bids and are not supposed to collude for one's own interest.

The number of bidders for any project may vary vastly depending on several factors, some of which have to do with the project itself[4] and the owner. Other factors have to do with the situation of the bidder and also prevailing economic conditions. Public agencies also have the right not to award the contract to the lowest qualified bidder but to rebid the project instead. This is done sometimes if the agency believes that such action is in the best interest of the agency and public money. The rebid may occur with or without adjustments to the project design and/or contract terms.

---

3 https://www.construction.com/.
4 Mainly type, size, and location of the project, which define the contractor's ability, level of comfort, and risk.

Table 3.1 Contract type versus potential number of bidders.

Contractor procurement type	Public advertisement?	Owner requests bids from	Competition
Public/competitive bidding	Yes	Many contractors, anyone qualified and interested	High
Selective bidding	No	A few selected contractors, by invitation only	Moderate
Direct award/negotiation	No	One contractor (at least at a time)	Low

Some public agencies sign "unit-price" contracts[5] with contractors. These contracts allow the agency to award small projects (usually up to $100,000 to $200,000) without a bidding process. These contracts usually cover items such as sidewalks, pavement, speedbumps, and road repair. Such contracts are usually valid for a year or two years, and can then be renewed, likely after adjustments. Even with such an arrangement, the public agency will award small projects that fit within this category in rotation among contractors with valid unit-price contracts with the agency. Thus, the distribution of projects is fair, and no single contractor gets several projects in the same period of time, while others do not get any.

Selective Bidding: Selective bidding is when an owner invites only a selected group of contractors to bid on a project. This choice can be based on objective or subjective considerations such as certain criteria that are important to the owner (high quality, speed, and deep expertise in a particular area), previous experience, and/or reputation. Government agencies use this procurement method when the situation is justifiable, such as in cases of national security. Public corporations may also do this when there are a few contractors that can be trusted with this type of project.

Negotiation/Direct Award: When an owner likes to deal with a specific contractor and the rules allow such arrangement, it is done through direct award with or even without negotiation. The owner contacts the contractor in this situation to inform him/her of the project and express interest in signing a contract with the contractor. This may take some negotiation which will result either in signing a contract or not. In the latter case, the owner may look for another contractor. It is unusual and unprofessional for an owner to be negotiating separately and simultaneously with more than one contractor, unless it is a case of competitive bidding. It is perfectly fine though if the owner has another contractor in mind in case the negotiations with the first one fail.

Similar to the previous type, this arrangement is not permissible for public agencies and most publicly owned private organizations, except in justifiable cases (Table 3.1).

Selection Criteria:

Public agencies and most publicly owned private organizations are usually required to choose the lowest qualified bidder with some exceptions. Private owners have the discretion to choose contractors as they like. The choice of the successful contractor can be based on:

A. Lowest price
B. A combination of price and speed
C. Best qualifications
D. A mixture of price and qualifications

---

5 Unit-price contracts will be discussed in detail later.

Options C and D can be somewhat subjective and dependent on the assessment of the owner. In some cases, the owner forms a "Selection Committee" to help pick the successful contractor. The committee members must understand well the owner's demands, priorities, and constraints. Furthermore, the committee members must together set clear, objective, and measurable criteria for evaluating candidates. Most of these situations require presentations from the candidates and involve intensive interviews with them. The most important question the owner will ask the candidate is: Why should I hire you and not others? The author witnessed situations in which a candidate, who was initially ranked third or even lower, managed to jump to top place after the presentation and interview. It is not a simple process because it requires the owner's persuasion and impression, using all professional means, including past projects/experience and the technical team's qualifications. Creative and innovative ideas will always help.

> Although the contractor's bid or offer amount is important, contractors are not always selected solely on the basis of price, even when they are qualified.

One of the most important concepts in dealing with a client is not to focus merely on your own qualifications, but rather on how these qualifications can best serve the client. Owners' criteria include cost, time (schedule), quality, and others, but the priorities differ not only from one owner to another but also from one project to another.

Qualifying Bidders

Owners require certain qualifications bidding contractors must satisfy. Therefore, advertisements in public bidding usually say, "The project will be awarded to the lowest qualified bidder." The owner needs to inspect the bid and make sure the bidder satisfies all conditions set by the owner in the invitation to bid. This process can be either pre- or post-qualification.

Pre-bidding qualification is when every bid is inspected and vetted for qualification upon submission. In this case, only qualified bids go to the next step, and the successful contractor will be the lowest bidder in this pool of qualified bidders. Post-bidding qualification is when bids are not vetted for qualification upon submission. After the submission period expires and the pool includes all bidders, the lowest bidder is picked and then vetted. If qualified, he/she will be the successful bidder. Otherwise, the next lowest bidder is picked and vetted, and so on.

The pre-qualification method is usually used if the number of bidders is small, but when the number is large, it will be more time-efficient to use post-qualification.

Bid Documents

Bid documents include two groups of documents:

1. The detailed design, which includes the drawings (plans), specifications, and any other supplement issued by the designer. They may include documents created by a third party (such as a manufacturer) for components in the design such as pre-engineered assemblies.

2. Documents created for the bidding process such as the invitation for bids, instructions to bidders, bid data sheet (BDS), evaluation and qualification criteria, bidding forms, and the proposed Contract Documents.

Addenda: When the owner makes a change, correction, or clarification in the design, during the bidding process, in a way that may affect the bids in any way, the owner must issue a memorandum to all bidders informing them of such change, correction, or clarification. This memorandum is called addendum (plural is addenda). If this addendum requires the bidders to do additional work and/or redo their work, the owner must extend the bid submission date to allow the bidders to implement and reflect the new information.

Updating forms and formats: Many organizations, public and private, that do projects frequently, have standard bidding forms. It is important for the organization to periodically review and update these forms. There are many reasons that compel such updating. The construction industry is always advancing and producing new concepts, materials, equipment, and other things. For example, the Construction Specification Institute (CSI), MasterFormat is being updated constantly, with a major change happened in 2004, going from 16 to 50 divisions[6]. Professional organizations, such as the Project Management Institute (PMI), AACE International, The Construction Management Association of America (CMAA), and others, continuously update their standards and references. Many software companies are being acquired by other companies, thus changing the name of software programs. Many products are emerging, while others are disappearing or becoming obsolete. Even local building codes may be updated, so any reference to such codes must be current. Implementing a mandatory review, annual or other, in the organization's policy is a necessary and important matter.

Digital Procurement

The digital procurement method is rapidly emerging. It saves time and money on both sides: the owner's agency and bidders. There are various websites and applications that provide electronic bidding and related services. It also provides a speedy and efficient method of communication for questions (request for information [RFI]) and others. Bids can also be submitted electronically using both fillable documents (PDF) and attachments. This method also has the advantage of having a real-time stamp, so there will not be a dispute about the time of the contractor's submission.

Bids for Design–Build (DB) contracts

Solicitation for design–build (DB) contracts follows the same procedure, except that there will be no detailed design available for bidders. The owner will likely provide as much information about the project design as possible and available. Scope definition, project charter, conceptual design, and other documents can be helpful in the initial definition of the project for better contracting terms.

In normal situations, professional entities specialize either in design or construction, not both. Construction project management can be done either by the owner's own team or through third-party

---

6 There will be more details on the CSI MasterFormat in the cost estimating chapter.

specialized companies. The bottom line is that the entity that manages and supervises the construction process on behalf of the owner must be different from the one doing it.

However, in the case of DB contracts, the owner signs only one contract with one entity that will do both design and construction for the project. In almost all these cases, the DB entity is either a design or construction firm. If it is a design firm, it will acquire the services of a construction firm to do the construction services. Inversely, if it is a construction firm, it will acquire the services of a design firm to do the design services. In all cases, the owner will deal with only one entity that will provide both services.

Joint Ventures: A joint venture (JV) is a business entity created by two or more parties, generally characterized by shared ownership, shared returns and risks, and shared governance. Companies typically pursue JVs for one of four reasons: to access a new market, particularly emerging market; to gain scale efficiencies by combining assets and operations; to share risk for major investments or projects; or to access skills and capabilities[7].

JVs, by nature, are temporary entities. It is usually stated in the JV creation contract that the JV will automatically be dissolved by the end of the said project. The parent companies may extend/renew the JV for other project(s).

The advantages to those partnering in the JV are:

1. A JV is usually an entity, legally independent from the parent companies. This limits their liability.
2. The partnering companies can combine and complement their expertise and resources to reach a level of competency higher than either one of them. The JV will be able to take on large projects that are likely beyond the capabilities of the individual parent companies.
3. The JV opens the door for future cooperation between the entities.

> There is no such thing as the "best type of contracts." Every situation is different, and the choice of the type of contract must be based on the merit of the situation.

The JV has an advantage to the owner in that the JV as an entity is bigger and more capable than the individual parent companies. The owner signs only one contract with the JV, which is a better arrangement than signing two or more individual contracts.

When Time Matters

In certain projects, when time matters, such as in major highway projects, the government agency likes to minimize the period when the highway shuts down or encourages the contractor to work during off-peak hours for the convenience of the public. The owner or government agency may resort to traditional or other more creative contracting methods to pressure the contractor to complete the project on the completion date stipulated in the contract or expedite it as much as possible. The traditional method adds a clause in the contract, imposing a penalty (called liquidated damages, or LDs for short) on the contractor for every day of delay beyond the contractually specified project duration. This penalty can be fixed

---

7 Roos, Alexander; Khanna, Dinesh; Verma, Sharad; Lang, Nikolaus; Dolya, Alex; Nath, Gaurav; Hammoud, Tawfik. "Getting More Value from Joint Ventures". Transaction Advisors. ISSN 2329-9134.

(e.g., $5000/day) or progressive, such as $2000/day for the first week, $2500/day for the second week, and $3000 for every day after two weeks. The penalty may be coupled with a bonus for every day of completion earlier than the contractual completion date.

Usually, but not always, there is a cap on the LDs and/or the bonus. LDs, in nature, are not supposed to be punitive to the contractor. Instead, LDs should represent the owner's actual/potential losses as a result of the delay. In public projects such as highways, bridges, and government buildings, it is extremely difficult to quantify such losses, so public agencies estimate the LDs based on the importance of the project to the public and their inconvenience as a result of inability to use the facility or reduction in its capacity. There are always legal challenges to the amount of these LDs.

Nontraditional and more creative methods include $A + B$ contracting and lane renting. The first one is a bidding/contracting method that factors time plus cost to determine the low bid. Under the $A + B$ method, the owner, usually a public agency, sets a hypothetical cost per day for the project. Each bidder submits a bid with two components:

- A: dollar amount for contract items (i.e., cost estimate).
- B: days required to complete the project.

Bid days are multiplied by the set "cost per day" and added to the "A component" to obtain the total bid. $A + (B \times \text{cost per day}) = \text{total bid}$. This is also called cost plus time or bi-parameter bidding/contracting method.

For example, there are three bidders for a project. The owner sets a hypothetical value of $10,000/day of construction:

Table 3.2 Example of bid analysis using the $A + B$ contracting method.

Bidder	Cost of items (A)	Days	Days × multiplier (B)	Total bid (A) + (B)
Bidder 1	$4,200,000	130	$1,300,000	$5,500,000
Bidder 2	$4,350,000	120	$1,200,000	$5,550,000
Bidder 3	$3,990,000	155	$1,550,000	$5,540,000

From Table 3.2, we find that Bidder 1 is the lowest, taking into consideration the committed number of days for construction, even though he/she is not the lowest in terms of dollar amount. The value of the contract is part "A" (cost of items), only though $4,200,000 in the above case.

The lane-renting method started in the United Kingdom, and later some government agencies in the United States started using it. Like the $A + B$ method, it adds a hypothetical monetary amount to the contractor's bid amount for the closing of road and highway lanes during construction using varying rates calculated by a formula. The main purpose is to minimize the complete or partial closure of these lanes during construction work, especially in peak hours, and thus reduce the public's inconvenience and losses.

LDs can still apply with either of the two methods mentioned in case the contractor took more days than originally committed to finish the project. In all cases ($A + B$ and lane renting contracting methods; with or without LDs), the contractor factors all possibilities into its bid amount. Delays may and do happen, so the cost estimate has to take into consideration these possibilities and their consequences, in order to be realistic.

In addition, in some situations, the contractor may have to accelerate (compress/crash) the project schedule to meet the owner's contractual requirements. More than likely, there will be a cost impact to this acceleration that must be taken into consideration when estimating or updating the project costs. The cost estimator may become a key player in providing the management with information that is necessary for making decisions regarding this acceleration[8].

Bidding with Options

In some situations, the owner may not be sure about an item or items in the project such as:

- Adding a swimming pool or children's playground
- Using carpet or ceramic tile
- Adding solar panels on the roof

The owner will have the bid based on the most likely scenario, but add possible scenarios as options:

- If the swimming pool is unlikely, then the base bid will be without the pool but will have the pool as an "add option."
- If carpet is the likely scenario, then the owner will have the base bid with carpet and add an option of installing ceramic tile (with certain specifications) in lieu of the carpet.
- If the owner is likely to have solar panels installed, then they should be included in the base bid with an option to delete them.

In this case, the bidding contractor will give a price for the base bid and separate prices for each option, in positive or negative amounts. These options may also impact the project time (schedule), increasing or decreasing the duration time. Later, the owner has the liberty of settling on the base bid only or with one or more of the options, using the numbers and conditions the contractor submitted in his/her bid.

To Bid or Not to Bid?

Bidding on a project is costly for the contractor. It takes many hours to read the project documentation, do the detailed estimate, prepare other required documents (a detailed schedule, for example), and finalize the estimate. The contractor must decide to bid or not to bid, based on several factors:

1. Type of project/construction involved compared to the contractor's area of work specialty. The contractor may, of course, subcontract any portion of the job that is outside its area of expertise (or practice) or that it feels can be performed more efficiently by a specialty subcontractor. The contractor may also decide to enter into a new specialty.
2. Location of the project: Contractors usually like to work on projects within their region or where they can provide efficient services. Taking on a job far away from headquarters or regional offices may

8 Selah Mubarak, *How to Estimate with RS Means, Basic Skills for Building Construction*, 5th ed., (Hoboken, NJ: John Wiley & Sons, 2020), chapter 3, pp. 45–47.

create logistical problems and result in higher costs, in addition to the possibility of being unfamiliar with that specific local market (vendors, suppliers, subcontractors, and labor). Sometimes, a contractor decides to expand into new territories based on organizational strategic goals.

3. The size of the project: Most contractors have operational and financial limitations to the volume of projects they can take on. This may be dictated by their resources, bonding capacity, or both. It is possible sometimes to team up with someone else to reach the required capacity.

4. The project's owner and the architect/engineer (A/E): Reputation and previous experience with a particular owner and/or designer may influence the contractor's decision to bid or not to bid. Alternatively, it might be reflected in the contractor's price tag: lowering or raising.

5. The amount of work currently being carried out or planned as compared to the capacity and strength of the contractor's organization. When economic conditions are favorable, contractors are usually busy, and the demand exceeds supply. Contractors in this case are less likely to bid on projects that are not their favorites. Remember that, unlike other industries, the contractor does not have an "on/off" switch when economic conditions abruptly change. Contracting and construction is a process that takes a long time, and in most cases, the contractor must finish what already started, regardless of the changes in the market.

6. Other bidders (the competition): Most contractors know competitors' strong and weak points, so they can have a feeling of their odds of winning the project. This may encourage or discourage him/her to bid, that is, if he/she knows the names of the competitors. Of course, there is no guarantee here and surprises may and do happen.

7. The availability of equipment, qualified personnel, and subcontractors required compared to those available to the contractor: this is a matter of both feasibility and economy to the contractor. Theoretically, the contractor can acquire any item, but the cost must be acceptable to keep the contractor's overall cost reasonable.

8. Future strategic plans: The project may be in a geographic area that is outside the contractor's base or even outside his/her area of expertise/preference, but he/she chooses to bid on the project to expand the business and fulfill the organization's strategic objectives.

9. How badly does the contractor want this project? Although the answer to this question has to do with current economic conditions as well as current and committed work, there are other factors that influence the contractor's desire to take on this project, such as plans to expand in a new region, establish a relationship, or other.

We can assign these factors into three groups:

I. Factors that have to do with the *contractor's* situation: Financial, resources, current ongoing projects, company's strategic plans

II. Factors that have to do with the *project*: Size, type, location, designer, owner, other bidders

III. *Other* factors such as prevailing economic conditions, future expectations (wars and labor strikes)

Preparing a bid costs the contractor money and time. Therefore, contractors must be selective in what projects they bid on, and where they see a realistic chance of winning and making profit or achieving other goals.

Adjusting or Withdrawing Bids: Bidders can modify or withdraw their bids before bid opening but once bids are opened, bidders cannot modify or adjust their bids. Withdrawing a bid after bid opening can subject the contractor to consequences, including possibly losing the bid bond or part of it. Sometimes, an exception is made if the contractor shows a major mishap such as a gross mathematical error or omission of a major item. In fact, after bid opening, if the owner finds out that there is a significant difference between the lowest and the next lowest bidders, it is a common and ethical practice to ask the lowest bidder to check his/her numbers in suspicion of a mistake. If this bidder finds out there was a major error in his/her calculations, he/she may be allowed to withdraw, sometimes with a penalty.

Bidding Tricks/Unethical Practices

1. *Bid Shopping* is the practice by which a general contractor (GC), after their bid is submitted with certain subcontractors' and material suppliers' quotes included, attempts to obtain prices from other subcontractors and material suppliers that are lower than those included in the GC's original estimate on which their bids are based. The practice may occur after a contract is awarded, pressuring subcontractors to reduce their prices included in the bid.

   It is an unethical practice because the GC has negotiated initially with a subcontractor and reached an agreement, even if not formal, but later wants to pressure the subcontractor to lower his/her price or else the GC will use another sub. The GC in this case will be deceiving subs and materials suppliers, and wasting their time and effort or reducing their expected profit.

2. *Bid Peddling* is bid shopping in reverse. Bid peddling occurs when a subcontractor who is not selected for a construction project contacts the prime contractor to substitute his or her company for a subcontractor on the original bid by offering to reduce its price. Sometimes this is done purposely by not submitting a bid during the bidding period and waiting for another subcontractor to submit a bid, and then offering to perform the work at a lower price. This allows the subcontractor (practicing bid peddling) to save the effort of estimating costs and use the numbers produced by the other subcontractor.

   In addition to being unethical, most local laws outlaw this practice.

3. *Courtesy Bidding* occurs when a contractor agrees to submit a bid that is intended to be unsuccessful (purposely high) so that another conspirator can win the contract. This happens typically when the owner requires a minimum number of qualified bidders, usually three, but there are only two of them. One of these two bidders may seek a courtesy bid from a colleague. Courtesy bidding is also known as complementary bidding or cover bidding. It is a form of bid rigging, which is illegal.

4. *Bid Unbalancing* is redistributing costs among the work items involved without changing the overall total bid price.

   Some contractors undertake this practice in fixed-price contracts (lump sum and unit price) to increase the profit (by lowering unit price on items with likely less than expected quantity and increasing the unit price on items with likely more than expected quantity) and/or shift cash flow to their advantage.

   It is an unethical and possibly illegal practice. It is also a risky practice that sometimes backfires on the contractor.

Table 3.3 shows two examples of unbalancing bids.

> Doing things professionally and legally is not only ethically better, but it is also better for the contractor's business interests in the long run.

Table 3.3 Bid unbalancing.

Item #	Description of work	Unit	Unit price	Total quantity	Scheduled value	Unbalanced numbers (1)			Unbalanced numbers (2)		
						Unit price	Scheduled value	Price difference	Unit price	Scheduled value	Price difference
1110	Excavation, common earth	CY	$4.00	3,300	$13,200.00	$8.00	$26,400.00	$13,200.00	$4.82	$15,911.37	$2,711.37
1120	Excavation, sand	CY	$2.50	6,200	$15,500.00	$4.87	$30,187.50	$14,687.50	$4.82	$29,894.09	$14,394.09
1130	Excavation, rock	CY	$13.00	1,150	$14,950.00	$5.00	$5,750.00	$(9,200.00)	$4.82	$5,544.87	$(9,405.13)
1140	Excavation. clay	CY	$7.50	2,875	$21,562.50	$1.00	$2,875.00	$(18,687.50)	$4.82	$13,862.18	$(7,700.32)
Totals	—	—	—	—	$65,212.50	—	$65,212.50	$—	—	$65,212.50	—

Types of Construction Contracts:

A. Fixed-price contracts: This group of contracts includes two types:
    a. Lump-sum contract is where a fixed price is quoted for an entire project. It represents the price tag the owner will pay the contractor for building the project. This type of contract requires that the contractor has possession of a full and complete set of plans and specifications, so he/she will be able to obtain an accurate estimate in order to give the lump-sum price. This fixed price is good unless there is a change order, which may change that price. Sometimes, the contract has some provisions that entitle the contractor to price adjustment under certain conditions.
    b. Unit price contract is when the nature of the work is known, and the quantity is also known within a certain range but not exactly. The contractor puts a price tag per unit and will be paid a total of this unit price multiplied by actual, not estimated, quantity. This type of contract is appropriate for work such as excavation, sidewalk, road paving, or resurfacing, and even for subcontracts for painting or floor covering. However, if the actual quantity turned out to be significantly different from the estimated one at the time of contract signing, the two sides may re-negotiate the unit price.

        Both types mentioned above put most or all the risk on the contractor's side, but there are certain provisions the contractor can put in the contract to shift some of that risk back to the owner. For example, if the contractor is concerned about a spike in the price of a major commodity such as lumber, steel, or asphalt, he/she may add a provision that any price escalation in that commodity higher than a certain annual percentage, say 5%, will be absorbed by the owner. In fact, it may be in the owner's interest to have such a provision because, without it, the contractor has to protect himself/herself by hiking the price.

B. Cost-plus-fee contracts: In this type of contract, the owner takes most of the risk because, unlike lump-sum contracts, the final cost to the owner will not be known in advance. Instead, the contractor will charge the owner an amount equal to his/her actual direct cost (labor and materials), in addition to a fee that covers indirect costs and profit. This group includes a few types:

    -   Cost plus a fixed (lump-sum) fee: In this type, the contractor is paid by the owner his/her actual direct cost (labor + materials) plus a lump-sum fee. This is not a common type of contract.
    -   Cost plus a percentage fee: In this type, the contractor is paid by the owner his/her actual direct cost (labor + materials) plus a fee that is a percentage of the direct cost.
    -   Cost plus a sliding-scale fee: In this type of contract, the contractor is paid by the owner the actual direct cost (labor + materials) plus a fee, inversely variable according to the increase or decrease of the actual cost. Thus, it is in the best interest of both parties to keep the total cost low.
    -   Cost plus a fixed fee with a bonus/penalty: This type is somewhat similar to the one before. The contract provides incentives to the contractor to keep the total cost as low as possible. A number is assigned in the contract as the final total cost; if the actual cost gets below (or above) this number, the contractor gets a bonus (or penalty) proportional to the difference.
    -   Cost plus fee with a guaranteed maximum price (GMP): This type of contract stipulates a maximum dollar amount that the project cannot exceed. Once it exceeds this value, the contractor covers the difference, and the owner is locked into that GMP. This is a very common type of contract.

The bottom three types, cost plus a sliding-scale fee, cost plus a fixed fee with a bonus/penalty, and cost plus fee with a GMP, have an incentive for the contractor to prevent the cost from escalating out of control and thus, shift back some of the risk to the contractor.

It is possible, however, that in a cost-plus-fee contract between the GC and the owner, the GC requires lump-sum contracts from some of his/her own subcontractors. Such a decision depends on the risk or uncertainty involved in the specific subcontract scope of work. For example, the project may be a stadium with a unique, never-built-before dome, using cutting-edge technology. The risk in the project is high, which justifies a cost-plus-fee contract between the GC and the owner. The risk is mostly in the dome and the opening/closing mechanism. So, packages such as concrete foundation, seats/bleachers, and plumbing, have well-defined scope with little uncertainty, which allows them to be offered to subcontractors on a fixed price (mostly lump-sum) basis. This helps both the GC and owner restrict the risk to certain components of the project.

> The most important difference among types of contracts is the assignment of risk between the owner and the contractor.

The above mentioned variety of contracts differs in several aspects, mainly in risk allocation. Risk is one of the most important factors in terms of the contract, as well as the contractor's project planning including cost and schedule.

The project risk is subject to several factors such as the type and duration of the project and events (local and worldwide).

Typically, for any owner, the question is: What contract type is most appropriate for me? There is no one answer that fits all situations, but there are guidelines:

> There is no such thing as the "best type of contracts." Every situation is different, and the choice of the type of contract must be based on the merit of the situation.

1. Lump-sum contract is not recommended if the construction contract is to be signed while the detailed design is not finished. How can the contractor give a fixed accurate number for the cost of the project without knowing exactly what some of the items are? If the contractor is pressured into signing a lump-sum contract in this situation, he/she will be compelled to assume the worst, consequently raising the price tag to a level that is unfair to the owner.

2. Also, a lump-sum contract is not recommended if the design is complicated, has not been done before, or contains a new technology that does not make the contractor confident in his/her numbers. The same argument as above: If the contractor is pressured into signing a lump-sum contract in this situation, he/she will be compelled to assume the worst, consequently raising the price tag to a level that is unfair to the owner.

3. If the project is relatively homogeneous (such as a huge task, for example, excavation, road paving, and sidewalk), it may lend itself to a unit-price contract. However, determining the approximate total quantity is important for the calculation of indirect and other costs.

4. If a cost-plus-fee contract is to be used, it is recommended for the owner to have safeguards by invoking a clause restricting the cost escalation or sharing them with the contractor in case the cost goes above a certain level.

5. Fairness and professional practices are pillars in business dealings among parties. Best results are obtained when parties are honest about their knowledge, capability, and willingness to take responsibility. A situation may happen that puts one of the two contracting parties at a severe disadvantage due to unpredictable and uncontrollable circumstances. For example, the price of a major commodity spikes beyond expectations in a fixed-price contract. The owner is encouraged to be fair and understanding toward the other side, and not take advantage of the situation. The last thing an owner wants to do is work with a disgruntled contractor who knows this is a money-losing project.

6. The contractor always adds profit and contingency allowances[9]. These two items are directly related to the risk taken. So, it is logical and expected that the more risk the contractor assumes, the higher the cost will be on the owners' side.

7. There is no such thing as shifting the risk for no extra cost. Shifting the risk always comes at a cost, visible or hidden.

8. If there is a high degree of uncertainty/risk in the project, it is unfair to both the owner and the contractor to sign a lump-sum contract. The contractor will raise the project price tag in proportion to the risk to protect himself/herself. It will be better for both parties to lower the project price tag in exchange for shifting the risk to the owner, perhaps by signing a cost-plus-fee contract (Figure 3.1).

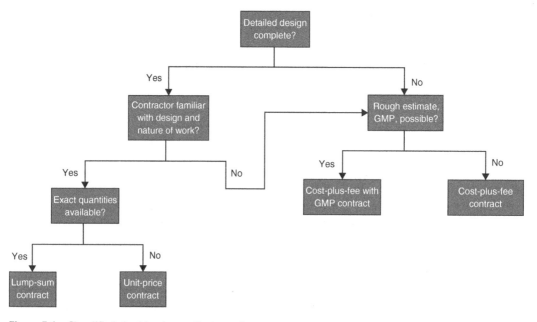

Figure 3.1 Simplified chart to choose the type of contract.

---

9 These two terms will be discussed in detail in the cost estimating chapter.

The guidelines of the Construction Management Association of America[10]:

1. Risk should be assigned to the party who can best control it.
2. Risk should be assigned to the party who can bear the risk at the lowest cost.
3. Risk should be assigned to the Owner when no other party can control the risk or bear the loss.
4. Assumption of risk by the other parties to the construction process results in increases in cost (visible or hidden) to the Owner.

Figure 3.2 illustrates the risk taken by the owner and contractor based on the type of contract. In general, fixed-price contracts assign the highest risk to the GC, while cost-plus-fee contracts assign the highest risk to the owner. There are modifications in each type of contract to shift some of the risk back to the other party.

Contract Templates

When it comes to choosing the contract form, the owner has two choices: use a template well-known in the industry or write his/her own contract. The author strongly recommends the first choice as these templates were written by professionals, vetted by lawyers, and used thousands of times over the years. Besides, every template allows the user to customize it and make any modifications needed. Also, in the case of a dispute resolution, the mediator, arbitrator, or judge is usually familiar with these templates and does not have to spend time understanding and searching the contract.

Some organizations that frequently contract to do projects, including some government agencies, have their own templates that they use for their projects. Also, many contractors and home builders who build and sell homes on a large scale have their own contract forms.

The most common and popular construction contract templates in the United States are produced by the American Institute for Architects (AIA). They issue many templates for different types of agreements and situations[11]. There are also ConsensusDocs forms[12]. In Europe, the Middle East, and North Africa, the most popular templates are issued by FIDIC[13] and NEC[14].

Project Delivery Methods

A delivery method is an approach the owner of a project uses to assign responsibility and authority among the project team members in order to manage the entire lifecycle of the project process from inception to final completion. A delivery method is *not* a type of contract, although there is often a correlation between the delivery method and the type of contract.

A project can be packaged in a variety of ways:

A. Design–Bid–Build (DBB) (also known as Design–Bid–Construct or the traditional method). In this method, the owner first contracts with a designer who completes the design documents (the drawings and specifications). Then, the owner procures a contractor to build the project in another

---

10 Capstone: The History of Construction Management Practice and Procedures, 2003 Edition, Construction Management Association of America, Inc. (CMAA).

11 https://www.aiacontracts.org/.

12 https://www.consensusdocs.org/contracts/.

13 https://fidic.org/sites/default/files/FIDIC_Suite_of_Contracts_0.pdf.

14 https://www.neccontract.com/products/contracts.

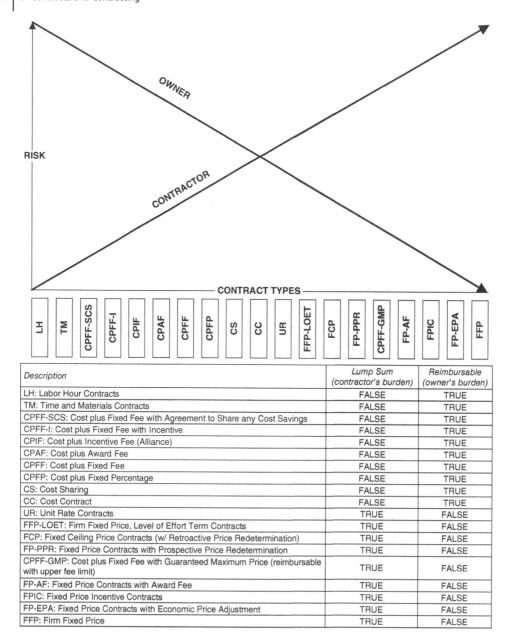

Description	Lump Sum (contractor's burden)	Reimbursable (owner's burden)
LH: Labor Hour Contracts	FALSE	TRUE
TM: Time and Materials Contracts	FALSE	TRUE
CPFF-SCS: Cost plus Fixed Fee with Agreement to Share any Cost Savings	FALSE	TRUE
CPFF-I: Cost plus Fixed Fee with Incentive	FALSE	TRUE
CPIF: Cost plus Incentive Fee (Alliance)	FALSE	TRUE
CPAF: Cost plus Award Fee	FALSE	TRUE
CPFF: Cost plus Fixed Fee	FALSE	TRUE
CPFP: Cost plus Fixed Percentage	FALSE	TRUE
CS: Cost Sharing	FALSE	TRUE
CC: Cost Contract	FALSE	TRUE
UR: Unit Rate Contracts	TRUE	FALSE
FFP-LOET: Firm Fixed Price, Level of Effort Term Contracts	TRUE	FALSE
FCP: Fixed Ceiling Price Contracts (w/ Retroactive Price Redetermination)	TRUE	FALSE
FP-PPR: Fixed Price Contracts with Prospective Price Redetermination	TRUE	FALSE
CPFF-GMP: Cost plus Fixed Fee with Guaranteed Maximum Price (reimbursable with upper fee limit)	TRUE	FALSE
FP-AF: Fixed Price Contracts with Award Fee	TRUE	FALSE
FPIC: Fixed Price Incentive Contracts	TRUE	FALSE
FP-EPA: Fixed Price Contracts with Economic Price Adjustment	TRUE	FALSE
FFP: Firm Fixed Price	TRUE	FALSE

Figure 3.2 Allocation of risk between owners and contractors.
Source: AACE International, Recommended Practice 67R-11, Contract Risk Allocation - As Applied in Engineering, Procurement, and Construction. Revised January 14, 2014.

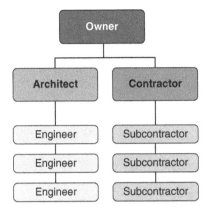

Figure 3.3 The traditional way of procuring construction.

process. The contractor has to examine the design documents before signing the construction contract. Figure 3.3 illustrates the relationships among contracting parties.

Characteristics of the Design–Bid–Build delivery method

1. The owner typically has two contracts (designer and contractor). There may be another contract if the owner decides to hire a third-party construction management consultant.
2. Best understood in the construction industry because it is simple and has a linear sequence of work (longest delivery). The operations proceed in sequence, as shown in Figure 3.4.

Main advantages of the Design–Bid–Build delivery method:

1. The owner retains control of the design and can make any changes before committing to a construction contract.
2. Procurement laws are well defined and known.
3. Owner gets the lowest possible cost through competitive bidding, as the detailed design documents are available to bidders.

Figure 3.4 Sequence of project operations.

Disadvantages of the Design–Bid–Build delivery method:

1. Owner's changes have higher costs due to the involvement of two separate entities (designer and contractor) and the fact that the design has been completed.
2. Most litigious because in most cases, the designer and contractor will blame each other.
3. Contractor does not have input during the design process.

B. Design–Build (also called Design–Construct): In this method, the owner signs one contract only for the design and construction of the project. In most cases, the entity that the owner contracts with is

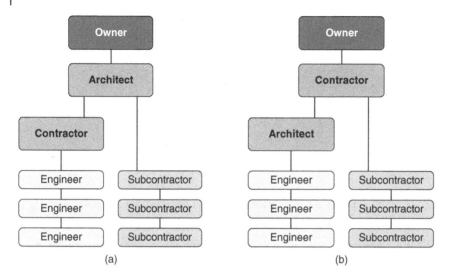

Figure 3.5 The Design–Build contracting method.

either a design or construction firm. In the first case, as shown in Figure 3.5a, the design firm will acquire a construction firm to build the project, while in the second case, as shown in Figure 3.5b, the construction firm will acquire a design firm to design the project. However, in both cases, the owner will have one contract and will not be responsible for the internal communication and issues between the two firms.

Characteristics of the Design–Build delivery method:

1. The owner has a single point of contact which means less headache.
2. Often is the fastest delivery method because construction can start before the design is completed, and there is no bidding/negotiation process for the construction phase.
3. Most cost-effective, as some studies suggest. While the DBB (traditional) method allows for competitive bidding, leading to the lowest cost for construction, the cost-effectiveness of this method comes from a few factors, including the early involvement of the contractor.
4. This method needs a well-defined scope when fast-tracking because the design and construction overlap. So, changes, especially if major, may be costly or even infeasible.
5. It is important to make timely decisions. Otherwise, we will lose the advantage of being the fastest delivery method, in addition to increasing the cost.
6. The owner must effectively administer the DB process.
7. The construction management consultant, representing the owners' interest, must be either part of the owner's organization or a third party.

Advantages of the Design–Build delivery method:

1. Construction input occurs during the design process, including the constructability study.
2. Overlaps and gaps in scope are identified during pre-construction.
3. Cost benefit of procuring the construction directly from the trades.

4. There are no markups on subcontracts or changes.
5. Improved schedule due to early resolution of design and construction issues.
6. Packaging work can allow for construction to start early.

Disadvantages of the Design–Build delivery method:

1. Owner responsible for changes, overlaps, and gaps in scope.
2. Subcontractors may be brought into the project late in the process (no input during the design).
3. Need up-front programs and performance criteria.
4. Owner is pushed for early decisions.

C. Construction Management: In this method, there is no GC, but another party is playing that role. It has two main models:

  – Agency CM (also called Pure-Agency CM, CM-for-Fee, or CM-not-at-Risk): The owner contracts directly with subcontractors. The role of the CM will be management and supervision but in advisory, not authority, role. This is illustrated in Figure 3.6.
  – CM-at-Risk: Combines Agency CM and GC role: The CM here contracts with subcontractors and will be managing, supervising, and taking responsibility for the work. This is illustrated in Figure 3.7.

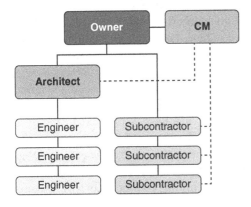

Figure 3.6 The pure agency CM contracting method.

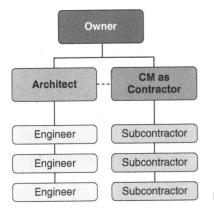

Figure 3.7 The CM-at-Risk contracting method.

Difference between Design–Build and Turnkey

In both types of projects, one entity (the contractor) takes responsibility for both design and construction of the project. In DB projects, the owner is fully involved in the design and construction from the start till the completion of the project. In turnkey projects, the contractor has full control of the project's design and construction. The owner's role is almost like buying a finished product. This puts the responsibility of financing the project on the contractor. In most arrangements, the owner pays a small security deposit when signing the contract, and then pays the entire balance at the time of closing (when taking possession of the project). This type of contract is popular with developers.

General Contractor's Acquisition: Contractual Obligations Between the General Contractor and Subcontractors

While owners have the contractor's acquisition options mentioned earlier, GCs have their own acquisition process and guidelines for selecting subcontractors and vendors. Even in competitive bidding cases, the GC usually is not obligated to go through a bidding process to select the subs. However, the owner may still have criteria and restrictions on the subcontractors and the vendors the GC can hire or buy from. Furthermore, many owners require the bidding GCs to have their subcontractors get approved by the owner, so the names of these subs must be included in the original bid. Owners may also require that a minimum percentage of the work must go to specific minority subcontractors, or that the GC must perform a minimum percentage of the work, for example, 50%, by own forces.GCs usually have a database for subcontractors by specialty, geographic location, and other criteria. Vendors are also organized in a database by type of business, location, and other criteria. These databases must contain contact information. All information must be updated continually so the GC may be able to select subcontractors and/or vendors for any project, based on certain set criteria. In the general case, subcontractors sign contracts with the GC, but, in addition to the terms of these contracts, they are still subject to the terms of the contract between the owner and the GC. The GC will be responsible for managing their own subcontractors and for their performance and work quality before the owner. Responsibility and liability of the GC extends also to include subcontractors' payments to their employees and vendors, as well as the safety of the entire job site. This probably answers, at least partially, the question some owners ask, "Why don't I just hire subs directly and save the GC's markup?" Owners, especially when lacking a dedicated professional(s) in their organization, get these benefits from signing only one contract with a GC as compared to signing contracts with multiple subcontractors:

1. The GC will be responsible for coordination among the different subs and trades.
2. The GC will assume the responsibility and liability for all the work, worksite, and workmanship.
3. Having peace of mind in case of a quality or warranty issue: The GC is the only party responsible before the owner. Otherwise, there will likely be "finger pointing" among the subs to avoid liability.

The argument above applies to the CM-at-Risk delivery method, as the CM plays the role of the GC, as there is only one party that takes all the responsibility and liability before the owner.

> Subcontractors: You better read the contract between the owner and the GC. The terms will likely apply to you!

Contract Ethics

The National Contract Management Association (NMCA) has a code of ethics. The code is intended to create public trust and confidence in the integrity of the contract management process. The guiding principles include the following[15]:

1. Accountability: Members accept responsibility for their own conduct and hold each other accountable.
2. Compliance with Laws: Members follow all laws and regulations.
3. Confidentiality: Members protect confidential information concerning the business affairs of any present or former employer, governmental agency, business partner, organization, or public body on which they serve.
4. Good Faith: Members conduct all business in good faith, make any required disclosures, and avoid actual or perceived conflicts of interest (whether by reason of financial interest, family relationship, or any other circumstances). Members strive to advance the profession and the Association without compensation.
5. Integrity: Members fulfill their duties without deception or misleading practices. Members actively support and encourage others in adhering to this Code.
6. Professionalism: Members educate themselves in all aspects of the contract management profession and apply this knowledge to the best of their ability to serve their employers, customers, clients, business partners, and the public. Members stay up to date on developments in the contract management field to maintain their knowledge, skills, and level of professional competence. Members make only truthful claims concerning their professional qualifications, education, certification status, or experience.
7. Respect: Members acknowledge the importance of diversity, equity, and inclusion across the Association and contract management profession. Members are always respectful of others and their opinions. Members do not make statements that could reflect poorly on another contract management professional, the reputation of the Association, or the contract management profession.
8. Trust: Members conduct themselves in a manner to establish and maintain trust and confidence in the integrity of the contract management profession and processes.
9. Violations: Members must report any suspected violations to an official of NCMA at either the chapter or national level. Violations of the Code are subject to disciplinary action at the discretion of NCMA up to, and including, revocation of membership and certification. Members shall not retaliate against anyone who raises a valid concern under this Code.

Conflict of Interest

A conflict of interest occurs when an entity or individual becomes unreliable because of a clash between personal (or self-serving) interests and professional duties or responsibilities. Such a conflict occurs when a company or person has a vested interest—such as money, status, knowledge, relationships, or reputation—which puts into question whether their actions, judgment, or decision-making can be

---

15 https://ncmahq.org/Web/Web/About/Leadership/Code-of-Ethics.aspx.

unbiased[16]. Conflict of interest is unethical and, in many cases, illegal. Some examples of a conflict of interest in the construction contracts are:

a. The contractor, in a cost-plus-fee contract, buys materials from a company that he/she has a vested interest, without shopping competitively.
b. The owner hires a construction management consultant to oversee the contractor's work, but this consultant is affiliated with the contractor.
c. An owner seeks mediation in a dispute case with the contractor, but the mediator has current business dealings with the contractor.
d. The contractor's project manager is planning to start own business, so he is trying to cut deals with his employer's clients.

We may have a situation that poses conflict of interest in its entirety such as in the examples mentioned above, or it can be a temporary situation. When such a situation arises, the party with the conflict of interest is usually asked to remove themselves, as it is often legally required of them. An example of that is if the board of trustees of an organization is investigating allegations against one of its members. This member must recuse himself/herself from this meeting. Also, if a bidder on a project for an organization is related to a member of the selection committee. This member must recuse himself/herself when opening the bids and selecting the winning bidder.

In general, clarity of the rules and the transparency in applying them are required for honest and legal contracting.

Exercises

3.1 What is a brief definition of a contract?

3.2 What are the requirements for any contract to be legally enforceable?

3.3 What are the different ways for an owner to procure the services of a contractor? Briefly describe each and compare them.

3.4 Can a public agency hire a contractor using the direct award method? Explain.

3.5 In addition to the lowest price, the owner may have other selection criteria, such as what?

3.6 What is the difference between bidders' pre-qualification and post-qualification? Under what circumstances do you expect each method?

3.7 What is an addendum? Explain and mention under what circumstances it is issued.

---

16 https://www.investopedia.com/terms/c/conflict-of-interest.asp.

3.8 What are the advantages and disadvantages of the design–build delivery method?

3.9 Can a bidder modify or withdraw their bids after submitting their bid? Be specific on when they can and when they cannot.

3.10 What is bidding with options?

3.11 What methods may an owner use when time matters?

3.12 Mention the factors that influence the contractor's decision to bid on a project.

3.13 What is bid shopping? Why is it unethical?

3.14 What is courtesy bidding?

3.15 What is bid unbalancing? Why do some contractors do it? What are the risks and potential problems with this practice?

3.16 What are the types of fixed-price contracts? Under what circumstances would you recommend each?

3.17 Which party bears most of the risk in lump-sum contracts? Why? Are there ways to shift some risk to the other party?

3.18 Which party bears most of the risk in cost-plus-fee contracts? Why? Are there ways to shift some risk to the other party?

3.19 Give an example of a project where you recommend:
a. Lump-sum contract
b. Unit price contract
c. Cost-plus-fee contract

3.20 Under what circumstances would you recommend a Pure-Agency CM delivery method to an owner?

3.21 Give two examples of "conflict of interest" in construction contracts.

3.22 Your friend is thinking of building an office building. He wants to save money by hiring subcontractors directly, rather than hiring a GC. What would you advise him? Your answer must make assumptions about the owner; mention these assumptions.

3.23 What is the difference between Design–Build and Turnkey delivery methods?

3.24 Private owners can award bids without competitive bidding. Government agencies, in most cases, cannot. Why? Are there exceptions? Explain.

3.25 Which type of contract do you recommend in each of these situations: LS (lump-sum), UP (unit price), or CF (cost-plus-fee)
   a. Single-family home with a design familiar to the contractor.
   b. Single-family home with a new design that is not familiar to the contractor.
   c. Several sidewalks with the total quantity approximately known.
   d. A hospital with incomplete detailed design.
   e. Residential complex, like another one built earlier.
   f. Oil refinery using a new technology.
   g. Excavation of several ponds. Quantities and types of soil roughly but not exactly known.
   h. A villa with a contractor who is experienced in this type of building.
   i. A tower building with the design incomplete.
   j. A stadium that has new innovative features that were never tried before.

4

The Planning Phase

Introduction

Carpenter's rule has been "Measure twice and cut once," which can be expanded to everyone in a layman language: take your time for adequate planning before you execute.

Planning, in general terms, has been defined as the comprehensive process of *thinking of* and *preparing for* the activities and actions[1] needed to successfully complete a project[2]. We can customize this definition according to the role of the party doing the planning. Keep in mind that planning, in this context, is distinct from scheduling. The relationship between the two areas is discussed in Chapter 7, Time Management.

Project planning should be practiced by all project participants, most importantly the project owner and contractor. However, it may also apply to other project parties: focusing on their own work and aligning it with the project's activities, scope, and constraints.

> Planning for any project or piece of work must have two components: "thinking of" and "preparing for" the project or work we like to do.

The dilemma of planning is that, in most cases, it requires preparation that needs effort and money for events that may or may not happen or may happen differently. On the other hand, if you do not do such planning, there may be costly consequences or lost opportunities. In many cases, the decision boils down to benefit-to-cost ratio. For example, if you learned that a hurricane is developing in the ocean and has a likelihood to hit your area in 7 to 8 days, what would you do? Preparation requires:

1. Buying plywood boards and other wood members to support windows and doors
2. Buying flashlights, candles, and batteries
3. Storing enough water for all purposes
4. Obtaining sandbags

---

1 The difference between activities and actions is simple. Activities take time such as inspection of the site or negotiating the contract. Actions are events, such as making a decision, that do not consume time, such as signing the contract. This will be explained in Chapter 7, Time Management.
2 Construction Project Scheduling and Control by Saleh Mubarak, 4th edition, Wiley, page 4.

Project Management in the Construction Industry: From Concept to Completion, First Edition. Saleh Mubarak.
© 2024 John Wiley & Sons, Inc. Published 2024 by John Wiley & Sons, Inc.
Companion Website: www.wiley.com/go/nextgencpm

5. Perhaps buying items such as a compact gas stove, electric generator (with fuel), and/or solar power generating system
6. Any other supplies such as nonperishable food items (that do not need refrigeration), board games, and ham radio
7. An evacuation plan if the hurricane is likely to be a major one

The question now is: Which among the above items, and how much/many are justified and recommended, and at what stage? If you delay your decision by a few days, the prediction of the hurricane's path and strength will be more accurate; however, you will have less time to prepare. Besides, many of these items will no longer be available. You need to estimate the likelihood of each event and the consequences and act accordingly.

> Measure twice and cut once!

The Project Management Plan[3]

The Project Management Plan is the document that describes how the project will be executed, monitored, controlled, and closed[4]. This plan represents the roadmap that shows how the owner intends to execute the project. It documents the approved scope, schedule and cost baselines, along with project planning assumptions, constraints, and alternatives.

The project management plan is the point of reference for measuring progress. It guides project execution and control. It has to be specific but dynamic. This requires the plan to have a clear and solid base but also leaves a flexibility for potential adjustments as needed.

The plan can take different shapes and have different contents depending on:

- The purpose of the plan
- The information available
- The timing of the plan
- The level of detail needed

So, it is normal to start with an abstract plan that could be one page or even less, outlining the scope of the project in a loosely-defined way, along with major constraints. As time passes and the "project lens" zooms closer on the project, more information will be available that will allow the plan to get clearer and more detailed. Planning starts when the project is just a concept and continues throughout the project stages until the final completion.

Planning may cost time and money, but it is an investment with a potentially high return. Many studies have shown that good planning saves time and money, and reduces headaches.

---

3 Must not be abbreviated as PMP, as this acronym is reserved for the PMI's PMP, project management professional, certification.
4 PMI-PMBOK, 7th edition, 2021.

> Those who rush to execution without adequate planning, allegedly because "we do not have time to plan," are likely to spend more time and money during the execution, not to mention the headache and frustration!

Terminology Internationally

An important note here, especially for our international readers outside the United States of America: the term planning in some other countries or regions may have different connotation. In the nomenclature of many countries around the world, planning may include the time aspect only, so it is synonymous with or even an alternative to the term scheduling. Professionals whose job is to create and manage schedules are given the title "planners" in these countries. In the UK, scheduling is called planning, and the schedule is called the "programme." The author is not taking sides here in terms of right versus wrong, but to put all readers of this book on the same page, we adopt the US terminology for the terms planning and scheduling. Planning is discussed in this chapter while scheduling is discussed in Chapter 8.

Planning for Different Project Parties

Planning must be done by each project party, but the plan timing and content differ depending not only on the role of the party but also on what each party's project is. For the owner, the project starts from the moment it is approved, while for the contractor, it starts from the moment he/she signs the contract. The arrow showing the contractor's project in Figure 4.1 is dashed on the left side (during the bidding/negotiation process) as the contractor can do some planning while not sure if he/she will get the project for certain. The end of the project is the same for both the contractor and the owner. After this, the owner will still be involved during the life service of the project; however, it will be another phase that is separate from the construction project.

Planning for a Project Owner

To an owner, planning includes but is not limited to defining project scope and constraints, performing feasibility studies (financial, legal, and other), and comparing alternative designs and execution methods. Most importantly, project planning helps the owner:

1. Make the decision whether to carry out the project, and if yes,
2. Execute the project in the best and most efficient manner.

The owner's plan can include elements that have to do with scope, design, alternate designs, cost, time, finance, land, procurement, operations, and so on.

Ironically, the owner who says, "I do not need planning" or "I do not have time for planning," is likely to really need planning. Take, as an example, a person who wants to take his family for a picnic in a

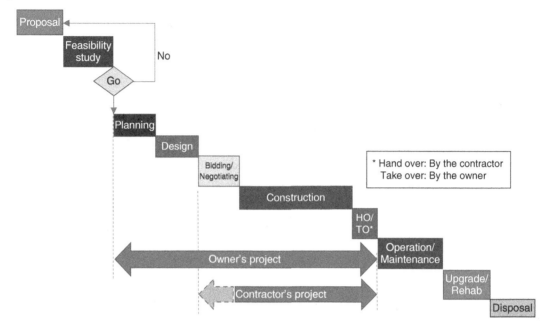

Figure 4.1 The owner's project versus the contractor's project.

place two-hours driving away. He may rush them without planning because "we do not have time!" As a result, he may discover – on their way or even after arriving at the destination – that they forgot to bring important things they needed for the picnic. This leads to extra cost and/or delay, in addition to frustration and dissatisfaction. Many owners do so because they are so excited and anxious to see their project finished; however, this feeling must be restrained by logical thinking and following essential steps.

One of the most important tasks for an owner, as part of the planning process, is to define the project's scope and major constraints: cost, timeframe, and quality/specifications. This is usually achieved during the design phase with the help of design professionals. One common dilemma in construction is the competition among project objectives and constraints. For example:

1. The owner likes to have an "above-average" finish, but the budget is limited and does not allow for such a wish.
2. The owner likes to accelerate the schedule so that the project can finish before the high season, but this acceleration will increase the cost.

What is important to the owner is to rank priorities because there will be situations where scope, specifications, time, cost, and probably other constraints will clash, and the owner must choose one over the other.

Figure 4.2 illustrates the relationship between project scope and constraints: cost, schedule (time), and quality (specifications). Any change in the scope or one of the constraints will likely impact the others. The project, typically, has additional constraints such as safety, environmental impact, public relations, customer satisfaction, employee satisfaction, and ultimately becoming profitable and supportive of the organization.

Figure 4.2 The "Golden Triangle": project scope and major constraints.

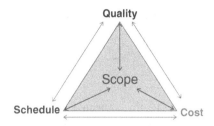

When designing a project, the design professional's main task is to represent the owner's main idea while reconciling the owner's "wish list," "need list," and "can list." This involves financial and legal challenges in the conversion of the original idea into a viable and feasible design. Many owners have ambitious or even unrealistic ideas that must go through the "reality filter" by the design professional. This may take the design through several iterations until it settles on the form that is both feasible and satisfactory to the owner. If later the owner cannot find a contractor to build the project within the constraints set in the design[5], the design professional may be held responsible.

Scope management before and during the construction phase is an extremely important topic that will be discussed in detail in the next chapter.

> Even though the project's parties are working on the same "project," each party plans differently based on its role and scope of work. However, plans of all parties have to mesh and integrate together to get the project completed efficiently.

Planning for a Contractor

For a construction contractor, planning helps prepare for and manage the project better and more efficiently. Planning is embodied in preparing the project management plan that will be the foundation for the main components of the project, such as cost, schedule, project control, quality control, risk management, and safety management. The plan outlines the resources needed to execute the project, the risks that may impact the outcomes of the project, and more. Its objective is to explore and define the most efficient way to execute the project.

It is important that planning starts as early as possible before the start of the project and continues till the end of the project. It predicts outcomes, events, and incidents, and creates mechanisms to manage them. This plan gets updated as the project progresses, with new items deleted, added, or adjusted.

How Far Does the Contractor's Planning Go?

Preparing a bid requires effort and money, but the contractor may not be the bid winner. So, the contractor typically estimates the likelihood of winning and the prize of winning (how much profit or other benefits are expected from that project) and then decides if it is worth the effort to bid.

---

5 Within certain reasonable margin. Also, unusual market changes may relieve the design professional from this liability.

In such situations, we may use the Expected Value theory combined with the benefit-to-cost ratio:

$$E(X) = \sum_i x_i \cdot p(x_i) \qquad\qquad (4.1)$$

where:

$E(X)$ is the "expected value" of an event, X,
i is the domain of all possible outcomes,
$p(x_i)$ is the probability of occurrence of outcome i
x_i is the value of consequence of outcome i

The contractor shall make an educated guess on the upcoming event, say a project may go for bidding. The contractor estimates the likelihood of the event happening (winning the project) based on certain observations and assumptions, and the consequences of all possible actions (not bidding, bidding but not winning, and bidding and winning), and then makes the decision that is expected to bring about the highest benefit-to-cost ratio. However, the decision is not purely scientific and objective. There are several factors involved such as work currently on hand, the market situation, other potential bidders, previous experience with this owner, how badly he/she needs this project, own company's financial stability and strength, organization's strategic plan, own attitude (conservative/liberal), own "gut feelings," and others. A good contractor may come out unsuccessful in one bidding experience but will not come empty-handed in several attempts. The likelihood of winning varies depending on several factors, but a savvy contractor should be able to survive and prosper even in a fluctuating market.

In such planning, information flows through the timeline but the available time and options get less, and their costs may increase. Figure 4.3 shows a project timeline for the contractor when contemplating bidding on a project. We can divide this timeline into four periods:

- Period 1 is when the contractor gets a tip – or expects – that a project may be available for bidding. Here are the uncertainties for the contractor:
  o Will the project go up for bidding?
  o If yes, will the contractor bid on it? and

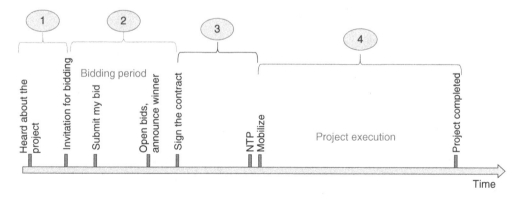

Figure 4.3 Project planning timeline for the contractor.

   o What are the chances for the contractor to get the project?
   o What is the potential profit of the project?

 – Period 2: Now the project is officially announced. The contractor needs to decide whether to bid on the project. The uncertainties for the contractor are:
   o If I bid on the project, what are the chances of getting the project? and
   o If I get the project, what is the expected net profit?
     Note that the two questions above are correlated: the contractor can lower his/her bid amount to improve the odds of winning, but this will reduce the expected net profit, and vice versa.

The decision to bid or not to bid depends on several factors mentioned earlier.

The contractor will assess the risk taken on the project, which will likely affect the bid markups, particularly profit and contingency allowances.

 – Period 3: Assuming the contractor submitted a bid, now it is a waiting period to find out if he/she will be the winning bidder.
 – Period 4: Now that our contractor has won the bid, planning will focus on the execution of the work.

> Time is a valuable resource that can work for or against you. With good planning, it should work for you.

For every period above, there are tasks the contractor can do to help plan better for the project such as:

1. Visit the site, not only to inspect it, but also to examine the surroundings.
2. Learn more about the project design from the available documents and other information.
3. Learn more about the project owner and the designer. Did the contractor have any previous experience with either?
4. Learn more about the competition (other bidders). This may be a factor to encourage or discourage the contractor from bidding.
5. Do an approximate cost estimate for the feasibility study. If the contractor decides to bid on the project, a detailed cost estimate and schedule will be needed later.
6. A second visit to the site may be helpful to identify some of the work items related to the site such as excavation, demolition, and removal. The visit also may help identify any legal challenges such as vegetation, historical monuments, wetlands, protected species, and hazardous or contaminated materials. Some of these items may need a permit that may take a long time.
7. Assess the risk taken in the project, which helps the contractor not only create a risk management plan, but also estimate the bid markup such as profit, contingency, and escalation allowances.
8. Estimate the project duration, perhaps utilizing a CPM[6] schedule. This helps the contractor plan for the start and finish of the project, as well as estimate the overhead expenses.
9. Do resource (labor, materials, and equipment) planning to make sure all required resources will be available. This must be accompanied by the cost of resources and alternative sources.

---

6 The critical path method (CPM) will be explained in Chapter 7, Time Management.

10. After signing the contract, the contractor may conduct another more investigative visit to the site to create a site and logistics plan, outlining locations of the project footprint, office trailers, storage, temporary utilities, parking, roads, and more. In many cases, a simple item, such as deciding the entry and exit points, can have an impact on the flow of work. This must also include a plan for emergencies. Any need for fencing, security, or night-lights?

11. Now that the contract is signed, and the project is about to start, the contractor shall revisit the resource acquisition plan in more detail. This includes labor, materials, equipment, staff, and other items. Special attention will be given to unique, custom, and long-lead items and major items with large quantities. A list of vendors and suppliers is a must.

12. The schedule may be resource-loaded and leveled for better efficiency.

13. The contractor shall form his/her own team. The team can be composed of dedicated (employed full-time), borrowed, and temporary staff. The team will also grow and shrink as work requires.

14. The general contractor must also have a list of all subcontractors to be involved in the project.

15. Risk management planning is extremely important for the contractor.

16. The contractor must generate a cash flow diagram, anticipating all expenses and incoming payments.

17. A documentation system has to be established already.

18. Any other task that helps the contractor better execute the project.

The concept we just illustrated applies also to the case when the owner is negotiating with the contractor without the bidding process. It starts with the contractor's uncertainty about whether he/she will get the project. It may take several rounds of negotiations and information exchange till the dust settles and the contractor knows the situation for sure. Again, the contractor will be spending time and effort with nothing secured, so he/she will use the previous guidelines in benefit-to-cost ratio. Keep in mind that the contractor's benefits may be beyond making a profit from this project. They may include establishing a relationship with this owner, entering a new business territory, or others.

So, project planning starts when nothing is certain and intensifies as project information increases and uncertainty decreases. The author calls the period between signing the contract and receiving the notice to proceed (NTP) issued the golden period because during this period, the contractor has secured the project, yet the time clock does not start ticking. The contract time does not officially start counting against the contractor till the NTP is issued. In many cases, this may not happen till days or even weeks after signing the contract.

> Planning costs time and money, sometimes for events that are uncertain, but the cost of a lack of planning, in case these events occur, could be much higher! Here is where the contractor needs to plan efficiently.

What does the Contractor's Project Management Plan include?

Typically, the contractor's project management plan includes the major elements of the project, with a focus on all the necessary information to successfully get the project off the ground and running till completion. It defines the following:

- What: A brief but clear definition of the project's scope
- Where: The location of the project

- Whose: The owner, design consultant, and other project participants, and their roles
- Who: The contractor's team that will be responsible for executing the project, including subcontractors
- How much: The project's budget
- When / for how long: The project's timeline
- Other: Any issue such as quality and risk management that may need to be planned

The plan may take several iterations as more information becomes available.

If you do not have a plan, you likely do not know:

1. Where you are going, and
2. How to measure progress.

If you do not know where you are going, then all roads go there!

The Relationship Between the Owner, the Designer, and the Contractor

The relationship between the owner and other project partners, mainly the designer and the contractor, takes the form of any relationship between a client and a professional: the client hires the professional for services in exchange for a set fee[7]. This relationship in construction is intricate and extends over a long period. The main challenge for both the designer and the contractor is to fulfill the owner's demands within the constraints of the law and their own profitability. This includes demands that may not be feasible, legal, ethical, or efficient. In addition, the owner's personality may be difficult to deal with or may micromanage and interfere with their professional work to the point of annoyance[8]. The dilemma for the designer and the contractor is that the owner is "the boss" who pays the bills. It will be great if they can satisfy the owner in every aspect, but the bottom line is to satisfy the written agreement with the owner.

The owner's demands have to go through the "professional filter," which is the job of the designer and contractor. They are supposed to respond to these demands with a professional answer that is in the best interest of the owner. In many situations, the answer to an owner's request is, "No, we cannot do this, but instead, we have a couple of other options." The designer or the contractor must provide responses to difficult or infeasible requests by offering practical and viable alternatives. They must provide the owner with the pros and cons of every alternative, along with the cost, and probably their recommendation. At this point, the owner may be in a better position to make an informed decision. This practice is not restricted to infeasible requests, but also includes requests that are feasible but not optimum or suitable for whatever reason. The design and construction professionals are legally, professionally, and ethically obligated to provide the owner with the expert opinion and best advice to fulfill his/her requirements. Such advice becomes more important in the planning stage because many decisions are hard and costly to

---

7 The fee can be a lump sum, a percentage, or a combination, with or without other stipulations.
8 This point is mentioned in detail in Chapter 16, Soft Skills for Project Management.

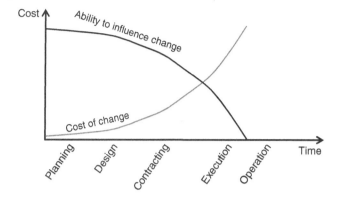

Figure 4.4 Failure to plan is planning to fail!

change later. Figure 4.4 shows the relationship between "ability to influence change" and "cost of change" on one side and time on the other side.

> Planning implies that work should be proactive, not reactive. People who do not plan, spend their time putting out fires!
> Good planners do not do so, but even if/when they have to put out a fire, the fire may be an opportunity for them!

Preconstruction Services

Preconstruction services can be part of the planning process for both the owner and the contractor. These services are offered by a construction company, usually the one that is likely to take on the project but before the construction work begins, and possibly during the design or specifically in its early stages. Preconstruction services have their own contract between the owner and the contractor, independently from the construction contract. More importantly, the owner has no commitment to sign a contract with that company later for the construction work.

This type of service explores the project, from a construction perspective, giving an assessment of its requirements and exploring ways to improve and optimize the construction process and work. The preconstruction services contract may also allow and enable the contractor to:

1. Explore more about options for the project's requirements, time, money, and others
2. Identify potential issues with the design
3. Make suggestions to improve the design from a constructability and value engineering (VE) point of view. This can be for the sake of saving time and/or money, improving the quality, functionality, or other aspects of the project.

By the end of the preconstruction services phase, the owner can make a better decision regarding the construction phase: retain the same construction company for the construction work, hire someone else, or even postpone or cancel the project. This service has the advantage of minimizing the owner's risk through the process of exploration. Also, there is a huge advantage to the contractor in getting involved in the project early, so he/she can be better prepared when construction starts, assuming the owner will retain the same company. Nowadays, building information modeling (BIM) is being utilized sometimes as part of preconstruction studies. It allows the owner and contractor to simulate the construction process, which allows them to detect clashes and inefficiencies, and improve the process in general. This is a benefit to both the owner and the contractor.

It is possible that the owners, who have their own technical teams in their organization and do frequent projects, to do the preconstruction study, or part of it, internally. Many large organizations that do projects frequently have their own design and construction teams that do preliminary studies before assigning the entire project, design and construction, to external parties.

Constructability and Value Engineering

Constructability (also called buildability) is "the optimum use of construction knowledge and experience in planning, design, procurement, and field operations to achieve overall project objectives" (Construction Industry Institute, 1986a, p. 2). It is the practice of reviewing the design (drawings and specifications) prior to construction[9], mostly in the preconstruction phase, with the intent of correcting errors, conflicts, delays, omissions, inconsistencies, or discrepancies.

Value engineering (VE) is a science that studies the relative value of various materials and construction techniques. VE considers the initial cost of construction, coupled with the estimated cost of maintenance, energy use, life expectancy, and replacement cost.

Constructability and VE are two different missions. They are not mutually exclusive, but they may have a little overlap between them. There are some subtle differences between them. Constructability must be involved in the early stages of the design. It is mostly a feasibility study that directly impacts the construction work. It looks for any potential problems when constructing the project. VE is an optimization study for the entire lifecycle of the project. It usually comes after substantial design decisions have been made. VE looks at the entire lifecycle of the project, not just the upfront cost, for the improvement of the project in every aspect, such as cost (upfront and operation), functionality, safety, and quality. The general contractor is usually who conducts the constructability study, obviously because it is the party that will perform the construction. VE, on the other hand, is usually done by brainstorming sessions with several parties involved such as the designer, the owner, the contractor, and likely a consultant. It may be performed under the management and leadership of a specialized VE consultant.

As we mentioned in an earlier chapter, the architect balances aesthetics with functionality, while the engineer (mainly the structural) assures safety, and this all together determines cost. In any constructability or VE study, the participants may joggle aesthetics, functionality, and cost, but safety must be a cornerstone that cannot be compromised.

> Investing a little time and money in constructability and VE studies can bring a high return in saving money and time, and improving other aspects.

---

9 It can be conducted prior to issuing for the bid, prior to signing the contract, or after signing the contract, depending on the timing of the study and the party doing it.

The Contractor's Organization's Strategic Planning

The contractor has to balance all projects under his/her organization, including finance and resources. But most importantly, aligning these projects, current and future, with the organization's strategic goals.

Although this is not a book on strategic planning, we will briefly mention the connection between planning at the project level and the contractor's organization's strategic planning, as shown in Figure 4.5.

Perhaps the most important aspect of the contractor's organization is cash flow, which has to be balanced among projects to make sure the organization keeps operating and in good health. For example, a strong, well-established contractor can take on a risky project, which comes with it high risk and high potential profit. The contractor makes such a decision based on the worst-case scenario that even if this project fails, other projects the contractor is carrying out will be sufficient for maintaining the running of the organization. This argument does not apply to another contractor who is new in the market or is financially weak.

Another point regarding strategic planning is expanding to:

1. New types of projects
2. New region/country/territory
3. Larger projects
4. New sector/type of clients
5. New role or a new type of contract/service

Expanding in any new direction, such as the five scenarios mentioned above, needs careful study to make sure that such a move will serve the organization's goals. Reaching the goals may – and usually does – take several steps till the results can be materialized. For example, the contractor may not achieve the usual amount of profit in the first project, but the focus of this project was to establish a "foothold" in this territory or with this client.

Strategic planning requires that different departments in the organization be totally open, sharing information and expectations with other departments and the leadership. The leadership must plan and act with emotional intelligence. Passion is great only if it keeps logic in control of emotions.

Strategic planning also must include a vision for how to view the organization 10, 20, or more years from now. This vision needs a plan that must address all aspects such as risks, cash flow, resources, logistics, and politics. Alternative plans must exist because the future may bring about unexpected changes and because sometimes things do not go exactly as planned.

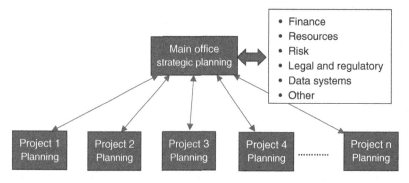

Figure 4.5 Strategic planning at the corporate level.

> A vision without a plan is just a dream. A plan without a vision is just drudgery. But a vision with a plan can change the world.
>
> — Joel Barker

Exercises

4.1 Define planning and give an example.

4.2 What is the main dilemma in a contractor's planning?

4.3 Define the Project Management Plan. What are the basic elements contained in the plan?

4.4 For a typical construction project, each of the involved parties, owner, designer, and contractor, needs to do their own planning. Explain this point, focusing on the differences among their planning.

4.5 What are the main benefits of the contractor's planning?

4.6 Explain the principle of benefit-to-cost ratio in contractor's planning.

4.7 Planning for an owner helps with two objectives, explain.

4.8 The contractor's planning depends on the stage he/she is in. Considering the four stages shown below, mention actions the contractor can do as part of his/her planning effort in each stage.

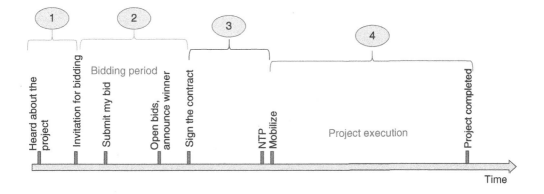

4.9 Why do we call the period between signing the contract and having the NTP issued the "golden period"?

4.10 What do preconstruction services include? What is the main purpose of such services?

4.11 Define constructability. Who usually performs it? Why?

4.12 Define value engineering. What benefits does it attain?

4.13 Why is performing constructability and value engineering important during the planning process?

4.14 How do project planning and organizational strategic planning work together?

4.15 How would you respond to those owners who say, "I don't have time for planning?"

5

Scope Management

Introduction

The Project Management Institute (PMI), defines the scope as "the sum of the products, services, and results to be provided as a project." For construction projects, scope simply describes the final product as defined in the contract documents. Scope management in construction goes through two phases: definition and maintenance. In the first phase, the project owner defines the project scope or outcome with the help of the design professional. The role of the designer is to carry out the owner's idea in a feasible design that complies with the legal and professional standards and fulfills the owner's requirements. This usually takes the owner's original idea into some "tweaking" and refining until it finally gets represented in the completed design. This design goes to the contractor to build it, which may bring about some changes requested by the owner or suggested by the contractor for a variety of reasons such as Value Engineering (VE)[1]. These changes may, and usually do, continue throughout the construction process[2] till the project gets completed. Changes during the transitions may be minor or major, and likely to impact project constraints such as cost and schedule. They are illustrated in Figure 5.1, with the emphasis that most or all of the scope definition happens during the design phase. However, changes may occur during the contractor's procurement and construction phases.

Scope Management

One of the major tasks for an owner is to define the project scope. Scope management is defining the project's final product within the set requirements and constraints, and then managing this scope during the execution phase until final completion. The difficulty lies in the conflict between the owner's requirements with legal and professional limits, as well as the conflict among the owner's own constraints.

---

1 A science that studies the relative value of various materials and construction techniques. Value engineering considers the initial cost of construction, coupled with the estimated cost of maintenance, energy use, life expectancy, and replacement cost.
2 We call them change orders, COs.

Project Management in the Construction Industry: From Concept to Completion, First Edition. Saleh Mubarak.
© 2024 John Wiley & Sons, Inc. Published 2024 by John Wiley & Sons, Inc.
Companion Website: www.wiley.com/go/nextgencpm

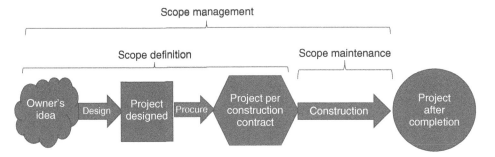

Figure 5.1 The project scope transitions from raw idea to completed product.

The owner starts with a product idea, as shown in Figure 5.1, but may also have financial, time, and other constraints. Figure 5.2, the "Golden Triangle," represents the traditional exchange among project scope and major constraints:

Figure 5.2 The "Golden Triangle": project scope and major constraints.

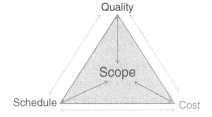

A. The heart of the triangle is the scope, which defines the finished product as demanded by the owner in the written contract. This scope is represented by a set of drawings and technical specifications (specs) which define quality. It has a cost estimate (definitive or approximate) and a timeline (schedule). Any change in the scope is likely to have an impact on these three elements: quality (specs), cost, and/or schedule.

B. Even if the scope does not change, but one of the three elements (specs, cost, and schedule) changes, the other two elements are likely to be impacted. This part will be discussed later, particularly under the "cost-time trade-off."

Scope Definition

The role of the design professional must be in helping the owner define and refine the scope, by reconciling the three lists (Figure 5.3): the wish list, the need list, and the can list.

The designer should go through these steps:

1. Obtain and grasp the owner's row idea of the needed project.
2. Understand the objective of the project, going beyond the owner's definition of the project.
3. Learn, directly and indirectly, the owner's time, money, and other constraints.
4. Work with the owner on prioritizing the requirements ("absolutely need," "better to have," and "nice to have"), along with alternatives. This may take intensive conversation with the owner. The outcome must be documented in a list, in itemized order based on priority and cost.

Figure 5.3 The client's "Three Lists" that must be reconciled.

What you like to have

What you need

What you can afford

5. If the designer can fulfill the owner's scope within the set constraints, it will be great, and the mission will be easy. However, in many cases, this may not be possible, and the scope and constraints compete, as outlined earlier. The mission then becomes "how to best accommodate the owner's requirements within the set constraints."

6. In many cases, the owner may not be sure or firm about the constraints or some of them such as the budget and timeline. In this case, the designer will make the project's scope based on the minimum level of constraints, and then add "options" that may be implemented if feasible. For example, an owner likes to add a swimming pool to the home he/she likes to build but is not sure he/she can afford it. The designer will prepare a plan without the pool, and then add an alternate design or an option with the pool[3].

> The owner must rationally define the project scope without emotional impulses or external biased influence. Most importantly, the owner must focus on needs first and then wishes, within the financial and other constraints.

7. If there is more than one person representing the owner, such as a business partner or a spouse, the designer must take input only from the real decision-maker, although listening to others is still a good idea.

In many cases, especially major projects, the designer starts with producing preliminary (also called conceptual) design. In building projects, this may include only some of the architectural drawings that show the elevations, plans, and sections. This gives the opportunity to the owner to make changes before the detailed design is made. Furthermore, the detailed design may come out in stages called design development. Each stage is named DD##, where "##" represents the percent of completion. So, DD30, DD60,

---

3 The earlier the decision is made, the less it will cost, in addition to other benefits. This applies in variable degrees from severely to no effect, depending on the item. A decision to add a swimming pool is greatly impacted by the timing of the decision, while a decision to switch from carpet to ceramic tile may come relatively late with little or no impact (see Figure 4.3).

and DD90 represent the stages at 30, 60, and 90% completion of the detailed design, respectively. Interestingly, DD100 is not automatically the same as the final design because although it represents 100% completion of the detailed design, it needs to be approved before it can be named "Final Design." This all gives the owner chances to review the design and make any necessary changes before it is finalized.

Implementing Building Information Modeling (BIM) also helps the owner visualize the finished product before anything happens on the ground. This gives an opportunity to examine different scenarios and alternatives and make changes early enough without negative consequences. BIM is a great tool for the contractor also. It allows the simulation of the construction process, providing an opportunity to detect clashes and optimize the process.

Again, the author likes to emphasize point #3 earlier, that the designer needs to learn, directly and indirectly, the owner's time, money, and other constraints. The design professional may make suggestions pertaining not only to the design of the project but also to any other related matter that are within the designer's expertise and is in the interest of the project and the owner. The architect may realize that the price of lumber is high, and the owner lives in a hurricane zone, so he/she may suggest having the planned house built with reinforced concrete or steel framing. The architect may also realize that for a 60-year-old client, even though has no special needs at that time, the design may be modified to accommodate possible future special needs. In brief, the designer must have a deep, long-term, and wide-angle vision of the owner's requirements.

Project scope must be:

1. Clear: So that everyone understands what is to be done,
2. Realistic: Feasible and within set constraints, and
3. Flexible: To accommodate minor changes[4] if / when they happen; either because of unexpected circumstances or if the owner decides so.

Despite all efforts, changes may and do happen. Such changes happen at different stages of the project and for many reasons, including:

1. The contractor conducted value engineering (VE) study with a focus on the project's lifecycle. As a result, there were several recommendations that can improved the construction process and/or the utilization of the project.
2. Unforeseen conditions forced the change. For example, after starting the excavation, the contractor discovered that the soil uncovered was different from what was specified in the soil report submitted previously by the owner. This led to a change in the foundation of the building.
3. Changes in the market such as new materials or new technology influenced a change.
4. The owner's financial situation changed, positively or negatively, so he/she wants to upgrade or downgrade.
5. The owner simply changed his/her mind about an item.
6. A change in the local building code was issued before the design was permitted.

---

4 The term "minor changes" here is in the context of impacting the scope, not necessarily the cost. As an example, if an owner decides to install marble tile instead of carpet, it is considered a minor change because it did not alter the scope even though the cost may be high. Generally, we do not allow changes to the project scope such as a major change to the size, type, or purpose of the project.

7. The design faced a legal challenge (e.g., zoning or wetland) and had to be modified to comply with the legal requirements.
8. The project's end users or their requirements changed after the design was completed.

> The more the decision for a change is delayed, the more likely it will increase the cost and duration.

Project Scope versus Product Scope

The term scope used in this chapter refers to the "final product" as described in the design and contract documents. In this context, we differentiate between the scope from the owner's and contractor's perspectives. Even though they both agree on the final product, the owner's focus is on the final product while the contractor's focus is on the entire process ending up with the finished product.

According to the PMI, project scope is the work performed to deliver a product, service, or result with specified features and functions, while product scope is the features and functions that characterize a product, service, or result[5]. The same concept applies in quality management (to be covered later) in defining quality control (product-oriented) and quality assurance (process-oriented).

The owner is recommended to have a scope management plan. It will be a component of the project management plan that describes how the scope will be defined, developed, monitored, controlled, and validated[6]. Having such a plan, which is prepared, studied, and documented, is good as a reference for scope change issues. It can be in the form of text or graphics (flow chart), but it has to be simple, clear, and broad to cover all potential situations.

Scope Creep

Even though some changes may have a positive impact on the project, too many changes, especially major ones, are a sign of poor scope management. One of the common symptoms of poor scope management is scope creep. It is defined as the continuous and gradual expansion of the scope of a project (size, area, design, materials, etc.) after the contract has been signed because of multiple and successive owner-issued change orders[7]. Scope creep usually results from the owner's lack of vision, lack of appreciation for the impact of changes on the cost and schedule, unrealistic expectations, emotional spurts, decentralized decision-making, and/or other factors. It usually results in negative consequences for the owner, including the reduction of the value obtained in the project. The PMI defines scope creep as the uncontrolled expansion to product or project scope without adjustments to time, cost, and resources[8]. Scope creep can happen in other cases such as when someone wants to buy a simple car, and then listens to the salesperson mentioning options and upgrades. It can even happen when a person goes to a supermarket to buy certain items but gets enticed by marketing ploys to buy more items, scattered strategically in the store. However, it can be more serious in construction because of its potential magnitude, and also because it may happen unnoticed over time in small successive steps.

---

5 PMI-PMBOK, 7th edition, 2021.
6 PMI-PMBOK, 7th edition, 2021.
7 In some parts of the world, a change order is called variation order, VO.
8 PMI-PMBOK, 7th edition, 2021.

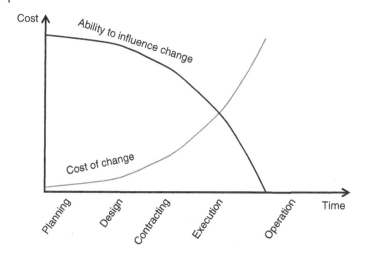

Figure 5.4 Impact of time on changes and their cost.

There are several problems with scope creep:

1. With many changes resulting from scope creep, the owner may suffer a major increase in cost but may not feel it because it is happening in small increments.
2. The changes put the owner in a noncompetitive cost situation with the contractor.
3. It will likely have a negative impact on the schedule.
4. It may have a negative impact on the morale of the contractor and workers, and cause frustration.

Figure 5.4 shows the importance of making decisions regarding the scope, as early as possible. The earlier we make the decision, the greater the ability to make such changes, and the less it will cost.

The author was involved as the contractor's cost estimator in a project to build a project for a city government. The city wanted to build a split-level parking garage on a piece of land it owned in a downtown location of the city. It had a budget limit of $2.5 million[9]. The initial estimate was just under the $2.5 million threshold, but the client kept making changes, mostly additions and upgrades till the estimate exceeded $4.2 million through at least 13 successive changes, with each change adding "just a little" to the cost estimate! Some changes triggered other additions. For example, as the parking garage was located on a corner of two streets. It was suggested to convert all parking spaces on the ground floor facing these two streets, to retail stores that can be rented out. It was a great idea, but it decreased the garage capacity by around 40 spaces. This led to the addition of one-half level of parking to compensate for these lost spaces. Then, the architect decided to add pavers on the sidewalk near the retail stores. As the final design was approved, the city's upper management was "surprised" by the $4.2 million price tag and did not like it. It took another round between the client and the designer to settle on a design with a price tag of around $3 million. The process of adding and then deleting took time, effort, and cost but fortunately, it was done during the design, and not the construction, process.

---

9 This project took place in 1998–99. The budget mentioned reflects prices at that time.

Another example of the importance of timing change: in an office building, the owner decides to change a few doors from 2'-8" wide to 3'-0"[10] wide. All walls are made of concrete masonry units (blocks). Let us consider the following six different scenarios for when the change was issued:

1. At the start of the design phase.
2. After the design is completed.
3. During the bidding phase.
4. After signing the construction contract but before the doors were delivered or the openings were framed.
5. After the door openings were framed but before the walls were finished.
6. After doors were installed and painted (everything was completed.)

The cost of the change for the six scenarios is as follows (not including the difference in price between the old and new doors):

1. Scenario 1: No extra charge.
2. Scenario 2: The architect's fee in redoing some of the drawings and specifications.
3. Scenario 3: The architect's fee as in scenario #2, plus the cost of issuing an addendum to all bidders.
4. Scenario 4: The architect's fee as in scenario #2, plus the cost of a change order.

Scenarios 5 and 6 also include the cost of the architect's fee as in scenario #2, plus a change order but the cost of the change order increases drastically from scenario #4 to scenario #5, and then from scenario #5 to scenario #6. In addition, scenarios #5 and 6 are likely to require time extension.

It is important to distinguish between scope creep and progressive elaboration. We already defined scope creep, which is generally looked at negatively and must be minimized and controlled. Progressive elaboration on the other hand refers to the ongoing improvement of a project plan based on new learnings and insights obtained during the project lifecycle[11]. This happens particularly in cases where construction starts before the final design is issued. An example of this situation is fast-tracking, where the project is divided into phases, and construction of a phase proceeds when the design of the subsequent phase is still underway, as shown in Figure 5.5.

However, in the case of fast-tracking the project, it is important to create the concept design, which identifies the major components and concepts of the project. After that, the detailed design can overlap with the construction, as illustrated in Figure 5.6.

In progressive elaboration, the contractor needs to make assumptions due to a lack of design details, but later when more design details become available, the newly revealed information may replace the contractor's assumptions, so a change is warranted.

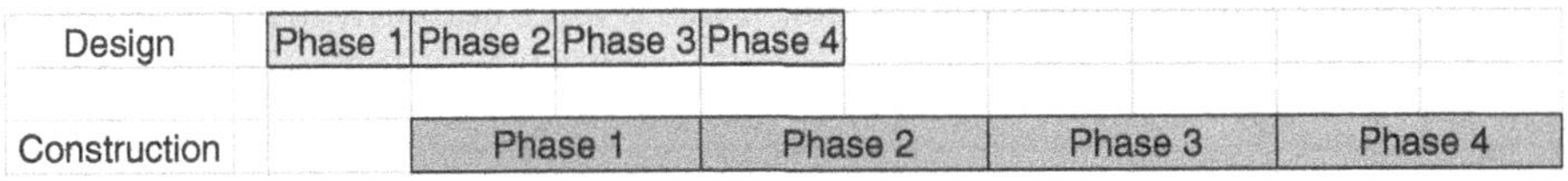

Design	Phase 1	Phase 2	Phase 3	Phase 4				
Construction		Phase 1		Phase 2		Phase 3		Phase 4

Figure 5.5 Fast-tracking design and construction.

---

10 " symbol means inches and the ' symbol means feet.
11 https://project-management.info/progressive-elaboration-in-project-management/.

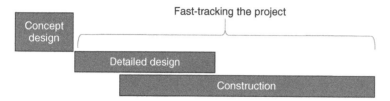

Figure 5.6 In fast-tracking detailed design and construction, concept design must be done first.

The PMI defines progressive elaboration as the iterative process of increasing the level of detail in a project management plan as greater amounts of information and more accurate estimates become available[12]. So, unlike scope creep, changes due to progressive elaboration are normal and expected.

Scope Maintenance

Scope maintenance during the construction phase is another important part of scope management. It covers the topic of change orders, which is directly related to many components of project management, such as cost management, scheduling, project controls, procurement, accounting, and others.

As discussed earlier, changes can happen at any time, both during the design and construction phases[13]. The term change simply means a departure from what was specified in the design and other contract documents, no matter what type, size, or impact. Sometimes, the change is not by choice such as in the case of subsurface or unforeseen conditions. At other times, it happens based on the order of the owner. Sometimes, the change happens because of an error or omission in the design documents. Changes can also happen for reasons that are not the fault of any of the contracting parties. Changes may have an impact on the project's cost, schedule, or both. The topic will be discussed in subsequent chapters.

Formalizing the Change Process

Creating and following a simple system to direct the process of decision-making is a great help. The following can be a foundation for such a system:

1. Is the change required or discretionary?
   - If mandatory, you may not need to look for alternatives and further "what if" investigation unless the alternative achieves the same objective in a better way.
2. What is the estimated cost impact?
   - The owner can set a threshold, say $1000 for a simple residence, where any change above that amount has to go through more investigative steps and/or needs more authority to approve.

---

12 PMI-PMBOK, 7th edition, 2021.

13 Although the official name of the change document differs. During the design phase, it is just instructions to the designer. During the bidding period, a change is issued in the form of an addendum. After signing the construction contract, it becomes a change order, CO.

- You can divide the cost into the upfront cost (construction) and lifetime cost (operation and maintenance).
- You may create a cost versus value assessment for the change. The value is assessed in terms of benefits to the owner.
3. Does this change have an impact on the schedule?
   - The assessment of such impact depends on the criticality of the schedule.
4. Would there be consequences for delaying the decision? Negative or positive?
   - For example, higher costs and more delays.
5. Is it a technical matter where expertise is needed?
6. Do you have previous experience with such an issue?
7. Did you check for alternatives?
8. It is a good idea, especially for those doing frequent projects, to keep a feedback record. This can help guide future decisions for similar items or situations.

> The owner's relationship with the designer and contractor is important and delicate. The owner is the party that makes decisions, pays the bills, and will own (and likely use) the project when completed. On the other hand, the designer and contractor are professionals and experts in their fields, so it is their professional and ethical duty to lay down all the facts and options to the owner, so he/she can make the best decision.

This process may be implemented in a programmable flow chart producing output that helps guide the decision-maker.

The Decision-making Process

In defining, designing, executing, and even operating the project, it helps tremendously when the owner has an efficient decision-making process. This process depends on several issues such as:

1. The type of owner: individual, family, or organization; public or private?
   - If a corporation or public entity, the formalities and bureaucracy of the owner's organization.
   - If the owner is individual, the type of personality / character.
2. The owner's previous knowledge and experience with this type of project.
3. The availability of design and construction experts in the owner's team. If unavailable, does the owner have access to independent and competent experts?
4. The volume and impact of the matter where the decision is to be made.
5. Time constraints: Is there a deadline for the decision?

The type of owner's personality/character can be branched into several factors. One of them is the owner's degree of dependency on experts. Decisions vary tremendously in type and volume (impact). Some of them are technical or lean to be so, while others are subjective and a matter of personal opinion or preference. It is important to realize the difference between the two cases, and thus use technical experts when needed.

There are many books and articles on improving the decision-making process, but here are some general tips. Keep in mind, these tips are from the project owner's perspective:

1. If there is more than one decision-maker, one of them must be the designated "authority." This may be clarified, especially in government agencies and well-structured corporations, but may not be so clear in other cases.
   - If the decision-making process and/or the authority of the decision-maker are not so clear, the mechanism for making the decision must be clearly defined in the contract.
2. If the matter is technical, an independent expert may be consulted. It is important to hear the contractor's opinion but getting an independent assessment is necessary, especially for major items.
3. If the investigation into the matter takes time, do not wait till the last minute, and then feel pushed to make a rushed decision.
4. If this decision is to be made in a meeting with the contractor or another consultant, educate yourself as much as possible before the meeting. If no competent consultant is available in the meeting, ask the contractor to wait till you can consult and make the proper decision.
5. Breaking down/dissecting complicated matter into simple components always helps.
6. The decision-maker must balance between hesitation and rushing. Some decisions may or even must be made instantly. Hesitation may increase the cost or even cause harm.

Exercises

5.1 Define project scope.

5.2 What is scope creep? What are its main causes?

5.3 What is scope management? What are its two main components?

5.4 Who takes primary responsibility for scope management?

5.5 How would the designer help the owner in defining the project scope?

5.6 The project scope must be well defined yet flexible. Please explain.

5.7 You are an architect. An owner comes to you requesting the design of a house that costs, according to the client's specifications and your initial calculations, $1.5 million, but his/her budget is limited to $800,000. How would you handle the situation?

5.8 Too many changes during the construction phase are an indication of poor scope management, yet not all changes are necessarily bad. Explain this statement.

5.9 Give an example of a change that the more we wait to make the decision, the more it costs.

5.10 What is progressive elaboration?

5.11   An owner organization with complex bureaucratic decision-making process, is planning to have a major construction project. You need to work with the organization as an advisor to simplify the decision-making process to define the project scope clearly early enough, so they can minimize the likelihood of changes. Describe the steps taken.

5.12   One of the recommendations in Question No. 11 is "The decision-maker must balance between hesitation and rushing." Explain with some examples.

6

Project Budgeting/Cost Management

Introduction

Cost management is arguably the most important aspect of project management because almost all other aspects will ultimately be expressed in monetary value. Construction companies are in the business of building projects to make profit, and this is the bottom line. Of course, doing so must come within professional, legal, and ethical boundaries.

Cost management is a cycle that starts when the project is just an idea and continues till the physical completion and fiscal closing of the project.

As explained earlier in Chapter 2, the project starts with an idea from the owner with an approximate "budget number." This number represents the owner's initial financial capacity or willingness to pay for the project. The ball is now in the designer's court to design the project within the owner's scope and constraints (money, time, and other.) The original budget number may have to be adjusted with the reconciliation of the owner's scope and reality according to the designer's professional judgment.

Next step is signing a contract with a contractor to build the project, by means of bidding or negotiation. This step may force another adjustment in the budget number but this time, it will make it more definitive. After that, the construction process starts. Many events and actions may force yet more adjustments in the budget number. The owner, along with the contractor, will monitor the cost during construction, always comparing actual cost to planned cost, and detecting any variances. This step is very important, particularly in cost-plus-fee contracts, where the owner bears higher responsibility and risk.

In the end, the owner would like to finish the project within set constraints (cost, time, quality, and other), and the contractor would like also to finish the project with a satisfied owner[1] and profit amount as expected.

---

1 Subtle but important note here is that "owner's satisfaction" may be a subjective term and is not legally binding while "fulfilling the contract" is an objective term and legally binding. In all cases, it will be great if both owner's satisfaction and fulfilling the contract come together.

Cost Definitions

Cost is the amount measured in money, cash expended, or liability incurred, in consideration of goods and/or services received. From a total cost management perspective, cost may include any investment of resources in strategic assets including time, monetary, human, and physical resources.[2]

Value: We can define value as the perceived benefits of a specific product or service, received by a customer, relative to the price paid by the customer.

$$\text{Value} = \frac{\text{Perceived benefits received}}{\text{Price paid}\left(\text{cost}\right)}$$

It is important to note that the price may be the same for all customers, but benefits differ from one person to another and from one situation to another, even for the same person. Realizing the difference between cost and value is important to owners so they can make the best decision based on their priorities. For example, the value of a small elevator in a two-story house may be little to none for one person who is young and healthy, and very high for another person who has special needs. The value of a swimming pool in the backyard of a house means a lot to someone who will be using it constantly, while it is a nuisance of little value to another person who will not use it.

As a generic example: An electric vacuum cleaner that has superb cleaning power on bare floors (mainly tile). The value of this vacuum cleaner is vastly different between a building owner where all or most of the floor is tile versus another owner where all the floor is carpet.

Furthermore, the term cost will likely imply the upfront cost of the item being considered, while the value is more connected with the use or the possession of the item, so it automatically extends through the life of the item.

The value can relate to cost items only, as in the previous example, or may relate to other considerations such as time, environment, public image (reputation), historical, emotional, or other consideration. For example, the value of a shopping center project is higher to the owner if completed in October rather than January. Using environmentally friendly materials or solutions even if it raises the upfront cost slightly, may add value to the owner. A person may choose a location, design, or materials that have symbolic or emotional meaning to them. Some people may pay a lot more for an item than its market worth because of a symbolic value or emotional attachment to the item. Often, it is difficult to put a numerical value on such factors. One of the challenges that the designer and contractor face with some owners is reconciling the owner's "wish list", "need list" and "budget/can afford list." This challenge can be resolved logically by prioritizing the owner's needs and then wishes, and then drawing the line based on the available budget. Emotional intelligence helps in these situations.[3]

Cost estimate is the approximation of the cost of a project, item, or service. It comes in the form of a single total number, but it is the product of the process of cost estimating. This process differs according to the type of estimate; approximate or detailed. The AACE International defines a cost estimate as a

---

2 AACE International Recommended Practice No. 10S-90.

3 The author created an Excel module called "The Optimization of the Selection Criteria," that helps in making a rational decision when multifactors with different weights are being considered.

compilation of all the possible costs of the elements of a project or effort included within an agreed upon scope.[4]

Cost estimating is the predictive process used to quantify, cost, and price the resources required by the scope of an investment option, activity, or project. Cost estimating is a process used to predict future costs.[5]

As a predictive process, estimating must address risks and uncertainties. The outputs of estimating are used primarily as inputs for budgeting, cost or value analysis, decision-making in business, asset, and project planning, or for project cost and schedule control processes.[6]

Quantity estimating/surveying is the process of measuring the work of the project in the form of a series of quantified work items.

Budget (or project budget) is the total amount of authorized financial resources allocated for the particular purpose(s) of the sponsored project for a specific period. It is the primary financial document that constitutes the necessary funds for implementing the project and producing the deliverables. It is also defined simply as a planned allocation of resources.

Baseline budget is the approved cost estimate of the project including approved changes. It represents the price tag the contractor puts on the project and the owner approves.

The main difference between a budget and a baseline budget is that the budget can be set individually. A baseline budget is a set number that the owner and general contractor agree on in the contract. For example, an owner may think of building a shopping center with a $5 million budget in mind. This number is based on the owner's discretion and financial ability or willingness to pay. When this owner goes through the design and contracting phases, this number may change based on many considerations. Let us assume that the owner and general contractor signed a $5,250,000 lump sum contract. This amount will represent the baseline budget.

> Owners: When it comes to elective options (upgrades or added items), do not just ask about the cost, ask also about its value to you!

Cost Control is the continuous practice of:

1. Tracking actual cost
2. Comparing actual cost to the baseline budget, and finding any cost variances
3. Analyzing cost variances to find the causes
4. Taking corrective action whenever and wherever needed

It is also called cost tracking. It is practiced by the contractor during the construction phase. The term (cost monitoring) usually goes to the owner, and it includes the first three steps above. The owner cannot directly take corrective action since the work site is "owned" technically by the general contractor. The owner, of course, has other means to influence the work. This will be discussed later.

---

4 AACE® International Recommended Practice No. 10S-90, Rev. November 14, 2014.
5 AACE® International Recommended Practice No. 10S-90, Rev. November 14, 2014.
6 AACE® International Recommended Practice No. 10S-90, Rev. November 14, 2014.

Figure 6.1 The project cost cycle.

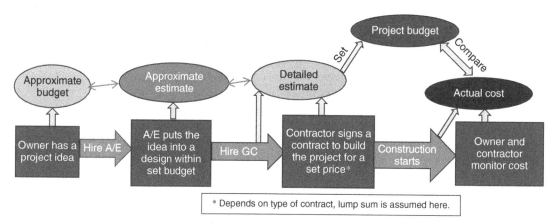

Figure 6.2 The project cost cycle.

Figure 6.1 gives a brief sequence of the project cost: It starts with an estimate, then a budget is set, and finally the practice of cost control during the construction process. Figure 6.2 introduces the same cycle in more detail.

Total cost management (TCM) is the effective application of professional and technical expertise to plan and control resources, costs, profitability, and risks. Simply stated, it is a systematic approach to managing cost throughout the life cycle of any enterprise, program, facility, project, product, or service. This is accomplished through the application of cost engineering and cost management principles, proven methodologies, and the latest technology in support of the management process. It can also be considered the sum of the practices and processes that an enterprise uses to manage the total life cycle cost investment in its portfolio of strategic assets.[7]

The concept of TCM was developed by AACE International to cover and optimize the entire project cost cycle. It is somewhat analogous to the Total Quality Management (TQM) that has been known and practiced for decades.

Schedule of values is a listing of elements, systems, items, or other subdivisions of the work, establishing a value for each, the total of which equals the contract sum. The schedule of values is used for establishing the cash flow of a project and serves as the basis for payment requests. The AACE International defines it as a detailed statement furnished by a construction contractor, builder, or other, apportioning the contract value into work packages. It is used as the basis for submitting and reviewing progress payments.[8]

---

7 AACE® International Recommended Practice No. 10S-90, Rev. November 14, 2014
8 AACE® International Recommended Practice No. 10S-90, Rev. November 14, 2014

Schedule of Contract Values

PROJECT NAME:
PROJECT #:
CONTRACTOR:

APPLICATION NO:
APPLICATION DATE:
PERIOD TO:
PERCENT COMPLETE TO DATE:

A	B	C	D	E	F
Item #	Description of Work	Quantity	Unit of Measure	Unit Price	Extension
	GRAND TOTAL				0
	(Alternatives, addition or subtraction)				

Figure 6.3 Schedule of values simplest template.

In its simplest form, as in Figure 6.3, it is a spreadsheet with a list of items; each with its code, description, total quantity, unit of measure, unit price, and total cost (quantity × unit price). The summation of all items' total cost must equal the contract sum. Such template is used for bidding and likely a part of the contract documents. It may contain, as per the bidder's or negotiator's request, alternatives to be added or subtracted.

It is possible to add more columns for project control purposes, showing quantities and prices for work completed (previous and this period), deductions (retainage + other), materials stored, cost to complete, cost at completion, and possibly other items in order to calculate progress payments and remaining work amounts and cost. Such schedule of values with these additional items is used with schedule updates (Chapter 8) and payment requests (Chapter 10).

The schedule of values is also called project cost breakdown or simply cost breakdown. In the UK and other countries, it is called Bill of Quantities.

Types of Cost Estimates from Accuracy Perspective

The process of estimating the cost of a project is somewhat like zooming the lens of a camera on a faraway object. We start with a fuzzy low-accuracy picture and then gradually zoom in. As we zoom in at the object, we see more clearly, and we know more about it. The picture becomes more accurate.

Experts and professional organizations divide cost estimates into several classes based on several criteria, most importantly the accuracy.[9] However, the author believes that cost estimates are basically two

9 AACE International five classes.

groups, approximate and detailed. The difference between the two groups is more than just the range of accuracy or effort level, it is the methodology of preparing the estimate. Approximate estimates are based on comparisons to previous similar projects with known actual costs, and then applying the unit cost produced by these past projects to the proposed project, after making appropriate adjustments. For example, a contractor who built several mid-rise office buildings with an average actual cost of \$185/SF for these projects. Now, he likes to estimate a newly proposed project of a similar nature with a total area of 42,000 SF. He can use the number \$185/SF after modifying it for inflation and project characteristics (such as the size and location of the project and the level of finish). So now, he suggests \$220/SF for the new project. Finally, he multiplies the modified unit cost by the total area of the project:

$$\text{Project approximate cost} = 42{,}000 \ \text{SF} \times \$220 \,/\, \text{SF} = \$9{,}240{,}000$$

The above number has low accuracy and is used to give an idea of the cost of the project. But this accuracy depends on several factors, most importantly the similarities between the proposed project and the previous ones that it is being compared to. Sometimes, the cost estimator realizes some differences, such as the level of finish or height of the ceiling,[10] so he/she take that into consideration when applying the adjustments. This helps improve the accuracy of the estimate.

This method is applied using one of several parameters, not necessarily the area unit (square feet or square meter) but it can be a pupil/student for a school, a bed for a hospital, an inmate for prison, or a "key" for a hotel. For industrial projects, it can be the unit of production such as a power plant (megawatt per hour), water or sewage treatment plant (million gallons per day, MGD), or oil refinery (barrels per day).

The advantage of such an approximate estimate is that it is quick and needs little effort. Also, it does not require the project's detailed design. Approximate estimates, also called analogous, rough or "back-of-the-envelop" estimates, are used for:

1. Feasibility studies, especially for the owner
2. To help decide to bid or not to bid (for a contractor)
3. Comparing alternatives (alternative projects or designs)
4. Making *initial* financial arrangements
5. It is possible for a contractor to use approximate estimates to come up with a Guaranteed Maximum Price (GMP), when signing a cost-plus-fee contract.

Detailed estimates are based on actual quantity takeoff from the detailed design of the project, pricing each item individually, adding indirect cost, and then putting the estimate together. This, of course, requires the availability of detailed design. Preparing detailed estimates is a time and cost-consuming task, and in many cases, requires the involvement of several professionals from different specialties. Detailed estimates are also called definitive or itemized estimates. They are usually used for:

1. Making *final* financial arrangements
2. Bidding or negotiating a project, especially when it is a lump-sum contract, and
3. Project/cost control (comparing actual cost to planned cost during the execution of the project).

---

10 Height of the ceiling impacts many items, not just the walls and partitions. For example, the capacity of the heating/air-conditioning system may have to be increased. If there is an elevator, it may have to be modified.

Table 6.1 AACE International classes of estimates.

ESTIMATE CLASS	Primary characteristic	Secondary characteristic			
	LEVEL OF PROJECT DEFINITION Expressed as % of complete definition	END USAGE Typical purpose of estimate	METHODOLOGY Typical estimating method	EXPECTED ACCURACY RANGE Typical variation in low and high ranges[a]	PREPARATION EFFORT Typical degree of effort relative to least cost index[b]
Class 5	0% to 2%	Concept screening	Capacity factored, parametric models, judgment, or analogy	L: −20% to −50% H: +30% to +100%	1
Class 4	1% to 15%	Study or feasibility	Equipment factored or parametric models	L: −15% to −30% H: +20% to +50%	2 to 4
Class 3	10% to 40%	Budget, authorization, or control	Semi-detailed unit costs with assembly level line items	L: −10% to −20% H: +10% to +30%	3 to 10
Class 2	30% to 70%	Control or bid/tender	Detailed unit cost with forced detailed take-off	L: −5% to −15% H: +5% to +20%	4 to 20
Class 1	50% to 100%	Check estimate or bid/tender	Detailed unit cost with detailed take-off	L: −3% to −10% H: +3% to +15%	5 to 100

[a] The state of process technology and availability of applicable reference cost data affect the range markedly. The +/- value represents typical percentage variation of actual costs from the cost estimate after application of contingency (typically at a 50% level of confidence) for given scope.

[b] If the range index value of "1" represents 0.005% of project costs, then an index value of 100 represents 0.5%. Estimate preparation effort is highly dependent upon the size of the project and the quality of estimating data and tools.

In some cases, the cost estimator needs to do an estimate when the detailed design is still being produced in stages; often called DDxx, where the xx represents the percentage of completion: for example, DD30, DD60, DD90, DD100, and final. In this case, the estimate is "hybrid" between the approximate and the detailed. Known items are estimated in detail and missing or unknown items are estimated based on suitable assumptions. As the detailed design advances, more information become available to the estimator, and more assumptions are replaced by definitive design information, thus, increasing the engineering (design) percentage completion (Table 6.1). This is shown in Figure 6.4 where the accuracy of the estimate improves with the increase in engineering (design) percent complete.

The process of quantity takeoff has been greatly improved in both effort needed and accuracy with the advancement of computer software and hardware tools. It is now possible to get the quantities directly from the CAD design and link them directly to building information modeling (BIM).

In addition to quantities, it is important for the estimator to know the specifications of the cost items. For example, for concrete items, the estimator needs to know the type and strength of the concrete and whether any additives are required. Knowing the brand and type of products is also

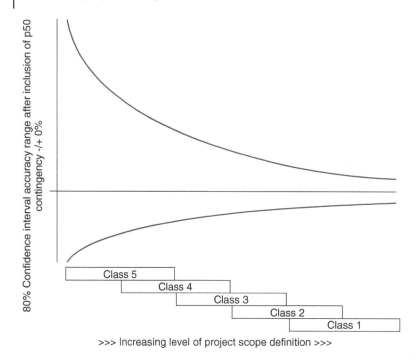

Figure 6.4 Estimate accuracy improves as the level of project definition improves[11].

important, such as plumbing, HVAC,[12] electrical, and finishing items. When getting a quote from a vendor, it is important to know what is included and what is not included in the price, such as taxes, delivery, setting up, warranty, disposal, and other items. It is important to know all the terms with the vendor's quote such as price details (if there is a discount for large quantity), the expiration date of the quote (the prices are good till when?), and the terms of payment. Such price quotes must be in writing, paper or electronic (email.) Sometimes, especially in a volatile economy, the terms of the price quote change, while the estimator may not pay attention to this change. For example, the standard price quote for most materials is 30 days but in the few months after the COVID-19 pandemic started in 2020, some prices were rising and fluctuating dramatically, so that period was reduced by many vendors to 10 days or less.

Estimating labor is critical because the unit price of any item is inversely proportional to the production rate. So, the same exact item will cost more or less, depending on crew production rate, which, in turn, is dependent on several factors such as weather conditions, skill level, and complexity of the job.

---

11 Guide to Cost Estimate Classification System, TCM Framework: 7.3 – Cost Estimating and Budgeting, Rev. August 29, 2022.
12 heating, ventilation, and air conditioning.

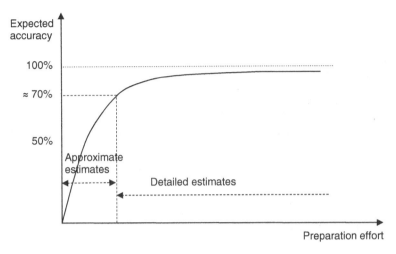

Figure 6.5 Accuracy versus effort.

Putting the estimate together is the final step that requires knowledge and experience. It involves adding other expenses such as mobilization, demobilization, indirect costs,[13] and making final adjustments and judgment.

The curve in Figure 6.5, which shows the relationship between estimate accuracy and preparation effort, demonstrates the point that it takes relatively small effort to do an approximate estimate, while doing a detailed estimate required much bigger effort.

Keep in mind that cost estimates, no matter how accurate, cannot reach 100% accuracy, simply because they are predictions of future expenses. So, when mathematical calculations and estimating software programs produce numbers such as $8,576,239.54; they are giving a false impression of high accuracy, but the number still has a margin of error that can be high or low. For this reason, final numbers should be rounded depending on the level of accuracy, purpose of the estimate, and the personal judgment of the estimator. For example, the number mentioned earlier can be written as $8,576,000 or $8,600,000.

Parametric estimating is an estimating method in which an algorithm to calculate cost or duration based on historical data and project parameters.[14]

Parametric estimating uses a statistical relationship between relevant historical data and other variables (e.g. square footage in construction) to calculate a cost estimate for project work. This technique can produce higher levels of accuracy depending upon the sophistication and underlying data built into the model. Parametric cost estimates can be applied to a total project or segments of a project, in conjunction with other estimating methods. Parametric estimating works well for industrial projects such as water treatment plants and oil refineries where there is a nonlinear relationship between the cost and the main product capacity (million gallon per day, MGD, thousand barrels per day, megawatts), for example:

$$C_i = C_m \left(Q_i / Q_m \right)^n$$

---

13 To be explained later in this chapter.
14 PMI-PMBOK, 7th edition.

where:

C_i is the cost of the new proposed project
C_m is the cost of the old built (known) project
Q_i is the capacity of the new proposed project
Q_m is the capacity of the old built (known) project
n is a parametric factor that is established for this type of projects

The term "3-Points Estimating" is also used sometimes as an estimating "method." However, it is not a method of estimating but rather a method of adjusting the estimate. It works by taking an average of three previously known numbers, optimistic, most likely, and pessimistic. This average may or may not be weighted:

$$\text{Triangular Distribution: } E = (O + M + P)/3$$
$$\text{Beta Distribution}(\text{PERT}): E = (O + 4M + P)/6$$

where:

E is the expected value for the estimate.
O is the optimistic estimate.
M is the most likely estimate.
P is the pessimistic estimate.

The triangular distribution gives equal weights to the three values. The second (beta or PERT) distribution gives heavier weight (four times) to the most likely value over either the optimistic or pessimistic, which seems to be more reasonable because the likelihood of the occurrence of the "most likely" scenario is more than either the optimistic or pessimistic.

Project (or Work) Breakdown and Format of the Estimate

When making a detailed estimate, the project needs to be broken down into elements that can be identified and estimated individually. This breakdown happens progressively, starting with level 1 (level 0 is the entire project) and going down to levels 4, 5, or even 6.

The way the cost estimator breaks down the project into items is different than the scheduler's breakdown. The cost estimator cares more about the type (nature) of the item while the scheduler cares more about its sequence, timing, and location. For example, in a multi-story concrete building, the cost estimator may combine all concrete beams or suspended slabs, but the scheduler cannot.

In the past, estimators used to organize their estimates based on their own preferences or their company's policy. In 1995, the Construction Specifications Institute (CSI), introduced an organization called MasterFormat that included 16 divisions. This format quickly gained popularity and became the standard for most estimates in the construction industry. It helped organize and standardize estimates, minimizing the likelihood of duplicating or omitting cost items. However, many items did not clearly fit any of these divisions, such as fiber optic cables and security systems. Besides, with the advancement of science and technology, many new items emerged that also did not fit any of the existing divisions. In 2004, the CSI changed and expanded its MasterFormat to include 50 divisions (starting with Division 00 and ending with Division 49), many of them unnamed but reserved for future use. This MasterFormat keeps getting updates but maintains the same structure. There will be more details on project breakdown in both chapters, scheduling and project controls.

The CSI MasterFormat

The following are the titles of the 50 divisions under the CSI MasterFormat. Note:

- The last division is No. 49, but when we count Division 00, they are 50.
- Several divisions are empty and reserved for future expansion.
- This MasterFormat is being updated periodically by CSI. This includes modifying some titles and naming empty divisions. However, the overall structure of the MasterFormat and the number of divisions are not expected to change soon.

Procurement and Contracting Requirements Group:

- Division 00 – Procurement and Contracting Requirements

Specifications Group
General Requirements Subgroup

- Division 01 – General Requirements

Facility Construction Subgroup

- Division 02 – Existing Conditions
- Division 03 – Concrete
- Division 04 – Masonry
- Division 05 – Metals
- Division 06 – Wood, Plastics, and Composites
- Division 07 – Thermal and Moisture Protection
- Division 08 – Openings
- Division 09 – Finishes
- Division 10 – Specialties
- Division 11 – Equipment
- Division 12 – Furnishings
- Division 13 – Special Construction
- Division 14 – Conveying Equipment
- Division 15 – Plumbing + HVAC
- Division 16 – Electrical + Lighting
- Division 17 – Reserved for Future Expansion
- Division 18 – Reserved for Future Expansion
- Division 19 – Reserved for Future Expansion

Facility Services Subgroup:

- Division 20 – Mechanical Support
- Division 21 – Fire Suppression
- Division 22 – Plumbing
- Division 23 – Heating Ventilating and Air Conditioning
- Division 24 – Reserved for Future Expansion
- Division 25 – Integrated Automation

- Division 26 – Electrical
- Division 27 – Communications
- Division 28 – Electronic Safety and Security
- Division 29 – Reserved for Future Expansion

Site and Infrastructure Subgroup:

- Division 30 – Reserved for Future Expansion
- Division 31 – Earthwork
- Division 32 – Exterior Improvements
- Division 33 – Utilities
- Division 34 – Transportation
- Division 35 – Waterways and Marine Construction
- Division 36 – Reserved for Future Expansion
- Division 37 – Reserved for Future Expansion
- Division 38 – Reserved for Future Expansion
- Division 39 – Reserved for Future Expansion

Process Equipment Subgroup:

- Division 40 – Process Interconnections
- Division 41 – Material Processing and Handling Equipment
- Division 42 – Process Heating, Cooling, and Drying Equipment
- Division 43 – Process Gas and Liquid Handling, Purification and Storage Equipment
- Division 44 – Pollution Control Equipment
- Division 45 – Industry-Specific Manufacturing Equipment
- Division 46 – Water and Wastewater Equipment
- Division 47 – Reserved for Future Expansion
- Division 48 – Electrical Power Generation
- Division 49 – Reserved for Future Expansion

The CSI MasterFormat helped organize and standardize estimates.

The CSI UniFormat

UniFormat is a method of arranging construction information based on functional elements, or parts of a facility characterized by their functions, without regard to the materials and methods used to accomplish them. These elements are often referred to as systems or assemblies.

1. Element A: Substructure
2. Element B: Shell
3. Element C: Interiors
4. Element D: Services

5. Element E: Equipment and Furnishings
6. Element F: Special Construction and Demolition
7. Element G: Sitework
8. Element Z: General

It almost follows the construction in chronological order.

Direct versus Indirect Costs

There are many cost items the estimator must account for in the estimate. These expenses are classified into direct and indirect. A direct expense is the one that can be linked to a specific cost item in the project. An indirect expense is an expense that cannot be linked to a specific cost item in the project. It can be linked to the project or the company (main office).

Direct expenses include labor, equipment, materials, subcontracts, and other specific expenses.

- Labor includes all crafts of labor that are directly involved in the construction work. It does not include salary employees like the project management team members or the administration. The estimator must also know if the labor is union or open shop because wages vary.
- Equipment includes all equipment that will be used in the construction effort and will not be permanently installed in the project.
- Materials include all materials that are either used as permanent part of the project (like concrete or steel) or consumed during the construction process (such as concrete formwork.)
- Subcontracts: Each subcontract sum is considered a direct expense to the general contractor, although it is divided into direct and indirect expenses to the subcontractor. The general contractor usually marks up this subcontract amount for own overhead and profit.
- Other direct expenses such as mobilization, bonds, and insurances (covering this project), and others such as an OSHA citation.[15]

> A direct expense is the one that can be linked to a specific cost line item.

Labor Cost

Employees pay part of their paycheck for taxes and other voluntary and nonvoluntary contributions. The employer also has to contribute additional amounts on behalf of its employees. Here is a list of some of the additional costs to the employer:

1. Workers' Compensation (WC) Insurance: This includes all employees; onsite, in the office, and others, so long as they are on the payroll and during work hours or performing official duties. This averages

---

15 Some contractors may treat this as an overhead expense, but the author prefers not to do so. It is a matter of own opinion and discretion.

17.1% of construction workers' paychecks for 2023[16], but it is much lower for office employees. It also differs from one contractor to another as discussed earlier.

2. Federal Insurance Contributions Act (FICA), and FICA medical (Social Security and Medicare tax): Currently, this is 7.65% of the employee's paycheck from the employer.

   This includes a 6.2% Social Security tax and 1.45% Medicare tax on earnings. In 2023, only the first $162,200 of earnings are subject to Social Security tax.

3. Unemployment tax/insurance: Most employers pay both a Federal and a state unemployment tax. Only the employer pays Federal Unemployment Tax Act (FUTA) tax; it is not deducted from the employee's wages.[17] The percentage and other terms differ from one state to another. For example, Florida calls it "Reemployment tax," where only the first $7000 of wages paid to each employee by their employer in a calendar year is taxable. Also, employers with stable employment records receive reduced tax rates after a qualifying period.[18]

4. Any other applicable taxes or fees.

5. Employers provided benefits such as paid vacation, medical insurance, and contributions to retirement funds.

Labor wages are also subject to Federal, state, and local laws, as well as collective bargaining agreements for union workers. Many laws require additional payment when working overtime, usually 1.5 to 2.0 times the regular rate. Overtime has been a debatable topic. There is no question that it has a negative impact on productivity but there is no agreement among experts on the magnitude of this impact through time. The author believes while overtime may be a solution for short-term need such as schedule acceleration, it is not a good idea as a long-term policy. In most cases of overtime, the employer pays more to get less per hour.

> Workers' Compensation insurance is a major expense but those contractors with good safety records pay less premiums than their counterparts who do not have good records.

Equipment Cost

Equipment cost is an essential part of any construction project budget. Equipment – ranging in size from a small air compressor to a huge tower crane – may be owned by the contractor, rented, or leased.[19] In all cases, equipment costs must be accounted for in the contractor's cost estimating and budgeting. It is important to keep in mind that equipment incurs costs whenever it is under the contractor's control, even when idle. Equipment expenses include:

1. Depreciation, which refers to a decrease in value due to deterioration, and sometimes obsolescence. The amount of depreciation according to the tax books may differ from actual depreciation (market value).

---

16 RS Means R0131 note, BCCD 2022.

17 https://www.irs.gov/individuals/international-taxpayers/federal-unemployment-tax#:~:text=Only%20the%20employer%20pays%20FUTA,the%20Instructions%20for%20Form%20940.

18 https://floridarevenue.com/taxes/taxesfees/Pages/reemployment.aspx.

19 Lease is a long-term rent possibly with other conditions, but, unlike short-term renting, the lessee may have the responsibility for the maintenance, insurance, and other expenses of the equipment.

2. Investment (interest): This expense represents the "lost opportunity" cost – that is, the potential profit that would have been made if the money invested in equipment had been used in a different investment. It is also called interest cost when the equipment is purchased using an interest-bearing loan.
3. Taxes, insurance, and storage: These expenses are taken either as a lump sum or a percentage of the average value of the equipment.
4. Maintenance: Usually taken as percentage (50 to 100) of the straight-line depreciation cost. Maintenance costs start low, then increase year after year as the equipment ages. However, spreading the maintenance cost uniformly over the useful life of the equipment is an acceptable practice for analysis.
5. Operating expenses, such as fuel, oil, other lubricants, tires, and other attachments such as excavator blades.

In the case of equipment on rent or lease, some of these expenses may be the responsibility of equipment vendor (lessor). It is to the advantage of the contractor who owns the equipment to keep it as busy as possible because the cost per hour decreases with more work hours per year. Equipment cost per hour is calculated using Eq. 6.1.

$$\text{Total cost / hour} = \frac{\text{Annual ownership and maintenance cost}}{\text{Operating hours / year}} + \text{Operating expenses / hour} \qquad (6.1)$$

Equipment replacement is mostly a matter of economics. New equipment usually needs little maintenance. As it ages, its value decreases and the cost of maintenance increases. At a certain point, it will be more economical to replace the old equipment with a new one.[20] In this case, the contractor will get new equipment with better productivity and less maintenance, and possibly technologically more advanced.

The decision whether to buy, lease, or rent equipment depends on several factors, most importantly the extent of time the equipment will be needed and the financial situation of the contractor.

> Some people say "It did not cost me anything, I used my own money (or asset). This is a myth because the money or asset could have been used to make money.

Materials Cost

Materials cost, in a typical construction project, is more than any other cost category. The management of materials is very important for optimizing the cost and making the construction process as smooth as possible, fulfilling cost, time, quality, safety, and other constraints.

Materials management is defined as the planning and controlling of all necessary efforts to ensure that the correct quality and quantity of materials and equipment are appropriately specified in a timely manner, are obtained at a reasonable cost, and are available when needed. A materials management system includes the major functions of *identifying*, *acquiring*, *distributing*, and *disposing* of materials needed for a construction project.[21]

---

20 More details can be found in engineering economy books.
21 Construction Industry Institute (CII) Publication No. 7-1, later updated in reports IR7-3 - Procurement and Materials Management: A Guide to Effective Project Execution, and Global Procurement & Materials Management (Best Practice) RT-257.

The objectives of materials management are to[22]:

1. Obtaining the best value for purchased materials.
2. Assuring supplies are on hand when and where required.
3. Reducing inventory to the lowest amount required.
4. Assuring quality requirements are met.
5. Providing efficient low-cost movement of supplies to the site and within site storage.

Cost of Materials Categories:

1. Purchase costs: This represents the cost of the item quoted/charged by the vendor. It may or may not include delivery costs. There is no standard for shipping cost, especially internationally. Sometimes, especially in major shipments, the vendor provides the cost FOB (acronym for "free on board" or "freight on board"). This defines the point when liability and ownership of goods are transferred from a seller to a buyer, including the responsibility and cost for shipping. It can be the vendor's warehouse, a train station, a seaport, or airport, or other. So, it is important for the contractor to know exactly what the vendor's delivery means.
2. Order costs: This represents the administrative cost to research, find, negotiate, and purchase the materials.
3. Holding costs: This represents the shipping and handling (not paid for by the vendor), storage, and any other handling costs before installation. Inspection may be considered as an additional cost, although it is usually part of quality assurance / quality control plan.
4. Shortage (unavailability) costs: This is a hypothetical cost that represents potential loss to the contractor if the material does not arrive on time. It can include the crew cost, if idle because of the late arrival of the materials, penalty for delay, acceleration as a result of this delay, or any other cost resulting from the delay in materials arrival.

Sales tax is another cost the contractor must be aware of. Most materials are subject to state and local taxes but there are some exceptions sometimes. Materials imported internationally are usually subject to customs fees.

The contractor must pay attention to the terms and conditions of the purchased materials contract. For example, the term *2/10 NET 30,* when mentioned, means a 2% discount if the bill is paid fully within 10 days of the delivery date or the full amount is due in 30 days. After 30 days, interest and/or a penalty may apply. This point underscores the advantage and importance of having positive cash flow.

The important thing for the cost estimator is to provide the information for the project manager for optimizing the entire process of the project construction. We must keep in mind that cost, even though very important, there are other aspects that must be considered, such as time.

Regarding the timing of the materials purchase, there are two "extreme theories", with reality always somewhere in between:

A. Just-in-time, which means that the material is delivered at the time of installation, so there will be no storage.

---

22 Construction Industry Institute (CII) Publication No. 7-1, later updated in reports IR7-3 - Procurement and Materials Management: A Guide to Effective Project Execution, and Global Procurement & Materials Management (Best Practice) RT-257.

B. Inventory buffer, which means that all materials are purchased and stored in advance, so the contractor will have it handy when needed.

Characteristics (advantages/disadvantages) of the Just-in-time theory:

1. No storage or extra handling cost.
2. Less likely to be subject to theft or vandalism.
3. High "unavailability" cost due to possibility of delivery delay.
4. Requires having vendors in close proximity in order to have quick and inexpensive delivery.
5. Higher purchase and order costs due to the large number of small orders.
6. Does not require large upfront cash flow for the contractor.

The characteristics (advantages/disadvantages) of the inventory buffer theory are almost the opposite:

1. Requires storage and extra handling costs. May also require electronic inventory (bar-code) system.
2. More likely to be subject to theft or vandalism and may require special security arrangements.
3. Low or no "unavailability" cost.
4. No need for vendors in close proximity. Shipments come pre-arranged and – most likely–in large amounts.
5. Requires high capital in the case when the contractor is not reimbursed for purchased materials by the owner till they are installed.
6. Possible volume discount but higher holding cost (frozen money), especially if the contract does not allow the contractor to be reimbursed for purchased but not installed materials.

In the end, the contractor has to balance between the two theories, depending on the project location and other relevant criteria (Figure 6.6).

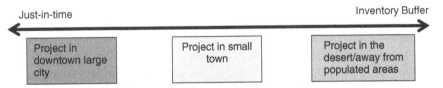

Figure 6.6 Materials management: Just-in-time versus inventory buffer theories.

Order Quantity

When ordering certain materials, the cost estimator needs to add waste allowance for certain materials:

Order quantity = Install quantity + waste allowance

Waste allowance is further divided into design waste and construction waste. Design waste comes from the difference between the material's unit dimensions and the dimensions of the place where it will be installed. For example, if we need wood joists, 13′ long, we must buy the 14′ long and cut one foot, because lumber is sold only in increments of 2 feet (8′, 10′, 12′,. . .) Another example is ceramic tile. If the tile size is 16″ by 16″ and the place is 6′–4″ wide, we need 5 tiles in a row, but the fifth one is to be cut to 12″ wide. Design waste for flooring (tile, carpet) increases when it flooring materials has a pattern, and also when installed inclined. Construction waste allowance, on the other hand, results from accidental damage, loss, or other unintended waste of the member. It may also include small leftovers (surplus) that are disposed at the end of the project because it is not practical to utilize it somewhere else. Defective materials may qualify for free replacement, but it depends on the purchase agreement terms and the vendor's or manufacturer's policy.

The amount or percentage of waste allowance added to the contractor's order depends on factors such as:

a. Type and cost of material
b. Location, dimensions, and position / direction of the member to be installed
c. Design pattern, if any
d. The cost of the material and the usability of the surplus
e. Labor skill (skill in cutting and handling a piece/tendency to break it). This may be influenced also by the weather, as some materials become more brittle in cold temperature. Other materials may become unusable with excessive moisture.
f. The criticality of the activity needing the material and the ease/speed of obtaining a replacement

For example, for concrete ready-mix, waste allowance is usually around 5% except for foundations such as footings, where it goes to 10% or more. Ceramic tile waste allowance depends on tile design, if it is plain or has pattern, the size of the tile, and the shape and size of the floor to be tiled. Design waste will increase if the tile is to be installed diagonally or in a way not parallel to the walls. For lumber, order quantity must specify the dimensions of the ordered member. So, for the joist example just given earlier, assume we need 50 joists: The order quantity includes, for example, 54 pieces of $2'' \times 10'' \times 14'$[23], which includes design waste (1 foot off every piece) and construction waste (4 extra pieces). Any piece that remains unused after the completion of the project, can likely be used later by the contractor on other projects, or returned to the vendor if the purchase agreement allows such practice. Lumber, in particular, is estimated, ordered, purchased, and paid for by nominal, not actual, size. The actual section dimensions of the $2'' \times 10''$ joists mentioned earlier are $1.5'' \times 9.25''$.

Another factor that affects the order quantity is the sales "package size." For example, bricks are typically sold in pallets of 534 bricks. So, if the contractor needs 18,500 bricks, including 3% waste allowance, the order size will be 35 pallets $\times$ 534 = 18,690 bricks. The amount of brick surplus after the completion of the project depends on how many bricks were wasted. The contractor may use them in a subsequent project.

Billing the client must be based on the installed quantity, but it is customary for many items that are cut or partial pieces to count as full pieces, such as concrete blocks, bricks, and tiles.

Designers, especially now with intelligent CAD systems, are encouraged to minimize design waste by making minor adjustments either to the building dimensions or in specifying the materials units to be used. In one project the author was involved in, the design required floor joists of length 14'–2'', which required purchasing 16' long joists. Moving the partition on the second floor by 2 inches allowed the use of 14 feet joists in lieu of 16 feet long. This not only resulted in savings on the cost of lumber, but also in cutting time as the new position of the partition required exactly 14 feet long joists, saving cutting time. In the cases when the pieces must be cut, the contractor may be creative in finding a use for the cut pieces such as blocking between studs.

---

23 " symbol means inches and the ' symbol means feet.

Pay attention to units! Construction cost estimates involve many units, some of which are unknown outside of the construction industry or even the craft. Here are some tips:

1. Certain materials have their own units such as lumber, which is measured by board foot (BF), which is $1' \times 1' \times 1''$, but since the unit is relatively small, we use MBF,[24] which means 1000 BF. Also, roofers use the square (SQ), which is 100 square feet. Interestingly, this unit is used for roof shingles, tile, and insulation (felt), but we still measure by square feet for plywood sheathing for the same roof.
2. Several items can be measured by unit (each) or area. For example, concrete masonry units (CMU, also known as concrete blocks) and bricks, as these materials come in different sizes. We may price them by each; block or brick, or the area in square feet. Also, for flooring, we measure tile by square foot or each, while we measure carpet by square yard. The baseboard is usually measured by linear foot (LF). Note that we may not measure the accessories, such as mortar, grout, or minor reinforcement, but take it in proportion to the quantity of blocks, bricks, or tiles.
3. For soil, we measure by cubic yard, however, we have different measurements:
   - For excavating undisturbed earth, we measure by bank cubic yards (BCY).
   - For hauling earth, we measure by loose cubic yards (LCY).
   - For compaction, we measure by compact cubic yards (CCY), although this may also depend on the depth of each compaction layer.
4. Some materials are measured by the unit it is packaged and sold in such as some liquids that are sold in gallons. The estimator needs to know the proportion between the gallon of the liquid and the area it covers. Also, some admixtures that are sold in bags of specific weight, so the estimator needs to know the conversion factor to calculate the number of bags needed.

For this reason, it is strongly recommended that the estimator always writes the unit next to the number. Also, when writing the price, it is mandatory to write the currency symbol, $ for example. If this is an international project or involves another currency, the dollar symbol must be preceded by the country it represents, US$ or USD for example, since several countries use their own dollar.

Comparing the three major categories of direct cost (see Table 6.2).

Table 6.2 Comparing labor, equipment, and materials.

Comparison category	Labor	Equipment	Materials
Subject to productivity fluctuations	Yes	Yes	No
Add waste allowance	No	No	Yes
Cost money while on site? (idle)	Yes	Yes	No[a]
Overtime cost	> normal	≤ normal[b]	NA
Delivery charge?	No	Yes	Yes

(Continued)

---

24 M here is in Roman numerals, which equals one thousand. This symbol is also used with bricks: M Bricks means 1,000 bricks. For other trades, such as electricians and electronic professionals, M means one million, which comes from the Greek alphabet mega.

Table 6.2 (Continued)

Comparison category	Labor	Equipment	Materials
Mobilization/demobilization charge?	No	Likely	No
Add taxes to cost	Yes[c]	Yes[c]	Yes[c]
Add fringe benefits to cost	Yes	No	No
May need resource leveling	Yes	Yes	No
May need safe storage	No	Yes	Yes
May be subject to damage, theft, or vandalism	No	Yes	Yes
Protected by Workers' Compensation Insurance	Yes	No	No
Protected by Builder's Risk Insurance	No	Yes	Yes

[a] There is a small cost for storage and insurance, but it covers all materials. Also, you may add frozen capital cost.
[b] Utilizing the rented equipment for extended hours per day and/or per week, may result in reduced cost per hour, but not always.
[c] Taxes differ among the three categories and depend on local laws and regulations.

Materials management considerations:

Not all construction materials are treated the same. Here are some points that must be considered when ordering materials:

1. Some materials can be stored but others, mainly concrete ready mix, cannot. If the materials cannot be stored, they have to be either delivered at the time of installation or made on-site.
2. Some materials are heavy and require hoisting equipment. Transporting, storing, and handling this type of materials must be planned carefully.
3. Some materials can be stored outdoors such as sand, concrete blocks, bricks, and some types of ceramic and marble tile. Other materials need to be stored indoors, such as doors and windows, sacks of cement, paint, and wood flooring. Some of these materials need to be stored within a certain temperature or moisture range, so the storage may need to be air-conditioned. Keep in mind that some materials can be adversely impacted by extreme temperatures (high or low), moisture, or dust. So, talking about "indoor storage" may not automatically satisfy the requirements for all materials.
4. Some materials have to be carefully handled and securely stored such as:
   a. Dangerous and toxic/chemical materials such as explosives, acids, and other chemicals.
   b. Expensive materials that can be subject to damage or theft.
   c. Fragile materials such as glass and ornamental work.
   d. Small pieces that need a storage system with identification criteria such as screws and bolts.
   e. Organic materials such as wood may be subject to decay by insects or fungi.
   f. Materials that have to be stored separately such as chlorine and iron/metal pieces, because chlorine causes oxidation.
5. In deciding the order quantity versus installed quantity, several factors must be considered, such as the cost of the unit item, the expected percentage of waste, ability to use surplus in other projects or return to vendor, storage capacity and restrictions (for large items), and shortage cost.

6. Some materials are custom-made or custom-ordered. Those have to be ordered according to the exact quantity needed, especially when expensive, because the surplus cannot be used in other projects, nor can it be returned to the vendor or manufacturer. The waste allowance in this case has to be balanced between the costs of surplus versus the shortage cost.

7. There are sophisticated computer systems for managing materials. Such system can show:
   a. The description of the item (along with specifications and/or dimensions)
   b. Any special comments such as fragile, toxic, and dangerous
   c. Total quantity ordered along with the order(s) information. This may include quantity identified as defective, and action taken
   d. Total quantity used and quantity available in storage
   e. A barcode and barcode reader may be used
   f. The system may also have information on vendors available for each type of material, along with their contact and other information

   Using such a system is very useful for the project management team, in order to know what is available in storage and order any needed materials on time.

8. In all cases, the cost of storage has to consider all expenses related to storage including, but not limited to, storage structure, handling, utilities, security, frozen capital, and insurance.

Indirect Expenses

Indirect expenses include:

a. Overhead: Any expense incurred to support the business while not being directly related to a specific product or service. It includes project and main office overhead.
   (i) Project overhead includes items such as salaried staff and vehicles dedicated to the project, even when working out of the main office. Project overhead also includes expenses such as job trailers with their furniture, equipment, and utilities. Also, any project expense that cannot be linked to a specific cost item, such as site fence, signage, night lighting, security system, and guard.
   Project overhead expenses continue throughout the life of the project, although they fluctuate depending on the volume and work needs. For this reason, the contractor needs to estimate the duration of the project in order to estimate the amount of its overhead.
   (ii) Main office overhead that includes items serving the entire corporation and not a specific project(s). This includes staff working in the main office (unless they are dedicated to specific project), the main office itself with its operating expenses, and professional fees such as accounting or legal, if serving the entire organization. Main office expenses include general expenses such as advertising, marketing, and bidding projects.

b. Profit allowance

c. Contingency allowance

d. Inflation/escalation allowance

e. Cost of finance: The contractor, in most cases, will have a negative cash flow throughout the project. This requires him/her to either borrow money or use their own funds. This represents an additional cost. Even when using their own funds, the contractor must add the cost of "lost opportunity" which represents the potential profit if these funds were invested.

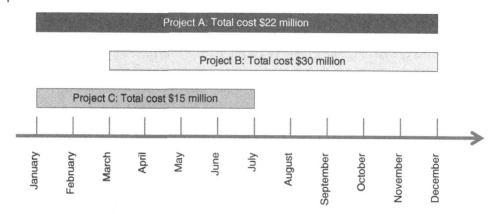

Figure 6.7 Distribution of an overhead item over several projects.

If a resource, considered overhead expense, is shared among all or some projects, the organization has two options to account for it:

A. Consider it as general overhead (main office expense), or
B. Consider it as project overhead and distribute it over the projects it serves.

If option B is chosen, the resource cost may be distributed among the sharing projects in proportion to the shared time and/or amount of the resource dedicated to each project. For example, a construction company bought a $10,000 software to be shared among 3 out of the 10 projects the company is executing. Its cost will be allocated among these projects in proportion to both their volume (cost) and length of use,[25] as shown in Figure 6.7.

Both project and main office overhead expenses may be planned and unplanned. They can be:

A. Continuous such as salaries and utilities,
B. Periodical such as taxes, legal fees, office supplies, maintenance, and
C. Accidental/incidental such as unexpected expenses for whatever reason.

Let us take a few examples of expenses and how they can be classified:

1. Granite countertop for a kitchen in the project building: Direct expense (materials).
2. Cold water for the workers in a specific project: Project overhead.
3. New furniture for meeting room in the job trailer: Project overhead.
4. A receptionist for the company's main office: Main office overhead.
5. A car and driver for clients/potential investors: Main office overhead.
6. Mortar for the block needed for the roof parapets: Direct expense (materials).
7. Bonus for project team for finishing ahead of schedule: Project overhead.
8. Fire extinguisher in the project trailer: Project overhead.
9. Fees paid to the state to renew the general contractor's license: Main office overhead.
10. Diesel generator to be permanently installed in a hospital project: Direct expense (materials).

---

25 Unlike human and equipment resources, which can be charged to different projects based on actual hours of utilization.

11. Diesel generator for generating power for night work: Project overhead.
12. Builder's risk insurance: Project overhead.
13. Cost estimating software: Main office overhead, unless the software is dedicated to a certain project, then it is project overhead.
14. Concrete forms: Direct expense (materials).
15. Subscription to Time magazine: Main office or project overhead, depending on where the magazine will be used.
16. Utilities bills: Main office bills are main office overhead, and project site bills are project overhead.
17. Safety officer salary: Project overhead if dedicated to one or few but not all projects, or main office overhead if covers all projects.
18. The CEO's airline ticket to check on the project: Can be either project or main office overhead.

Part of the expenses of the main office covers the expenses of searching, bidding, and negotiating for new projects. This is not a small item. In fact, this is so important to bring new business and keep the company running. The typical company will get only a percentage of those chased projects. This percentage varies depending on many factors including prevailing economic conditions and market situation. Typically, the contractor tries to maximize this percentage.

The general contractor's overhead (main office and project) varies widely among companies. According to RS Means, the percentage (of overhead) can range from 35% for a small contractor doing less than $500,000 (annually) to 5% for a large contractor with sales in excess of $100 million[26] with the percentage practically falls around 15–20% for most contractor. It is extremely important for the contractor to keep the overhead expenses under control because these expenses often become "invisible" expenses, and may get out of control, which can be a major reason for the failure of the company. On the flip side, there are contractors who look at almost every office employee as "overhead" in negative connotation, implying that it is an unnecessary expense. Both are extreme views that must be balanced in the best interests of the organization. Good contractors stay lean by acquiring the needed and justified resources that are necessary to run the business in the most efficient manner.

Sales Taxes

In most locations, the contractor has to add sales taxes to materials and possibly other expenses. Sometimes, there are taxes by the state, county, and city. In some cases, there may be exceptions and/or exemptions. In some cases, the public agency may be exempt from paying sales tax for materials in its own projects if it buys these materials directly. Thus, it may elect to buy and deliver expensive items to save on sales tax. This puts responsibility and liability on the owner's side for on-time delivery of these materials to the project, in coordination with the GC. Importing from another country is usually subject to customs fees.

Profit

Profit probably has more than one definition, but the author likes the simple definition: A return for taking risk. So, employees of a company receive salary but not profit because they are not taking risk but the company that hired them is taking a risk, so it should receive profit (or loss).

---

26 RS Means 2022 BCCD, R013113-50.

The contractor adds a profit allowance to the estimate, usually as a percentage of the direct or total cost, with the usual range of 5–10%. This percentage depends on factors such as:

1. Risk taken in this project. Profit percentage usually, but not always, is proportional to the risk taken.
2. Prevailing economic conditions and how badly the contractor wants the job.
3. Previous experience with the owner of the project.
4. Other bidders (competition).

It is not usual during hard economic times that a contractor bids or takes a job for 0% profit if the company's overhead is covered. This, at least, keeps the contractor's organization in business and pays expenses. In general, the contractor's attitude toward a potential project can be summarized as shown in Figure 6.8. It is not clear-cut (yes or no). Sometimes the contractor may have low interest in the project but shows interest with a higher profit margin.

Profit versus salary: To illustrate the difference between profit and salary, let us imagine that Joe decided to be an entrepreneur, selling sandwiches and hot dogs. He rents a mobile cart for $50/day. Consumable supplies (buns, food items, condiments, napkins, etc.) cost $100/day. After a full day of work (10 hours), Joe makes an average of $400 in sales. How much is his profit? The answer is *not* $250! Joe needs to pay himself a salary equivalent to what he would have made if he worked as an employee for someone else. Let us assume that the salary of an employee preparing and selling sandwiches is $180 for 10 hours/day. The net profit then is:

$$\text{Profit} = \text{Revenue} - \text{expenses} - \text{salary}$$

$$\text{Profit} = \$400 - \$50 - \$100 - \$180 = \$70$$

What if total sales per day are $300? The profit would be −$30, which is a loss. Does this automatically imply that it is better for Joe to quit his own business and work for someone else? Perhaps, but not necessarily. Many businesses start at a slow pace and then pick up gradually. However, some people do not have entrepreneurial and managerial skills, and they are better off working as employees.

For simplification, our example ignored many factors such as employer-provided benefits, if any. What if Joe got sick for a few days? As an entrepreneur, he will not get any income. In fact, he may still need to pay the cart rent and take other risks. So, it is not as simple as looking at the numbers, but this simplified example is only to explain the concept of profit.

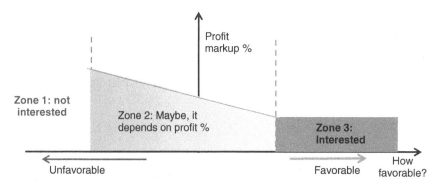

Figure 6.8 Contractor's interest in the project.

There is also the "It did not cost me anything" misconception. This is a situation that happens sometimes, especially with small contractors when using personal resources (equipment, money, or even humans) without compensation. The contractor (or any businessperson) must separate personal and business matters. So, if he uses his personal pickup truck (on a regular or frequent basis), he must charge the cost of using the truck to his business.[27] If his wife or teenage son/daughter is helping him (on a regular or frequent basis), he must pay them or at least count their effort as an expense. Even when using own funds to finance operations, he must charge his business interest or "lost opportunity" fee.

Calculating business expenses, including profit/loss, may be different from the official tax accounting. For example, a company may be making profit but acquiring new assets, so it shows zero profit for tax purposes. On the other side, a company may suffer big losses in one year, so it distributes this loss over several years to minimize the taxes in the upcoming profitable years. Even in subjects such as asset depreciation, business decisions may be different from what a contractor may do in his/her income tax forms. In this book, we do not involve detailed income tax laws and their loopholes. The book addresses issues from a pure business perspective.

Contingency Allowance

Contingency allowance is an amount of money set aside for events that are likely to happen but not exactly known in extent and timing. It is an important and delicate issue that needs to be balanced. If the contractor does not add a reasonable allowance for contingency, unexpected events may—and usually do—happen, which will be then taken out of his/her profit. However, the contractor must not overestimate or abuse the contingency allowance. Overestimating it will raise the contract price tag, which may reduce the chances of winning the project in the case of bidding. Abusing the concept of contingency may and does happen by using this fund for either expected expenditures or to cover for own faults and lack of planning and/or performance.

So, how much should the contractor estimate the contingency allowance? The amount—usually, as a percentage of the total construction cost amount, is directly proportional to the risk involved. This includes expenses from events such as accidents (natural or not) not covered by insurance and other expenses that either were not expected or happened to be more than originally estimated. The most important thing is not to allow contingency allowance to be used as an excuse for poor estimating or as a "general fund" when money is needed but cannot find other legitimate fund source.

Since risk is a major factor in estimating the percentages of both profit and contingency allowances, the author suggests that they are estimated separately but at the same time, taking each one into consideration when estimating the other. The overall estimate for both must be balanced, taking into consideration risk, and other influencing factors.

Having defined contingency, the interesting question is: What happens to the contingency fund by the end of the project in normal situations? The term "normal" here means things happened as expected, not better and not worse. Let us clarify this concept with an example. A contractor added to his estimate a

---

27 It is possible that a personal vehicle, home, or other property is being used partially for work, and this is declared officially in the person's income tax.

profit allowance of 5% and a contingency allowance of 6%. What would be his net profit after the completion of the project if:

A. Things occurred on a normal scale and pace,
B. Things went better than expected, or
C. Things went worse than expected.

The answers are:

A. Net profit ≈5%
B. Net profit > 5% but < 11%
C. Net profit < 5% but if things went much worse than expected, profit may be zero or even negative (loss)

This is further illustrated in Figure 6.9.

The term contingency allowance has to be linked to a specific project or other expense category. For any project, there must be a contingency allowance for the GC and perhaps the owner has its own contingency allowance, with the amounts varying according to the risk each party takes. The GC, or any

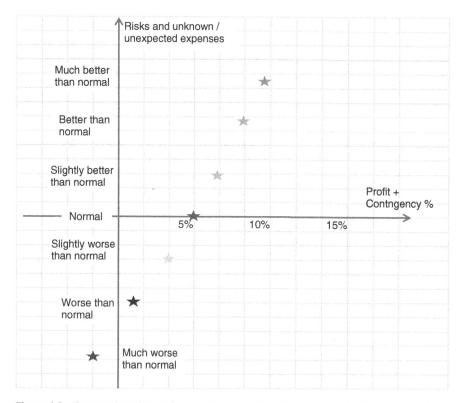

Figure 6.9 Expectations for profit + contingency with different project outcome scenarios.

Table 6.3 Allocating indirect expenses.

	Unallocated	Total allocation	Partial allocation
Labor	$375,211	$487,105	$456,564
Materials	$422,735	$548,801	$514,392
Equipment	$55,000	$71,402	$66,925
Subcontractor	$165,000	$214,206	$200,775
Subtotal	**$1,017,946**	**$1,321,513**	**$1,238,657**
Performance Bond	$11,600	$0	$11,600
Insurance	$30,538	$0	$30,538
Job overhead	$55,987	$0	$0
General Overhead	$50,897	$0	$0
Inflation/escalation	$40,718	$0	$40,718
Contingency	$50,897	$0	$0
Profit	$62,929	$0	$0
Grand total	**$1,321,513**	**$1,321,513**	**$1,321,513**

owner of a business, must also keep a general contingency allowance with the administrative budget, independent from any project budget.

Allocating Indirect Expenses:

Contractors may not like to show the details of their indirect expenses, so they may allocate them by adding them covertly to the direct expenses. Table 6.3 shows an example of an estimate for a small project in three formats: Unallocated, Total Allocation, and Partial Allocation.

Sources of Cost Estimating Databases

The best database for a cost estimator professional or construction company is their own, if such database exists and provided it is well-organized and updated regularly. In many situations, the estimator may not know the price of an item either because it is a new item, it is a known item but in a different location or country, the market has changed since the estimator last priced it, or for any other reason. The estimator can use commercial third-party databases such as RSMeans (from Gordian), ProEst (from Autodesk) in the United States. Davis Langdon's Spon's database is popular in the United Kingdom and other parts of the world. Compass International has commercial and industrial databases in the US and numerous other countries. NODOC—Nomitech Database is used for the oil and gas industry.

Software vendors also create their own databases but allow users to either build their own database or tie it into a third-party commercial database such as RSMeans and ProEst. With the advancement of technology, software users can update databases (prices and other data) online.

Equipment manuals and manufacturer's provided information help in calculating production rate, which helps the cost estimator calculate the unit price, as shown in Eq. (6.3).

Building your own database can be a difficult and time-consuming task. It will most likely require contributions from several specialists and a dedicated administrator who must be responsible for maintaining and updating it. It must be kept up to date in terms of item availability, prices, production rates, and other information. Items may be taken out of the database while new items are added. Databases can be resident on the hard drive of a personal computer in the simplest form or can be on the server and may be accessed online with proper security procedures. Security and backup of databases are extremely important.

Most databases, especially commercial ones, allow only the administrator to make changes to the database. So, when the cost estimator needs to make an adjustment to a unit price, production rate, or other, the item has to be pulled from the database to the estimate and then apply the adjustment.

In many instances, the cost estimator receives a quote from a vendor. The estimator must be careful in making sure the quote includes all the important information such as item detailed description, minimum quantity (especially when there is a volume discount), unit of measure, price currency (if international), price restrictions if any, price extras (taxes, shipping, delivery, etc.), validation of quote (period and conditions while this quote is valid), name of the salesperson providing the quote, return policy and restrictions. This quote must be in writing (email, fax) and must be kept as part of the estimate/project document.

Bonds and Insurance

A bond, also called surety bond, is an agreement whereby the bonding company (usually a specialty insurance company) guarantees to a property owner that a contractor will perform according to the contract documents. Construction surety bonds protect property owners against financial losses that could result from the default of a contractor. Surety bonds are three-party agreements between a surety, a contractor, and an owner, as illustrated in Figure 6.10.

Surety's obligations, however, are generally limited to the amount of the bond. If the surety incurs any liabilities in correcting the work or paying subcontractors and suppliers, that liability is usually

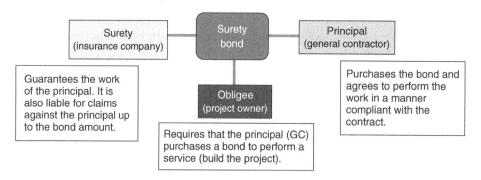

Figure 6.10 Surety bonds explained.

recovered from the contractor pursuant to a general indemnity agreement between the contractor and the surety, which includes personal indemnities from key parties with the contractor.[28,29]

The most used types are bid bonds, performance bonds, and payment bonds. There is also a warranty bond that sometimes the contractor has to provide.

A bid bond is usually required as part of the bidding process from all bidders to the project owner, to provide a guarantee that the winning bidder will undertake (sign) the contract under the terms at which they bid. The bid bond amount usually ranges between 5% and 10% of the bid amount, although it can go above or below this range. It is also possible that the owner sets the value of the bid bond, $100,000 for example. Some contractors, especially in small projects, provide a cashier's check as a "bid bond."

After the owner processes the bids and announces the winning bid, this contractor must comply with the requirements of the bid invitation and sign a contract with the owner, usually providing performance and payment bonds. Once this step is done, all other bidders receive back their bid bonds. So, if things happened according to the bidding and contracting rules, there will be no cost for the bid bond to the bidders. However, if the winning bidder did not comply with the requirements or refused to sign a contract for no legitimate reason, then this bidder will lose the bid bond sum or the difference between his/her bid and next bid; whichever is less.

Example for bid bond: A project owner sets the bid bond at $100,000. After opening the bids, the lowest qualified bidder's bid was $8,500,000, however, he refrained from signing a contract with the owner for no valid reason. Does he lose the bond?

A. If the next lowest qualified bidder's bid was $8,600,000 or more, then the first bidder loses all his bond.
B. If the next lowest qualified bidder's bid was under $8,600,000 say $8,570,000, then the first bidder loses the difference between the two bids, $70,000, and gets the rest of the bond money.

Performance Bond

Performance bond protects the owner from financial loss resulting from failure or default of the general contractor to perform the job according to the terms and conditions of the contract. The contractor usually has to provide performance and payment bonds as part of the contract requirements. Both bonds are usually obtained from an insurance company as a "surety." In case of a default or serious violation of the contract, the surety will be obligated to find another contractor to fulfill the contract terms. This, most likely, will cost the insurance company expenses, so it will try to recover it later from the defaulting contractor.

The cost of performance bond is a percentage of the entire bid amount, including itself. It ranges from 0.6% to 2.5%, the higher the contract amount, the lower the percentage. It also differs with the type of construction. It is higher for building (vertical) construction than roads and other horizontal construction projects. Tables 6.4–6.6 help give an idea of the cost of performance bond.[30]

---

28 https://www.aiacontracts.org/articles/6260866-surety-bonds-what-owners-should-know.
29 https://www.suretybondsdirect.com/educate/what-is-surety-bond.
30 All three tables are taken from the 2022 RS Means Building Construction Cost Data, BDDC.

Table 6.4 Calculation of performance bond for highways and bridges resurfacing projects.

Contract amount	Performance bond for building construction
up to $100,000	2.50%
$100,001–500,000	$2,500 + 1.5% of amount over $100,000
$500,001–2,500,000	$8,500 + 1.0% of amount over $500,000
$2,500,001–5,000,000	$28,500 + 0.75% of amount over $2,500,000
$5,000,001–7,500,000	$47,250 + 0.7% of amount over $5,000,000
over $7,500,000	$64,750 + 0.6% of amount over $7,500,000

Table 6.5 Calculation of performance bond for building construction.

Contract amount	Performance bond for highways and bridges: new
up to $100,000	1.50%
$100,001–500,000	$1,500 + 1.0% of amount over $100,000
$500,001–2,500,000	$5,500 + 0.7% of amount over $500,000
$2,500,001–5,000,000	$19,500 + 0.55% of amount over $2,500,000
$5,000,001–7,500,000	$33,250 + 0.5% of amount over $5,000,000
over $7,500,000	$45,750 + 0.45% of amount over $7,500,000

Table 6.6 Calculation of performance bond for new highways and bridges projects.

Contract amount	Performance bond for highways and bridges: resurfacing
up to $100,000	0.94%
$100,001–500,000	$940 + 0.72% of amount over $100,000
$500,001–2,500,000	$3,820 + 0.5% of amount over $500,000
$2,500,001–7,500,000	$15,820 + 0.45% of amount over $2,500,000
over $7,500,000	$39,570 + 0.4% of amount over $7,500,000

Payment Bond

Payment bond is a surety bond posted by a contractor to guarantee that its subcontractors, laborers, and material suppliers on the project will be paid for their products and/or services, so they will not place a "mechanics lien" (or simply lien) against the property or make claims against the owner.

A mechanic's lien is a guarantee of payment to builders, contractors, and construction firms that build or repair structures. Mechanic's liens also extend to suppliers of materials and subcontractors and cover

Table 6.7 Comparison among the three types of bonds.

Bond Type	Bid bond	Performance bond	Payment bond
Amount	Varies as set by the owner, either as a lump sum or percentage, usually 5–10% of the bid amount	Based on several tiers of the value of the contract with a descending percentage.	Varies from 0.5 to 3% and is usually combined with performance bonds.
Required from	All bidders	Only the lowest qualified bidder	Only the lowest qualified bidder
Final cost	Nothing unless the contractor is selected as the lowest qualified bidder and then refrains from signing a contract with the owner for no valid reason	The amount paid for the bond is added to the project expenses as it is not recoverable	The amount paid for the bond is added to the project expenses as it is not recoverable
Guarantees	The bidder will sign a contract with the owner if selected as the lowest qualified bidder	The completion of the project as per the contract terms	The GC will pay all subs, vendors, and workers, so the owner's project will not be subject to a lien

building repairs as well. It occurs if the claimant party did not receive full payment for materials/services rendered in the project. While the lien is placed, the owner may not be able to sell without court permission.

So, the payment bond will give the owner an assurance that the general contractor will make sure all the subs, vendors, and any other party that contributed to the project, will be paid so none of them will later place a lien against the project. Typically, with every progress payment, the general contractor must provide the owner with "lien releases" (also called lien waivers) from every party that is receiving payment, including the GC itself.

When payment bonds are issued with a performance bond, it is estimated that the premium will be between 1% and 2%, although the actual cost may vary depending on the credit history and background check of the contractor requesting the bond.

All public agencies require performance and payment bonds. Most private owners do so as well but it is a choice for them. Waiving such contract requirements will save money but increase the owner's risk. The Construction Industry Institute (CII), published an interesting study on the cost-trust relationship between the project owner and contractor. The project cost decreases with the increase in trust level between project parties (mainly the owner and general contractor) to the point of "blind trust," where the trust must not go farther.[31]

Table 6.7 shows a comparison among the three types of construction bonds.

Insurance

Insurance is a promise of reimbursement – by the insurance company to the insured – in the case of a covered loss. In construction, there are many types of insurance. The most common types are Workers' Compensation, builder's risk, and public/general liability insurance.

---

31 https://www.construction-institute.org/resources/knowledgebase/knowledge-areas/project-organization-communication/topics/rt-024.

Workers' compensation insurance covers a contractor's employees in case of personal injury or death during work or work-related illness, regardless of whose fault it was. Between 1911 and 1948, all 50 states passed Worker's Compensation Law. Even though the details of the law differ from one state to another, the essence of this important insurance is the same. It protects both sides, eliminating potentially thousands of cases of lawsuits. It protects the employee because the insurance covers the employee while at work,[32] without the burden of proof of fault. It protects the employer because it assigns financial amounts for all types of losses.

Workers' Compensation (WC) insurance rates depend on three factors:

1. Employee's trade (carpenter, electrician, etc.). The rate for WC is directly related to the risk involved with that type of work. High-risk trades such as structural steel workers (erectors, painters, welders), roofers, and wreckers have the highest rates. Office staff have low rates.
2. State where the project is located. WC insurance is subject to state law and regulations. Insurance companies have the same rates within the same state, but rates differ from one state to another. For example, WC insurance rates in Indiana are much lower than those in the neighboring state of Illinois.
3. Individual employers' past safety record, represented by its experience modification rate (EMR). The EMR ratings are calculated by the insurance industry based on the record of three years prior to last year. For example, in 2023, the records of 2019, 2020, and 2021 are used. The average EMR for the industry is 1.0. An EMR below 1.0 indicates a good (better than average) safety record, which rewards the employer with a discount proportional to its EMR. On the contrary, an EMR greater than 1.0 represents a safety record worse than average, which increases the WC rates for that employer by a percentage proportional to its EMR.

U.S. workers' compensation insurance rates are published in the RSMeans database, divided by trade and state. They represent the contractor with an average safety record (EMR = 1.0). The U.S. average rates for different crafts are 17.1% of labor rates in 2023. For bidding purposes, apply the full value of WC directly to total labor costs, or if labor is 38%, materials 42%, and overhead and profit 20% of total cost, carry $38/80 \times 17.1\% = 8.12\%$ of direct cost, or $38/100 \times 17.1\% = 6.5\%$ of total cost.

The above argument underscores the importance of safety records. Let us imagine two contractors: Contractor A with EMR = 0.80 and Contractor B with EMR = 1.30. In a building project worth $20 million, labor is expected to be around $7.6 million. Worker's Compensation insurance (national average) = $17.1\% \times \$7.6$ million = $1,299,600. This represents the amount of WC insurance a contractor with average safety record would pay for that project (also national WC rates.) Now, Contractor A pays $0.80 \times \$1,299,600 = \$1,039,680$, and Contractor B pays $1.30 \times \$1,299,600 = \$1,689,480$. The difference between the two contractors, $649,800, is about 3.25% of the total cost of the project. This difference can get bigger in states with higher-than-average WC rates such as Georgia, California, and Illinois.

---

32 The employee is at work even if he/she is not physically in the work site, as long as the employee is within work paid time.

> There are scenarios where we may wonder if WC insurance applies. An employee had an auto accident while driving:
>
> A. To buy materials for the project while within work hours.
> B. To a restaurant during the lunch hour.
> C. Back home after finishing work.
>
> The answers are: Yes for A, and no for B (assuming the lunch hour is not paid), and no for C.

Public/general Liability insurance covers third-party claims for personal injury or death, but rules may differ in some states to include coverage for other third-party claims. Depending on the individual policy, it may cover the assets of a business when it is sued for something it did (or did not do) to cause an injury or property damage. Example of typical coverage is an injury to a visitor to the site, damages to property outside the worksite resulted from the collapse of a crane, or lawsuit claiming that dust or noise from the worksite caused injury or illness to people living near the site. The cost of public liability insurance is about 2% of total project construction cost but it varies depending on several factors.

Builder's Risk insurance protects the contractor's insurable interest in materials, fixtures, and/or equipment awaiting installation (or after installation) during the construction or renovation of a building or structure, should those items sustain physical loss or damage from a covered loss. Cost is between 0.24% to 0.64% of total project construction cost.

> WC insurance covers all employees (laborers and staff) during work hours. Public/general liability covers other people who are not the contractor's employees. Builder's risk covers materials, equipment, and structures (all but humans).

There may be more types of insurance such as auto and "hired and non-owned auto" (HNOA) insurance, that cover automobiles owned and/or used in the project. Sometimes, the general contractor also acquires "umbrella insurance" to cover any claims that may not be covered or not covered enough by other insurance policies. It is also important to note that laws differ from state to state and from country to country. Also, there may be multiple jurisdictions where several insurance companies are involved, and typically every insurance company tries to have other companies pick the tab. Most policies also have a set deductible, which is the amount that the contractor (or policy holder) will pay out of their own funds before the insurance company pays anything.

Most importantly, the contractor must have enough coverage in terms of both the type of accident/incident and amount of coverage. Also, it is very important to keep a good safety record. It pays off in so many ways.

Bonds versus Insurance

Even though sureties that issue bonds are, in most cases, insurance companies, there are major differences between bonds and insurance:

1. Bonds are a three-party agreement (surety, contractor, owner) while insurance is a two-party agreement (insurance company, contractor).

2. The contractor acquires the bond for the owner's peace of mind but acquires the insurance for their own protection and peace of mind.
3. The money an insurance company pays for a claim is not recoverable but if the surety incurs expenses as a result of the contractor's default, it has the right to and does collect these expenses from the contractor.

> Aside from legal and contract requirements, contractors buy insurance for their own protection while they buy bonds for the owner's protection!

Inflation/Escalation Allowance

Inflation refers to changes over time in the overall level of prices of goods and services throughout the economy. The U.S. government measures inflation by comparing the current prices of about 80,000 items in a fixed basket of goods and services representing what Americans buy in their everyday lives to previous prices at the same time every year.

Escalation usually refers to the change in the price of one particular item or group of items of goods and/or services over a period of time. Both inflation and escalation have to be considered while preparing an estimate. Inflation is important because it is one reflection of the economy. Escalation is important also because it focuses on the major item(s) that represent the largest percentage of the contractor's cost, such as labor, lumber, steel, and concrete.

Since the estimate is performed before the project starts while actual expenditures occur later, inflation has to be considered. We can use the equation of the single payment with compound interest:

$$F = P \times (1+i)^n \tag{6.2}$$

As shown in Figure 6.11, the period "n" in the equation starts from the date of the database till the expected "midpoint" of the project. In this example, $n = 5 + 4 + 9 = 18$ months.

The contractor may focus on major cost items. He/she must also use educated common sense judgment to estimate possible price changes where the project is located, taking into consideration current

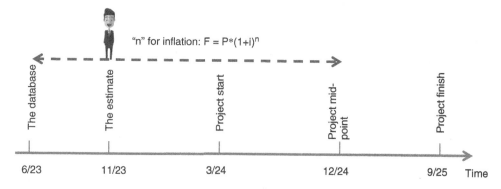

Figure 6.11 Adding the inflation allowance to the estimate.

and possible future events. The longer the forecasted period is, the higher the risk (uncertainty) will be. The contractor can also balance anticipated inflation/escalation with contingency allowances. It is good practice to be reasonably conservative and conservatively reasonable!

> Although inflation and escalation are important to the project's cost estimation, escalation of the cost of a particular item may be more important, especially when this item is a major part of the estimate.

Crew Productivity

Crews can be composed of laborers only or laborers with equipment. The production rate of the crew is very important for both the cost estimator and the scheduler. Crew production impacts both the activity's unit price and duration, as shown in Eqs. (6.3) and (6.4).

$$\text{Unit price} = \frac{\text{Crew daily cost}}{\text{Crew daily production}} \qquad\qquad (6.3)$$

$$\text{Activity duration} = \frac{\text{Total quantity}}{\text{Crew daily production}} \qquad\qquad (6.4)$$

As an example, we have an excavation crew that costs $1,945/day and has an average daily production of 760 cubic yards (CY). We need to excavate 11,000 CY.

$$\text{Unit price} = \frac{\$1,945/\text{day}}{760\,\text{CY}/\text{day}} = \$2.56/\text{CY}$$

And the total cost $= 11,000\,\text{CY} \times \$2.56/\text{CY} = \$28,151$

$$\text{Activity Duration} = \frac{11,000\,\text{CY}}{760\,\text{CY}/\text{day}} = 14.47 \approx 14\,\text{or}\,15\,\text{days}$$

Now imagine if the crew productivity (daily production) changes, increasing or decreasing, what will happen to the unit and total cost and to the duration? Assume productivity decreased by 20% due to adverse weather, so the crew daily production $= 760 \times (100\% - 20\%) = 608\,\text{CY/day}$

$$\text{Unit Price} = \frac{\$1,945}{608\,\text{CY}/\text{day}} = \$3.20/\text{CY}$$

And the total cost $= 11,000\,\text{CY} \times \$3.20/\text{CY} = \$35,189$

$$\text{Activity Duration} = \frac{11,000\,\text{CY}}{608\,\text{CY}/\text{day}} = 18.09 \approx 18\,\text{days}$$

The crew composition depends on several factors, but there are two cases:

A. Labor-dependent crew such as a crew laying blocks or bricks, building a roof, or painting. The number of laborers in the crew depends on the contractor's selection based on job requirements, available workers, workspace, and possibly other factors. In such a crew, the addition of a laborer will increase the production rate of the crew and vice versa. This is important when the contractor needs to accelerate the schedule by shortening the duration of critical activities and in resource allocation and leveling.

B. Equipment-dependent crew such as an excavation or concrete placement (by a pump) crew. In this case, the production rate is dependent on the equipment, so adding one laborer will not increase the production. Increasing production can happen either by replacing the machine with another higher capacity or adding a second crew.

> Remember that fluctuations in crew/labor productivity affect both the activity cost and duration.

Adjusting Estimates

Estimates must be adjusted for many factors, some of which have to do with productivity. When productivity changes, cost reacts inversely. There are also factors that affect cost but not productivity. Here is a list of some of them.

Factors that impact production rate:

1. Weather conditions: Temperature, humidity, wind, precipitation, visibility, lightning, dust, or others.
2. Labor skill level and type of labor (union or open shop).
3. Crew composition.
4. Type and condition of equipment and tools used.
5. Availability and readiness of materials.
6. Design complexity.
7. Location, local culture, and regulations.
8. Learning curves: This can be a factor in tasks with units of repetitive nature.
9. Fatigue/low morale: Fatigue may be caused by extreme weather, long-term overtime, or other factors. Stress or low morale may be caused by an accident on the job site, a looming strike, personal issues, or tension among the workforce members.
10. Management style and effectiveness.
11. Typical workday/workweek: Work hours per day and work days per week. Number, frequency, and length of breaks.
12. Quality assurance/quality control (QA/QC) requirements.
13. Communications system and protocol.
14. Multi-shift turnover.
15. Soil condition, especially for excavation.

16. Terrain/elevation/height of work item above the ground (if working on scaffolding for masonry, painting, etc.).
17. Safety and security requirements and conditions.
18. Job site logistics, conditions, and congestion.
19. Any other condition that may impact productivity.

Factors impacting price but not productivity:

1. Project location: This impacts the prices of resources, delivery charges, fees, taxes, competition, and other location-specific expenses.
2. Project size: When the project size is large, there is a likelihood of lower unit price for materials.
3. Expected inflation/escalation, taking into consideration prevailing economic conditions and expectations for world events.
4. Risk taken and desired profit.

Note that some factors mentioned earlier, especially weather, affect activities by variable degrees based on the activity's nature and location. For example, heavy rain hampers outdoor activities such as excavation a lot more than it does for indoor activities such as flooring.

Despite the fact that experts tried to derive equations and formulas for cost estimate adjustments based on some of the factors mentioned above, there is no such accurate formula. The adjustments must be made based on the experience and feelings of the cost estimator, after considering all the factors and obtaining as much information as possible.

One important consideration in adjusting estimates: When one aspect of the estimate changes, the estimator needs to look not only at that aspect by itself, but also any impact on other aspects of the estimate. For example, if a home builder knows the construction cost of a house, but the owner likes to raise the ceiling from 10 to 12 feet. Obviously, direct items such as the framing, walls, partitions, drywall, paint, stairs, etc., will be increased proportionally, but there are other items such as HVAC since the volume increases as well. It is also possible that window sizes may be affected. When this happens in a multi-story building, you need to consider the impact on the design and cost of the elevator. Raising the ceiling may also require the structural engineer to check the design, mainly the columns, and see if it needs to be strengthened.

Putting the Estimate Together

After completing the estimate, it has to go through the review process. This process can be a simple review by the contractor to the major items and the totals. Or, for large projects, the estimate will likely go through several reviews by a peer and the chief estimator, going up to the top person in the organization. It is very common for the top authority person to question, discuss, and even change some numbers until he/she is satisfied with the final number.

It is extremely important to reference and document everything involved in the estimate, so if the estimator or anyone else in the contractor's organization needs to verify or modify an item, they can do so easily and efficiently.

Cost Estimating Software Programs

Cost estimating software programs basically contain 3 parts:

1. The database where all items are stored with their description, cost, and other information,
2. The spreadsheet (the estimate) where the estimator can pull the needed items from the database, assign quantities to them,
3. The reports where the estimator can show the estimate in the desired manner. Also, the user can export the estimate to a third-party program such as Excel, where it can be adjusted and manipulated based on the user's requirements and desires.

Most commercial cost-estimating software programs control and protect their database, so the user may not be able to adjust prices, or productivities in the database, or add or delete items. But, once the items are pulled into the spreadsheet, they can be adjusted based on the estimator's requirements and judgment. The user can also create their own custom items that will be part of the user's profile and not the database.

The database must be updated periodically by its publisher (the owner company). If the contractor own his/her own database, then he/she can adjust it whenever desired, but when there are several users, care must be taken when allowing the users to change and modify items. It would be best if there is an administrator who controls the access to the database.

Repricing Estimates

In certain cases, after the estimate is completed, one or more factors change that trigger adjusting the estimate. Most software programs can handle such changes easily and efficiently. For example, assume that the contractor priced the ready-mix concrete, 4000 psi, at $125/CY, which is used by 120 items in the estimate. Shortly later, the price changed to $128/CY, so the contractor can make that change to the price of the concrete mix, and the program will make the change for all items that use that concrete mix in the estimate. If this change is global and applies to all upcoming estimates, the contractor needs to do it in the database, and then reprice the estimate. The "reprice" function takes all the items in the estimate through the updated database, adjusting the items that the unit price changed. If the contractor is estimating a project that is a duplicate of a previous project, but in a different location, he/she can reprice the project using the current database for the current location. Repricing estimate does not change or affect quantities.

One important estimate repricing may be needed if the entire project is postponed to a different time zone where productivity and/or unit prices differ. Let us consider a house project that was planned to start in a location that has a seasonal extreme weather, between June and September. The original plan called for construction to start in early December and finish in May, a total of 6 months. The owner had issues with financing the project, which delayed him and put the start in May. This slippage affects the estimate and the schedule, even if market prices do not change. As we saw in Eqs. (6.3) and (6.4), the unit price and duration of the work activity depend on crew productivity, which will decline during the project's new timeline. The contractor needs to reprice the estimate based on new production rates and unit prices. The schedule may need to be revisited too since production rates will decrease, and thus durations will increase.

The "Mathematical Intuition"

It is commonly believed that the natural ability of most people in doing mathematical operations has declined since the invention of calculators and computers. Although the author does not have the statistics to prove this claim, it is an observation by many of us, especially those who were born and raised before calculators and computers. What supports this observation is the known rule that every power or skill we have, mental or physical, increases by use and diminished by neglect. The author defines the mathematical intuition as the ability to:

1. Instantly/quickly perform mathematical operations and obtain reasonable and relatively accurate (approximate) results, and
2. Make a good commonsense judgment on answers, produced by self or others, and tell if the number is too low, too high, or about right.

Such intuition is important to the cost estimator and project manager since errors can still happen for several reasons. A failure to catch such mistakes can have a devastating effect on the contractor. In addition to the ability of mentally crunch numbers and know the rough answer, the speed of processing matters, sometimes considerably. A contractor may be in an unplanned meeting with a potential client and is asked about the cost of a project or work package, with no immediate access to a computer and other tools. The ability to visualize the project (or work package) and give an approximate estimate, even with a margin of error of up to 30%, is a big advantage. Similarly, if the cost estimator made a major error in the estimate (sometimes by either omitting or duplicating a digit when keyboarding), the contractor must be able to tell that such number is too low or too high. Discovering the error later may have negative consequences.

Is mathematical intuition inherited or acquired? Again, we do not have scientific data, but it is believed that it is both inherited and acquired. What is certain also is the fact that we can improve it by practice in our daily lives.

> We can generally say that the adjustments the contractor makes to the cost estimate are a combination of objective and subjective factors. It is based on past experience, investigation, expectations, and "gut feeling."

The "Good Cost Estimator's" Rules:

1. The good estimator must be able to produce an approximate cost (ballpark figure) for a project, with primitive or no tools available.
2. The good estimator must have the intuition to make mental judgment on numbers and tell if the number is too low or too high.
3. The good estimator never relies totally on machines (computers, calculators) without applying an educated commonsense judgment.
4. While the price of one item may be too high or too low, the overall estimate should be pretty accurate.

Exercises

6.1 Define cost and differentiate between cost and value. Give an example.

6.2 Define cost estimating and quantity estimating/surveying. Draw the differences between them.

6.3 To the quantity surveyor, doing quantity estimating for a project identical to a previous one is an easy task. This may not be the same for the cost estimator. Explain.

6.4 Define the baseline budget.

6.5 Moving the accuracy of the estimate from 70% to 90% takes a lot more effort than moving from 0% to 70%. Explain.

6.6 What is the Schedule of values? What elements does it contain?

6.7 Why is the schedule of values important even in lump-sum contracts?

6.8 An owner has complete design documents for a 20-story office building. He is asking for your opinion on the recommended type of contract and delivery method. Make recommendations along with justification.

6.9 The following expenses are direct expenses (D), job overhead (J), or general overhead (G)? (If expense is not clear, make an assumption and answer based on your assumption)
   a. A pickup truck for the project manager of a building project.
   b. Paint for the external walls of a building.
   c. A bonus paid to all company's employees at end of the year.
   d. Laptop for the cost estimator in the main office.
   e. A security guard to watch out at the project site.
   f. An airline ticket for the company's vice president to check the progress of the project.
   g. Kitchen cabinets for the new project.
   h. Safety equipment for workers in a project.
   i. Concrete retardant to slow the curing during summer.
   j. A software program to schedule the company's projects.
   k. Tile for floor seven of the building being built.
   l. A bonus paid to project employees and workers for finishing the project ahead of schedule.
   m. Scaffolding to paint the ceiling.
   n. Secretary in a project office (on-site).
   o. Formwork materials for a concrete suspended slab.
   p. Bulldozer to do excavation for the foundation.
   q. Advertisement to congratulate the CEO on his son's wedding.
   r. Printer for the project manager (on-site).

s. Overtime payment to workers to finish installing windows.

t. Party for the workers and staff to celebrate finishing the project without accident.

u. Marble countertop for the kitchens.

v. Safety violation for lack of temporary rail on the upper floors of the building.

w. Phone bills for the main office.

x. Cold water for the workers on site.

y. A safety officer who serves three projects out of the company's ten projects.

6.10 What are the major factors that influence the amount (percentage) of profit a contractor desires in a project?

6.11 What does the term overhead expenses mean and what does it include?

6.12 You are the CEO of a major construction company. You were told by the accountant that your overhead expenses are too high compared to companies of similar specialty and size. Create a plan to cut these expenses without harming the operations of the company.

6.13 In estimating the cost of a future project, which one is more important for the estimator: inflation or escalation? Why? Explain.

6.14 In balancing between Just-in-time theory and Inventory buffer, which theory you would lean toward if your project is:
  a. In the downtown of a major city.
  b. Away from the city; at least 100 km from the closest city or store.
  c. In a small town, not far from a major city.
  Justify your answer in each case.

6.15 Define cost control.

6.16 Is it okay to round numbers in cost control? What is the fundamental difference between cost estimating and cost control, regarding rounding numbers?

6.17 A general contractor based in New York would like to bid on a project in Arizona. He has never done work before in Arizona. Mention factors that may help him decide to bid or not to bid.

6.18 What is the "rule of thumb" that tells you if a cost item is direct or indirect expense?

6.19 What are the major factors that influence the amount (percentage) of contingency a contractor includes in the total cost for a project?

6.20 What discourages a contractor from bidding on too many projects?

6.21   Check the right boxes:

	Bonds			
No.	**Statement**	**Bid bond**	**Performance bond**	**Payment bond**
1	Required from all bidders			
2	Required from the lowest qualified bidder only			
3	It is refunded at the end of the process			
4	It guarantees that if chosen as the lowest qualified bidder, will sign a contract with the owner			
5	It assures the owner that the contractor will pay all his subcontractors, vendors, and laborers			
6	Protects the owner against liens against the project			
7	It assures the owner that the contractor will complete the project within the terms of the contract. Otherwise, the surety will get another contractor to complete the project within the terms of the contract			

6.22   Mention five factors that impact productivity.

6.23   Mention two factors that impact the price of an item but not productivity.

6.24   A crew of two laborers can install 300 SF of ceramic tile per day.
    a. What are the labor hours per SF?
    b. How many hours does the crew need to finish a 3700 SF floor?
    c. If productivity goes down by 20%, what are the daily production and the man-hours/SF?
    d. The productivity can improve by 10% if the contractor conducts training for his/her laborers at a cost of $300/laborer. Assume the contractor pays each worker $35/hour. Do you think, for this job, the training will pay for itself? (Ignore part c)

6.25   A crew of 5 laborers (3 masons and 2 helpers), must build a wall, 620 LF long and 5 ft high, using 8″ concrete blocks. The crew can do 520 blocks /day. The contract stipulates that the wall must finish in 7 days and pays $8.60/block. The contractor pays the mason $44/hour and the helper $32/hour. It costs the contractor $4.00/block including all accessories (materials only). The contractor's overhead cost is 12% of the total cost. Estimate:
    a. What is the "install quantity" and the "order quantity" with 5% waste allowance?
    b. What is the project's total baseline budget and duration?
    c. What is the contractor's cost per block and per SF?
    d. If everything runs as planned, what is the contractor's net profit?

  e. What are the total labor hours, labor hours per block, and labor hours per SF?

  f. Do you think the contractor will finish the project on schedule, if the crew keeps its production rate?

6.26 In Exercise No. 25, assume that adverse conditions resulted in a reduction of 15% of the production rate.

  a. What is the impact of this production rate reduction on the cost and duration?

  b. If the contractor needs to finish the wall on schedule and can have the crew work 10 hours/day (8 hours at regular pay and 2 hours @ 1.5 pay rate). Assume no loss of productivity due to overtime. How many overtime hours need to be rendered? What is the impact on the contractor's cost?

  c. If the owner allowed the contractor to finish in 8 days, what will the contractor's cost be?

6.27 An estimator needs to do an estimate for a project that is planned to start in January 2025 and lasts for 16 months. If (time now) is March 2024 and the database is 3-month-old. How long is the period for estimating the inflation/escalation? If the estimate, before adjusting for inflation, is $9,450,000, and the estimator needs to assume 5% annual inflation, how much the estimate will be after adjustment?

6.28 Assume that the lumber in Exercise No. 27 is estimated at $1.1 million, and the contractor would like to add a 10% annual escalation allowance (including inflation). How much would the estimate be?

6.29 How do you define "Mathematical Intuition"?

6.30 Some software programs have an option called "repricing estimates." Briefly describe this option and give an example.

6.31 What are the "Good Cost Estimator's" rules?

7

Time Management

Introduction

Time management in the general context is the ability to use one's time effectively and productively, especially at work. It is also defined as the process of organizing and planning how to utilize one's time among activities that must be completed. Time management, on a personal and professional level, is so important because if you can manage your time, you probably can get things done.

In this book, we are covering this important topic from a construction project management perspective. It is defined here simply as the ability of a project team to plan and finish the project by the contracted time. We will call it "project schedule management" because the main tool in managing time in projects is the schedule. In most cases, the challenge is not in getting extra time, but it is rather a matter of organizing and prioritizing work tasks.

Construction and other types of projects are almost always bound in the contract agreement by timeframe or finish date. The general contractor usually must submit a detailed schedule demonstrating the work sequence from start to finish, to complete the project on time. Preparing such a schedule is an important step; however, when project execution starts, the general contractor must update this schedule and make sure not to fall behind. This is easier said than done. The general contractor has the enormous and delicate responsibility of supervising, coordinating, and managing day-to-day work activities, so the project can be completed within budget, schedule, quality, and other contractual constraints.

Finishing on time is not just fulfilling a contractual commitment, it often has consequences. Many contracts have clauses for early finish bonuses and/or late-finish penalties in addition to have a likely impact on the contractor's upcoming projects. Add to that, indirect consequences such as the contractor's reputation and future opportunities. It was rightly said that time is money, sometimes literally!

In this chapter, we will cover project scheduling: how to create a schedule for a construction project, and how to manage it effectively throughout the project.

Basic Definitions

A *schedule* is a display of project activities and events, along with their timing, plotted chronologically on a calendar. Although the schedule thought process existed since humanity needed homes and other

Project Management in the Construction Industry: From Concept to Completion, First Edition. Saleh Mubarak.
© 2024 John Wiley & Sons, Inc. Published 2024 by John Wiley & Sons, Inc.
Companion Website: www.wiley.com/go/nextgencpm

projects, formalizing this process and making it more scientific, started in the twentieth century with the introduction of several scheduling methods, particularly the critical path method (CPM).

Project scheduling is defined as the determination of the *timing* and *sequence* of operations in the project and their assembly to give the overall completion time. The keywords here are timing and sequence, where time is expressed in form of the start and finish dates, while sequence is expressed in form of predecessors and successors.

Project scheduler is the professional whose basic duty is to create, critique, update, maintain, and manage a project schedule based on the CPM or other scientific scheduling methods. The scheduler is usually a part of the project management team, although it can be a third-party consultant.

An *activity* is a basic unit of work as part of the total project that is easily measured and controlled. The size of the activity varies, and it depends on the scheduler's philosophy but within certain contractual, professional, and practical guidelines. It is also called task. It consumes time, cost, and resources.

An *event* is a point in time that is usually the start or finish of a certain activity(s). It has no duration time. Important events are called milestones:

– Start milestones such as notice to proceed (NTP), and
– Finish milestones such as Substantial Completion.

> A schedule is simply a time estimate for the project. Just like cost estimates, we do our best to create accurate schedules, but things on the ground may happen differently. This is why updating the cost and schedule during the execution is of utmost importance.

History of Scheduling

When humanity needed houses, roads, temples, and other structures, there were "builders" who designed and built these projects. Knowing the composition of the work activities and their sequence was a matter of common sense, experience, and trial and error. Some of these projects were large and monumental such as the seven wonders of the world and other landmark structures. As mentioned in chapter two, the main objective in most of these ancient monuments was to please the ruler then, regardless of resource consumption, time, or other considerations. Today's projects, small and large, have limited financial and other resources and a tight schedule. Figure 7.1 shows the evolution of modern project scheduling.

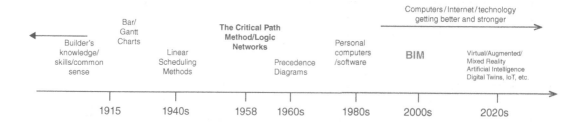

Figure 7.1 The evolution of scheduling methods.

It is not known exactly when bar charts started and who started them because several forms of such charts were used centuries ago, sometimes as making lines in the sand. However, modern bar charts and their professional use are credited to Henry Gantt, an American mechanical engineer (1861–1919), who advanced bar charts as a tool for routine operations, including measuring productivity levels of employees. This is where bar charts got the name Gantt charts. One of the first major applications of Gantt charts was by the United States during World War I, at the instigation of General William Crozier.

Bar Charts

Bar charts are defined as a graphic representation of project activities shown in time-scaled bar lines with no links shown between activities. Some people like to make Gantt charts as a distinctive type of bar charts, which is inaccurate. Bar charts are graphs with a wide variety of shapes and content, having one thing in common: representing work activities on a timescale. In the 1980s, personal computers allowed the creation of more complex and elaborate bar charts. At the same time, bar charts were getting more sophisticated with personal computers, the CPM was gaining strength and popularity with personal computers as well. As a result, CPM networks dominated project scheduling. However, bar charts remain as the most popular format for displaying schedules. So, CPM and bar charts worked together; the first has the brain (all calculations) and the second provided the body (the display/reporting.) Figures 7.2a and 7.2b show examples of simple bar charts.

In addition to the timeline, bar charts can display a host of data items such as the budget, man-hours, and project percent complete. Activities may also be represented by two bars: one representing actual performance and the other representing the baseline (as planned), as shown in Figures 7.3 and 7.4. However, the scheduler (or bar chart user) must not load the graph with too much data in a way that complicates the graph and makes it difficult to understand and interpret. The most important characteristic of bar charts is simplicity, so we should keep it this way.

Figures 7.2b, 7.3, and 7.4 belong to the same project; roof replacement. Figure 7.2b shows the simple bar chart, as planned. Figure 7.3 shows two bars per activity, representing its "as-built" and "as-planned" dates. Note that "as-built" bars are drawn only for activities or segments of activities that are already performed. Figure 7.4 is same as Figure 7.3 with the addition of project percent complete as a second Y-axis, and the S-curves (one for the "as-planned" and one for the as-built/forecasted). The second Y-axis can instead represent the budget or man-hours.

Advantages of bar charts:

1. Time-scaled: The length of the bar is proportional to the duration of the activity it represents.
2. Simple to prepare: No theories and/or science behind bar charts; just educated common sense.
3. Easy to understand and interpret by field crews. Therefore, it is still the most popular method of displaying schedules.
4. Can be more effective and accurate if based on CPM calculations.
5. The user can customize the shape and color of the bars to create a variety of bar chart graphs and to indicate work stoppage.
6. Can be loaded with other information (budget, man-hours, resources, percent complete, etc.) but as indicated earlier: we must not lose simplicity.

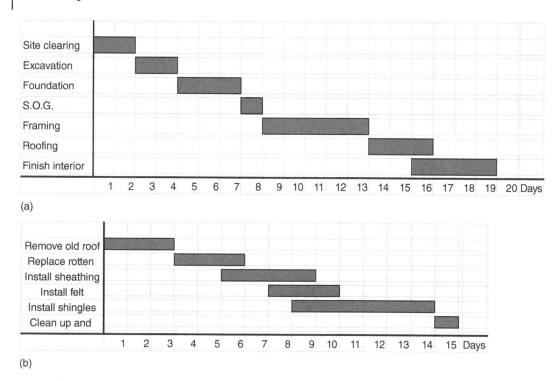

(a)

(b)

Figure 7.2 (a) Sample bar chart showing the construction of a small building. (b) Sample bar chart showing the replacement of a roof.

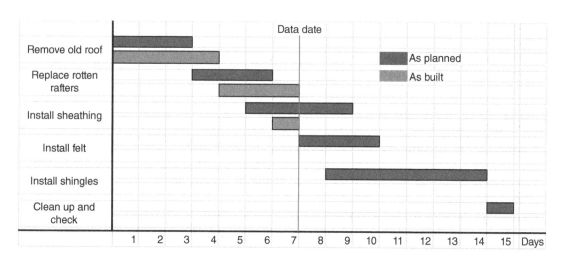

Figure 7.3 Sample bar chart showing activities' "as-built" and "as-planned" bars.

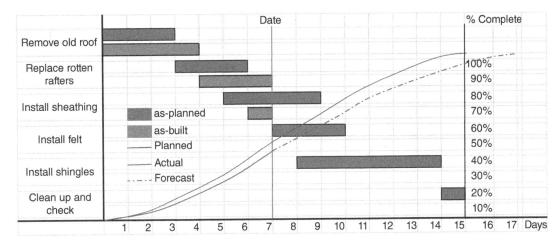

Figure 7.4 Sample bar chart with "as-built" and "as-planned" bars, percent complete, and the S-curve.

Shortcomings of bar charts:

1. Does not show logic (activities' interdependencies). For example, it does not tell why, in Figure 7.2a, we start roofing after framing: Is it a "logical" requirement, a resource restriction (we have only one crew), or other? Also, it does not explain why we start interior finish, in Figure 7.2a also, two days after roofing activity started and before its completion. Some computer programs can show logic links, but it may make the graph complicated and confusing.
2. Not practical for projects with too many activities. As a remedy, we can use bar charts effectively to show:

   a. A small selected group of the activities (subnet) such as those to be performed in a specific area of the project, by one subcontractor or crew, or during a specific period of time.
   b. Summary schedules where a bar represents a group of bars, such as "rolling up" (combining) activities by floor in a high-rise building.

> Perhaps the main advantage of bar charts is their simplicity and acceptability by field people. Do not lose that advantage by crowding too much information in the chart.

Logic Networks

A network is a logical and chronological, graphic representation of the activities (and events) composing a project. Network scheduling has revolutionized the management of construction projects. It has provided management with a more objective and scientific tool and methodology than simply relying on the project manager's experience and personal skills.

Network diagrams are basically two types: arrow diagrams and node diagrams, as shown in Figure 7.5a and b. Arrow diagrams were introduced first and became popular in the 1960s and 1970s. Node diagrams (in their simplest form) came out shortly after arrow diagrams but did not dominate till the late 1960s when precedence diagrams (networks) were introduced by Professor John Fondahl, with capabilities arrow diagrams did not have, as shown in Figure 7.6.

In arrow networks, the arrow represents an activity while a node represents an event or events of starting and/or finishing activities. In node networks, the node represents an activity, while an arrow between nodes represents a logical relationship (dependency) between these activities.

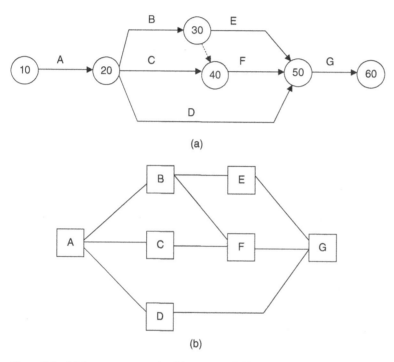

(a)

(b)

Figure 7.5 (a) An arrow network with seven activities. (b) A node network for the same project is shown in (a).

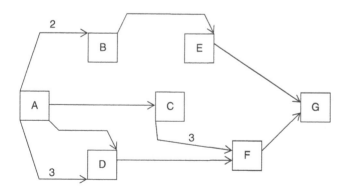

Figure 7.6 An example for precedence network.

Time-Scaled Logic Diagrams?

Some scheduling computer programs tried to combine the bar charts' advantage of being time-scaled along with the logic networks' advantage of showing activities dependencies. The result was called "time-scaled logic diagrams," but it did not work well in general because the too many intersecting lines and inability to maneuver the activities' size and location. In the author's opinion, this type of diagram may be fine for a limited number of activities, but it is hard to apply it to a network with hundreds of activities and hundreds of logical relationships. Oracle Primavera P6 has a feature called "Trace Logic" which allows the user to focus on an activity along with its predecessors and successors for up to three levels. Using the "spotlight principle" is always a good idea when we have a massive schedule. Use the software tools to choose a small number of activities and focus on them.

The Creation of the Project Schedule

The schedule starts with steps 1–4 below. These are the basic steps that without any of them, there will be no schedule.

1. *Breakdown the project into activities,*
2. *Determine activities' durations,*
3. *Define logical relationships (interdependencies), and*
4. *Create the network and perform CPM calculations.*

The 5th step is optional but strongly recommended. The 6th step is needed to assure the validity and accuracy of the schedule.

5. *Optional: resource/cost-load/level the schedule,*
6. *Review, finalize, and approve the schedule.*

Once construction starts, steps 7–9 must take place.

7. *Implement the schedule,*
8. *Monitor and update the schedule, and*
9. *Implement change orders.*

And finally, when the project is complete, the last step, 10, concludes the process and closes the circle.

10. *Closing out: feedback and archiving.*

The Project Management Institute (PMI) mentions seven project schedule management processes required to manage the timely completion of the project:

1. Plan schedule management
2. Define activities
3. Sequence activities
4. Estimate activity resources
5. Estimate activity durations

Title	Level
03 00 00 Concrete	1
03 10 00 Concrete Forming and Accessories	2
03 11 00 Concrete Forming	3
03 11 13 Structural Cast-in-Place Concrete Forming	4
03 11 13.13 Concrete Slip Forming	5
03 11 13.16 Concrete Shoring	5
03 11 13.19 Falsework	5
03 11 16 Architectural Cast-in Place Concrete Forming	4
03 11 16.13 Concrete Form Liners	5
03 11 19 Insulating Concrete Forming	4
03 11 23 Permanent Stair Forming	4

Figure 7.7 A partial breakdown from the CSI MasterFormat.

6. Develop schedule
7. Control schedule

The first six processes are considered "planning process" while the 7th step is "monitoring and controlling process."

Step #1: Breakdown the Project into Activities

The project must be broken down progressively into smaller and smaller components, with the final components called activities or tasks. The scheduler may use own judgement in this process, follow own organizational guidelines, or use industry-standard guidelines such as the CSI MasterFormat[1].

Looking at Figure 7.7, we see five levels of the work breakdown structure. We can add one more level with higher details, if needed: level 6, where units at level 5 are broken down to smaller components. Let us also not forget Level 0 which refers to the entire project before the breakdown. So, for example, we can say:

- At level 0, there is one component only: the project itself.
- At level 1, we may have a few (5–7) components,
- At level 2, we may have about 20–25 components,
- At level 3, we may have about 100–150 components,
- At level 4, we may have about 500–800 components,
- At level 5, we may have about 2000–3000 components, and
- At level 6, we may have about 8000–20,000 components.

Of course, the numbers above are not exclusive as the number of activities in projects varies significantly. In the past, before the availability of affordable personal computers and scheduling software, the scheduler kept the total number of activities to a reasonable limit for manual drawing and calculations of the CPM network. Nowadays, there is no such issue, so the number of activities in a project can

---

1 https://www.csiresources.org/home.

Table 7.1 Sample breakdown for concrete work in a 25-story building.

Work item	Level	No. of components
FRP Concrete work for building	2	1
FRP Concrete work, 3rd floor	3	25
FRP Concrete columns, 3rd floor	4	125
Formwork for conc. columns, 3rd floor	5	375
Formwork for conc. col C-4, 3rd floor	6	7500

be thousands and it is perfectly fine. However, the objective of the project breakdown is to attain control over the activities in terms of confidence in measurements: the duration before the start and the percent complete while in progress. The breakdown can stop when the scheduler feels confident in and control of the activity duration. Take, for example, a 25-floor concrete building with 20 columns. Each typical floor has five different concrete members (columns, beams, slabs, shear walls, and stairs). Assume that we start the project breakdown at level 1 by having a few major components, one of them is "Building Superstructure." At level 2, we take the component concrete work for building, combining the formwork, reinforcement, and concrete placement for all concrete members in the building, as shown in Table 7.1. We then break this one component by floor, producing 25 "activities." Next at level 3, we break these components by type (column, beam, . . .), producing 125 "activities." Next at level 4, we break each component to formwork, reinforcement, and concrete placement, producing 375 "activities." This may be a satisfying level to most schedulers because if we break activities further, say by individual column, we will have 7500 activities needlessly. Bear in mind that these are concrete activities only. If we add all activities from different specialties, we will end up with at least 4–5 times these numbers.

The number of activities depends on the size and complexity of the project as well as the philosophy and methodology of the scheduler. In general, the breakdown of the project into activities is influenced by factors such as:

1. Nature of the work/homogeneity: it is better for activities to be homogeneous:
   a. Having the same nature of the work throughout the activity,
   b. Performed by the same crew, and
   c. Measured by one unit.
   Using this principle, it is not recommended to combine concrete activities in one as FRP (formwork, reinforcement, placement).
   - Formwork is usually performed by carpentry crew and measure by contact area,
   - Reinforcement is usually performed by rodmen and measure by weight, and
   - Concrete placement is performed by the concrete placement crew and measure by volume.
2. Location/floor/segment: Even if activities have the same nature and being performed by the same crew, they must be separated if they are being performed at different locations. An example of this point is repetitive activity in multistory buildings. You may have the same concrete suspended slab in every typical floor, but they are different activities: different locations and different timing.

3. Size/duration: Sometimes, we have activities that are too large, even though they are of the same nature, performed by the same crew, and being performed contiguously. For better control, they may be divided into smaller and similar, but not necessarily equal, activities. For example, excavations of a large area may be divided by the grid, like a spreadsheet. A 600 linear feet (LF) long concrete masonry fence wall may also be divided by segments, 100 LF each, for example. Many contracts, especially public, restrict the duration of individual activities in the schedule to a maximum value, like 14 or 20 days.

4. Timing/chronology/phase: Sometimes, the contractor chooses to do the same activity in stages. For example, a contractor in some building projects likes to leave the slab-on-grade (SOG) on the ground floor till the end of the project, except one part that is completed early along with the walls around it and used as storage. In this case, the SOG, even though they have the same nature, is performed by the same crew, and being later contiguous, but the segments are not being performed at the same time.

5. Level of confidence in the duration: This is an interesting factor where we use the principle "divide and conquer." The scheduler's level of confidence in the activity's assigned duration may be subjective somewhat, but it has to do with the size and complexity of the activity, as well as the familiarity with it. Dividing a complex or large activity into smaller components will boost the scheduler's level of confidence in their duration.

In Figure 7.8, the *S*-curve is divided into three "zones." In zone 1 (breakdown levels 1 and 2), the confidence in the activity's duration is low. Confidence increases significantly with the breakdown to levels 3 and 4 in zone 2. In zone 3, the curve (confidence gain) slows down, so we get a small increase in confidence in exchange for more breakdowns (levels 5 and 6). The scheduler must keep breaking down the components till reaching a satisfactory level of confidence.

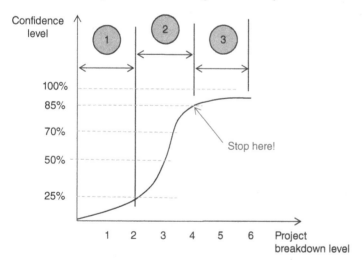

Figure 7.8 Project breakdown versus confidence level in activity's duration.

The breakdown level may not be the same for all activities. For example, if the scheduler feels confident about an activity's duration at level 4, but the confidence level is low for another activity at the same level, then the activity with less-than-satisfactory confidence only needs to be broken down into smaller activities, as shown in Figure 7.9.

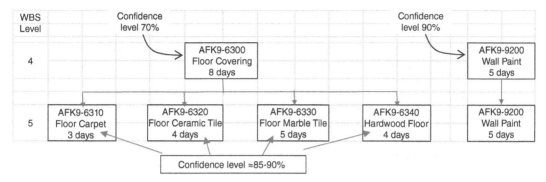

Figure 7.9 Project breakdown based on confidence.

6. Responsibility/trade: Activities must be performed by the same crew and be under one party (contractor and subcontractor). You may have two crews painting a building, but it is not a good idea to combine them in one activity because doing so may not accurately reflect their individual performance. One crew may be doing significantly better than the other one, so combining them may give misleading performance numbers.

The advantage of this progressive project breakdown system is the fact that the scheduler can roll the activities down or up as needed. For example, there may be over 3000 activities at level 5, but it can be rolled up to only around 100–200 at level 3. Specialized computer programs may allow such rolling using several methods for grouping activities. Also, during project execution, the project manager may practice project control at a mid-level, level 3 for example, and then drill down to level 4 or 5 if there is a problem that needs to be investigated.

It is interesting also to note the difference in mindset between the scheduler and cost estimator when it comes to project breakdown. While both pay attention to the type of work item and its nature, the estimator pays more attention to the materials and quantity. The scheduler focuses more on the sequence/chronology and location. So, an estimator may combine suspended concrete slabs on all floors in a multistory building, but the scheduler cannot . The scheduler may combine framing items or installing two different types of floor tile[2], while the estimator will not do so.

If we have to breakdown the project to units acceptable by both the scheduler and cost estimator, we may have to go to a higher level of details, say level 5 or 6, and allow the user to roll up items to lower level.

Step #2: Determine Activities' Durations

Determining activities' durations is one of the most important steps in creating the project schedule. It is important for calculating both the time and cost aspects of the work. But who determines durations? Some people incorrectly think it is the scheduler's job to do so. It is strongly recommended for the scheduler not to be responsible, at least single-handedly, for assigning durations without consulting with the project manager and teams' leaders. In the end, the project manager must review and approve the

---

2 If they are performed by the same crew and have roughly the same production rate.

schedule and make any needed adjustments. This also implies the acceptance of the schedule by all subs and teams' leaders. Assigning durations carries responsibility and liability with it, not only for the efficiency of the work plan and resource management but also for proving/disproving a delay claim dispute.

Determining duration requires, in most cases, the production rate in this activity. The source of production rates can be one of these:

1. Contractor's own records from past projects. The scheduler (or the person estimating duration) must consider how close the item being considered to the one compared to.
2. Ask crew leaders/subcontractors or an expert.
3. Equipment or vendor's technical information.
4. Commercial databases such as RS Means or Spon's price book.
5. Internet search but the user must carefully use these sources because internet search engines are not sources per se, but they provide the user with many references, some of which may not be reliable.

After getting the production rate from the source, it needs to be adjusted to the appropriate conditions. Activities vary in nature, crew composition, dependency on equipment, and productivity-impacting factors. So, there is no single method to determine durations. Some are calculated, while others are estimated or even "guesstimated." However, all these methods may have assumptions that require adjustments based on the conditions and merits of the activity and project. Production-impacting factors were mentioned and discussed in Chapter 6. Seeking duration accuracy is a must for the following reasons:

1. Knowing how long the crews are needed, so they are planned efficiently.
2. Calculating the cost of the activity because the crew's cost is calculated by time.
3. Determining the start dates for succeeding activities so their resources can be planned. This includes crews, materials, and any other requirements such as a crane and inspection.
4. Calculating the expected finish date for the project as well as important milestones.
5. Impact on cash flow and procurement.

Lack of accuracy in activities durations may lead to:

1. Delayed crews.
2. Subcontractors' conflict.
3. Shortage of materials/crews if the actual date for predecessor's finish happened early. Conversely, if the actual date of predecessor's finish is late, we may need to store materials that arrived as previously planned.
4. Possible problems with cash flow.
5. Disrupted plans and chaos.
6. Impact on succeeding activities as well as the entire project.
7. Wasted money.

Along with determining activities' durations, the scheduler needs to assign calendars to them. In many projects, not all crews work on the same calendar. Some may work five days a week, while others work six or even seven days a week. This matters because contracts usually count time in calendar, not work, days. So, a 10-day activity on a 5 workday/week may take more than another 11-day activity on a 6 workday/week. The calendar also includes non-workdays such as observed holidays and perhaps an allowance for "extreme weather days." There could also be special calendars for those activities that can only be performed on specific days based on the availability of a needed resource. In addition, activities will also have codes, such as area, floor, and responsibly, so they can be grouped, sorted, and filtered easily.

Step #3: Define Logical Relationships Among Activities

Setting up the activities' logic means defining the interdependency among them. It may seem an easy task, and it may be easy sometimes but not always. In many situations, there is more than one way to depict logic. It is a delicate task that needs to accurately represent the planned sequence of activities. It includes the following steps:

1. Assigning predecessor activities and the type of relationships: finish-to-start (FS), start-to-start (SS), finish-to-finish (FF), or start-to-finish (SF).
2. The amount of lag or lead[3], if any.
3. Imposed constraints, if any.

The scheduler has to pay attention to external relationships as well. This includes any activities that are not under the contractor's responsibility, but there is a dependency on them: either they depend on the contractor's work or the contractor's work depends on them. These external activities can be assigned to the owner, the designer, the professional construction management consultant, or another party.

Step #4: Create the Network and Perform CPM Calculations

After defining the activities, determining their durations, and sequencing them, we need to draw the logic network in order to do the critical path calculations. It is true that all construction projects networks nowadays are drawn and calculated by computer software, but the user must be familiar with the concept and be able to do it manually, including drawing the network and performing CPM calculations, for small projects.

During this stage, as explained later, we find out the project's calculated finish date, based on our assumptions of durations and logic. It is important to distinguish between this calculated finish date and the contract-imposed finish date. The contractor must make sure that the calculated finish date is no later than the contract-imposed finish date. In fact, it is better to have a time buffer between the two dates, we will later call it "time contingency."

Step #5: Resource-/Cost-Load/Level the Schedule

This step requires the user to allocate the cost and/or resources required for each activity. This requires cooperation with the cost-estimating department. Resource loading (allocation) is the assignment of the required resources to each activity, in the required amount and timing. Loading the cost can be done using one of two methods:

1. Loading the cost directly in the form of monetary amount.
2. Loading the resources (labor, materials, and equipment) in form of amounts (unit of measure for materials and unit of time, hour or day, for labor and equipment). The resources are usually priced

---

3 To be explained later in this chapter.

per unit along with other related information in the "resource dictionary" that belongs to the project. From this dictionary, each activity's cost is calculated. Keep in mind that crew (labor and equipment) cost depends on the activity's duration while the materials cost does not.

Loading resources has many advantages over cost-loading:

1. Ease of changing/updating the cost of a resource. This can save a lot of time. Take for example if the price of concrete mix changed from \$99/CY to \$104/CY. Assume also that there are 150 concrete activities in the schedule. Changing the price manually takes time and increases the likelihood for making errors. With resource-loading, the scheduler can change the unit price of the concrete mix in the resource dictionary. After that, the program will ask the user: "This change affects 150 items (activities), so do you want to execute this change?" With one keystroke, the scheduler updates the price of the concrete mix in 150 activities. Resource dictionaries can be edited and exported to other projects. There is one exception: when the activity is subcontracted, it may be resource-loaded by the subcontractor but will be cost-loaded by the general contractor.
2. Ability to integrate with accounting and procurement departments by providing the specific resource requirements for each activity: what it needs, how much/many, and when. The schedule is dynamic, so if there is a delay in one activity for whatever reason, it will instantly reflect the impact on this delay on succeeding and all other activities.
3. Ability to estimate and store (in database) productivities and man-hours, and to do cost analysis. This is important for estimating duration and cost of future projects. A direct correlation between the crew or resources and duration provides productivity. We can also find out the actual cost for the activity and cost per unit.
4. Ability to level the resources and to set upper limits on resource consumption. Leveling the resources (labor and equipment) minimizes the fluctuations in day-to-day resource usage throughout the project or for the duration of the resource need. It allows for more efficient use of resources by shifting activities within their available float, minimizing the peaks and valleys in the resource consumption profile.
5. Ability to use "resource-driven" activities. This is a feature in some computer programs for activities that need a fixed amount of a resource (usually labor-hours of certain craft) that allows the program to allocate this resource in varying amounts, based on availability, till the required amount (labor-hours) is fulfilled, regardless of duration.
6. Ability to create cash flow diagrams.

After the activities have been resource-loaded, the scheduler can level key resources. Resource leveling is an optimization technique that works by shifting noncritical activities within their available float.

The indirect cost (overhead and profit [O&P]) can be handled in several ways, depending on the preference of the contractor. For example, each activity can be loaded with its share of O&P as a separate item from its direct cost. Another way is to leave O&P outside individual activities and create one "hammock activity" that extends from the start of the project till the end and load it with indirect cost.

Step #6: Review, Finalize, and Approve the Schedule

In this step, the scheduler finalizes the schedule and makes sure there are no errors, omissions, redundancies, missing links, or any other inaccuracies. The schedule then goes to the project manager to review and make any necessary adjustments. The schedule also has to be reviewed and accepted by the teams' leaders. This is likely to be the baseline schedule that is used as a reference to measure progress. If so, it will be a part of the contract documents, so it better be accurate.

The scheduler is recommended to make a "health check" on the schedule. This health check may be based on criteria set by the scheduler, the company, or the DCMA-14[4].

Copies will be given to all project participants, starting with the owner, and going through the subcontractors. Each activity will have codes such as responsibility, area, or phase. So, it will be easy to produce a customized report such as a "sub-schedule" for a subcontractor.

Step #7: Implement the Schedule

Implementing the schedule means taking it from plan to action and from "theory" to application! The schedule is more than decoration on the trailer's walls. It is a roadmap for the project, so it must be followed as much as possible. This is primarily the responsibility of the general contractor's project manager and then the rest of the project management team members. The project manager will be responsible for assigning responsibilities and enforcing the duties. The upper management's commitment and support are crucial. We realize that the schedule is an estimate for future occurrences and that things may not happen exactly as forecasted, but we still have to:

A. Do our best in creating the schedule, as accurate and reflective of our plan as possible, and
B. Do our best on the ground to adhere to the schedule and take it seriously.

The schedule is more than decoration on the trailer's walls!

Step #8: Monitor and Update the Schedule

The schedule must be updated periodically, as the contract requires. This includes reflecting actual progress on the schedule and implementing any changes. It is a cycle where the scheduler is in the center: he/she receives the project progress report, implements it in the schedule for the update, and produces an updated schedule that goes to all project parties. The update is also involved in issues such as progress payments, which requires activities percent complete.

Step #9: Implement Change Orders

This step is practically part of step #8, as change orders are almost an inevitable part of construction projects. They happen for a variety of reasons, can be small or large, and can have an impact on cost, time, or both. They can involve addition, deletion, or substitution work. Change orders must be approved (by the owner and usually – but not always – the designer). The role of the scheduler is to implement

---

4 The DCMA-14 Point Assessment is a metric-evaluative tool intended to test the overall quality of a schedule. Developed in 2005 by the US Defense Contract Management Agency, the tool was designed for the Department of Defense as way to quantitatively evaluate project schedules.

approved change orders in the schedule which may result in modifying the baseline schedule. Sometimes, change orders come in form of "what if scenario," so the contractor can assess the impact of the change on cost and schedule. The owners can then approve, disapprove, or negotiate.

Step #10: Closing Out: Feedback and Archiving

After the project is physically completed and fiscally closed out, it has to be archived. Fiscal closing is an important step, financially and legally. It represents the settlement of all invoices, bills, payments, warranties, and other issues between the general contractor and the owner.

Archiving requires preserving all project documents, including copies of the schedule (baseline plus all updates including the final "as-built"). These documents may be needed to resolve any claim dispute. Also, the records will be used as a database for guiding and estimating future projects. Archived projects also include those canceled or postponed. All archived documents must be stored in their native format or a format that is easy to open.

It is recommended that all archived projects be stored separately from active projects, so the scheduler and the project management team members can see and focus only on active projects. If information is needed on an archived project, the user can activate it, extract needed information, and archive it again.

The Critical Path Method

The critical path method (CPM) is a scheduling technique using networks for graphic display of the work plan. The method is used to determine the length of a project and to identify the activities that are critical to the completion of the project.

The CPM made project scheduling more objective and scientific. Rather than leaving the entire process to the scheduler's subjective thinking and assessment, this method uses input from the scheduler (define activities, determine durations, and assign logic), and then calculates the output using known equations. Thus, network calculations determine when activities can be performed, the expected completion date of the project, and the critical path of the project.

The CPM uses logic networks for the calculations. These networks are composed of nodes, representing activities, and arrows, representing relationships (dependencies.) The network represents the project work expressed in a large group of sequenced activities through multiple paths. This is what distinguishes construction projects where there are usually many paths from start to finish, paralleling, overlapping, and crossing each other, unlike linear work (such as in assembly lines) where there is one main line of production, and everything else is subsidiary or contributary to it. The sample network shown in Figure 7.10 is just an example of a small construction project schedule. Real ones are presented in a much larger network.

The CPM calculates the length of all paths from start (activity A) to finish (milestone PF) using the given durations and relationships. It will also consider other impacting factors such as lags, calendars, and constraints[5]. The longest path in the network, from start to finish, is called the critical path, which determines

---

5 To be discussed later in this chapter.

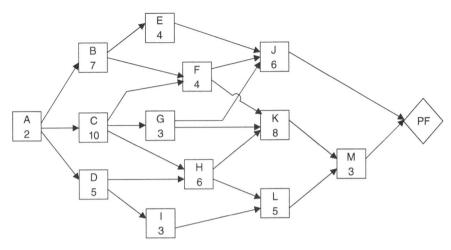

Figure 7.10 Sample logic network used with the Critical Path Method.

the length (duration) of the project. Thus, the critical path gets more attention than any other path because any delay in it, results in a delay to the project. This is the definition of the critical path in a nutshell, but there is a lot more to it that will be discussed in detail. The critical path calculations include the forward pass (from project start to project finish) and backward pass (from project finish to project start). The forward pass calculates the early dates (early start [ES] and early finish [EF]). Knowing the "early finish" of the last activity, we can tell the project's calculated finish date. The backward pass calculates the late dates (late start [LS] and late finish [LF]), which allows us to identify the critical path and calculate the float amounts for each activity.

One important clarification regarding dates: linguistically, if we say start activity A on day one, it may mean any time on that day. The same argument applies to the finish date. To eliminate confusion, we will adopt the "end of day" convention, which means that any date mentioned automatically means end of that day. Perhaps, we adopted this convention because when the contract says that the project must finish by a certain date, say 15 October, it legally means the end of business day of that date. So, all the start dates in the following examples are written as the end of day before. Activity A in Example 7.1 practically starts on the beginning of day 1, but we will write it as (end of) day 0, and so on. There is no ambiguity about finish dates since the date written is same as the finish date. Let us learn it more through examples.

Example 7.1   Draw the logic network for the 7-activity project, as shown in Table 7.2. Do the following:

1. Draw the logic network,
2. Calculate early dates for each activity and the completion time for the project,
3. Calculate late dates for each activity,
4. Identify the critical path and calculate the float amount for each activity.

(Continued)

Example 7.1   (Continued)

Table 7.2 A display of Example 7.1 activities along with their durations and predecessors.

Activity	Duration (days)	IPA[a]
A	5	—
B	8	A
C	6	A
D	9	B
E	6	B, C
F	3	C
G	1	D, E, F

[a] Immediately proceeding activities.

Solution:

1. Let us first draw the network:

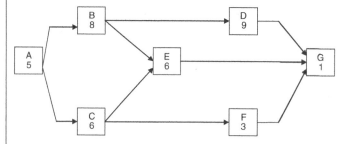

Figure 7.11a   The Logic network for Example 7.1.

2. The forward pass: we start with activity A, ES is on day 0 (practically beginning of day 1). Activity A takes 5 days of duration, so its EF is day 0 + 5 = 5. Next are both activities B and C with ES = 5, and then EF = ES + duration. Activity E cannot start till both its predecessors B and C are complete, so we pick the latter one, B. Activity E will have ES = 13, and EF = 13 + 6 = 19. We continue till the last activity, G, which will have ES = 22 and EF = 23.

   With the completion of the forward pass, we not only calculated the early dates for each activity but also the calculated finish date for the project, day 23.

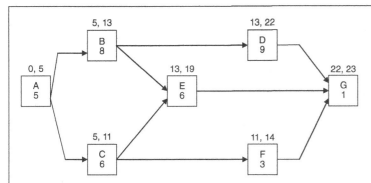

Figure 7.11b   The Logic network for Example 7.1 with the forward pass.

3. The backward pass: we start with activity G (the last activity or event). For now, we will start with the project-calculated finish date (23), which was calculated in the forward pass, and we enter it as the LF date for activity G. We then calculate the LS date = LF – Duration = 23 – 1 = 22. This date represents the LF date for the predecessors D, E, and F. We then calculate the LS date for the same three activities, using the equation: LS = LF – Duration. When we get to activity B, its LF date must be the earlier of the LS date for its predecessors D and E, which is 13, because activity B must be completed before either one starts (the earlier of 13 and 16). We apply the same argument to activities C and A: LF for C = 16 (earlier of 16 and 19), and LF for A = 5 (earlier of 5 and 10.)

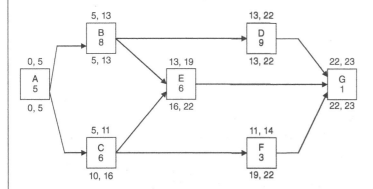

Figure 7.11c   The Logic network for Example 7.1 with both forward and backward passes.

4. Now, we identify path A-B-D-G as the longest path in the network, so it is the critical path. We realize that all activities on this path have late dates equal to their early dates, which means that they cannot be delayed; otherwise, the project completion date will be delayed. Other activities can be delayed by a limited amount of time called total float (TF).

(Continued)

Example 7.1 (Continued)

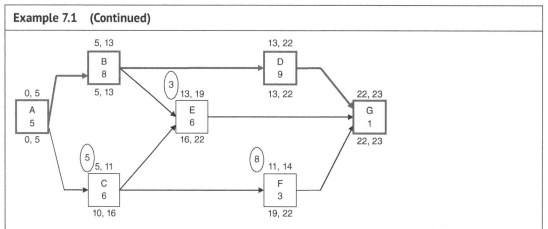

Figure 7.11d The final solution for Example 7.1.

5. Now, we calculate the amount of TF for each activity using the equations:

$$TF = LS - ES = LF - EF = LF - Duration - ES$$

$$TF \text{ for activity } E = 16 - 13 = 22 - 19 = 22 - 6 - 13 = 3 \text{ days}$$

The TF is shown in a bubble on the top left of activities C, E, and F. Activities A, B, D, and G are critical, so they have no float, or TF = 0, mathematically speaking.

CPM Definitions:

The critical path is the path on a logic network that takes longest calendar time than any other path on that network, from its start to finish. It takes into account the planned duration of activities, the type of logical relationships and lags among these activities, work calendars[6], imposed constraints, resource limitations and leveling, risk[7], and any other time-impacting factors.

In a brief definition, the critical path is the longest continuous path in a network from start to finish. It represents the shortest span of time to get the project done. It may sound like there is a contradiction between the two statements (longest versus shortest), but it is not. Here is a brief explanation: in Figure 7.12, we see six paths that go from project start to finish (this is NOT a bar chart. Bars here represent entire paths by their length). The longest path is path C which is 30 days. This 30-day path is the shortest period possible to complete the project.

More critical path definitions:

Total float, TF, is the maximum amount of time an activity can be delayed from its ES without delaying the entire project.

Forward pass: The process of going in a network, from start to finish, to calculate ES and EF dates for each activity.

---

6 Work calendars include all types of non-work time.
7 Risk is not factored in CPM calculations explicitly, but it may be taken into account when estimating the activities' duration. There are other methods that address risk in detailed calculations.

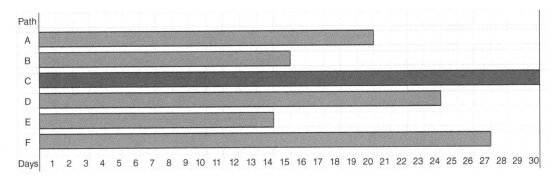

Figure 7.12 Graphic explanation of the critical path.

Backward pass: The process of going in a network, from finish to start, to calculate LS and LF dates for each activity.

We can tell the calculated finish date for the project after the forward pass, but we need both passes to identify the critical path and calculate the float amounts for activities.

ES: The earliest date the activity can start within project constraints.

EF: The earliest date the activity can finish within project constraints.

LS: The latest date the activity can start without delaying the completion of the project.

LF: The latest date the activity can finish without delaying the completion of the project.

An important point regarding early versus late dates is that a project manager needs to decide on dates within this range. Computer programs, by default, use early dates. Delaying is possible using constraints. The scheduler may print a report showing early and late dates for activities, however, such a report must not be given to field crews because it may confuse them or encourage them to delay activities.

Example 7.2 Draw the logic network for the 13-activity project, as shown in Table 7.3. Do the following:

1. Draw the logic network,
2. Calculate early dates for each activity and the completion time for the project,
3. Calculate late dates for each activity,
4. Identify the critical path and calculate the total float (TF) and free float (FF) amount for each activity.

Solution:

The solution is shown in Figure 7.13, but we would like to introduce a new term: free float, FF, which is part of TF. Let us look at:

1. Activity B: TF = 5 days. It can use all these 5 days without impacting the project calculated finish date (as the definition implies), but this will delay the ES of the successor E.
2. Activity F: TF = 2 days. If it uses any or all these 2 days, it will not impact the project calculated finish date, but it will delay the ES of the successor J.
3. Activity I: TF = 11 days. Using any or all of this float will not impact the project calculated finish date. However, using up to 8 out of these 11 days will not delay the ES of the successor L, but the remaining 3 days will.

(Continued)

Example 7.2   (Continued)

From the above discussion, we conclude that using the TF days for an activity, while it does not impact the project calculated finish date, may impact the ES of the successors.

Table 7.3   Information for Example 7.2.

Activity	Duration	IPA
A	2	—
B	7	A
C	10	A
D	5	A
E	4	B
F	4	B, C
G	3	C
H	6	C, D
I	3	D
J	6	E, F, G
K	8	F, G, H
L	5	H, I
M	3	K, L

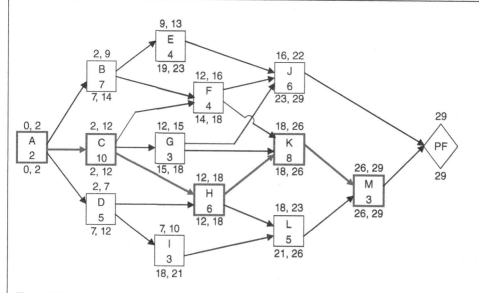

Figure 7.13   The final solution for Example 7.2.

Table 7.4 Numerical summary of Example 7.2 and the solution.

Activity	Duration	IPA	ES	EF	LS	LF	TF	FF
A	2	—	0	2	0	2	0	0
B	7	A	2	9	7	14	5	0
C	10	A	2	12	2	12	0	0
D	5	A	2	7	7	12	5	0
E	4	B	9	13	19	23	10	3
F	4	B, C	12	16	14	18	2	0
G	3	C	12	15	15	18	3	1
H	6	C, D	12	18	12	18	0	0
I	3	D	7	10	18	21	11	8
J	6	E, F, G	16	22	23	29	7	7
K	8	F, G, H	18	26	18	26	0	0
L	5	H, I	18	23	21	26	3	3
M	3	K, L	26	29	26	29	0	0

The numerical summary of Example 7.2 and its solution is shown in Table 7.4

Free Float (FF) is the maximum amount of time an activity can be delayed without delaying the ES of the succeeding activity(s).

- $FF_i = \min(ES_{i+1}) - EF_i$

Applying the above equation, let us calculate FF for activity E and G:

- $FF_E = \min(16) - 13 = 3$ days
- $FF_G = \min(16, 18) - 15 = 16 - 15 = 1$ day

Since FF is a portion of TF, then $FF \leq TF$. The author called FF the "unselfish portion" of TF. The other portion of TF that will impact the ES of succeeding activities, is called interfering float. There is another type of float, more restricted than FF, called independent float, which represents the amount of float that strictly belongs to the activity: not impacting succeeding activities and not being impacted by preceding activities. Float types are explained in Figure 7.14.

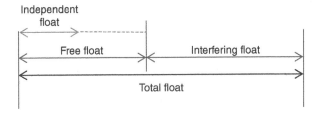

Figure 7.14 Types of float.

In general, here is the summary on float:

1. Total float, TF, even though given to each activity individually, it is shared by the activities on the path or some of them, in most cases. Using an activity float may result in reducing or even eliminating the float of its successors. For example, if activity B in Example 7.2 uses all its TF, activity F becomes critical and the TF of activities E and J is reduced by 5 and 2 days, respectively.
2. Float is one of the most important, interesting, and controversial issues in construction; technically and legally. Many contracts address the issue of "float ownership," which means the ability to use it without consequences. There will be further discussion of float in this chapter and other chapters.
3. Practically and professionally, we deal with TF and FF only. Also, most commercial scheduling software programs can calculate only these two types of float.

Tips on proper network drawing:

1. Activities in node networks are drawn as squares or rectangles, not circles. The reason for this is that relationships can be connected to the start or finish side of the activity, as shown in Figure 7.15a.

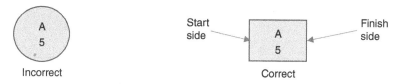

Figure 7.15a   Activity's node must be square or rectangular.

2. Connect activities only from their start or finish sides, as shown in Figure 7.15b.

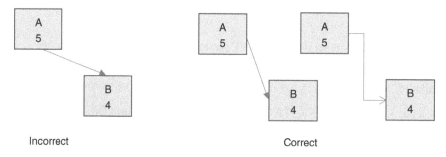

Figure 7.15b   Activities must be connected only from their start or finish sides.

3. Do not combine independent relationships, as shown in Figure 7.15c.

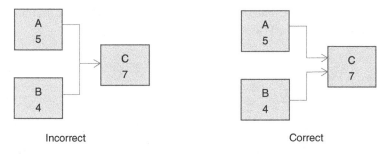

Figure 7.15c   Do not combine relationship lines.

4. Do not cross relationship lines, as shown in Figure 7.15d.

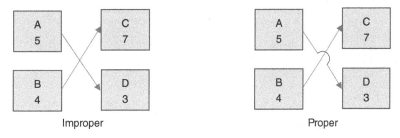

Figure 7.15d   Do not cross relationship lines.

5. Do not start the network with more than one activity. Instead, create a "project start" milestone as the single node to start the network, as shown in Figure 7.15e.

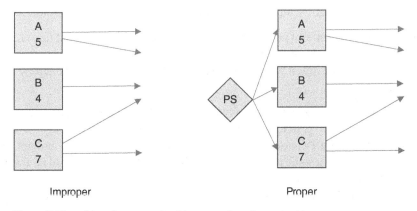

Figure 7.15e   Start the network with one node only.

6. Do not end the network with more than one end-activity. Instead, create a "project finish" milestone as the single node to end the network, as shown in Figure 7.15f.

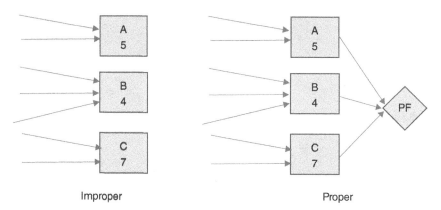

Improper   Proper

Figure 7.15f   End the network with one node only.

Constraints in the CPM schedule:

A constraint is an externally imposed restriction that may affect an activity's start and/or finish date. Constraints may conflict with logical relationships or delay activities, the path, or even the entire project. It is usually used to restrict the start and/or finish of an activity for a reason that cannot be represented by logical relationship with other activities. For example, an activity for installation of expensive equipment with TF = 30 days. The contractor may prefer to delay it as much as possible for cash flow or security reasons. A landscape activity may have large float but is delayed for reducing the maintenance of the landscape and also to protect it from construction work.

Let us consider these 3 scenarios for assigning a constraint on activity I in Example 7.2:

a. Start on Day 6: Such a constraint violates the logical relationship (ES = 7) and must not be allowed.
b. Start on or after Day 15: Such a constraint is valid and will reduce the TF from 11 to 3 days.
c. Start on Day 20: Such a constraint will result in a 2-day delay to the project, in addition to delays to the successors L and M.

Constraints that are tied to a date (start or finish on/before/after a fixed date) are static. This means that if a change happens in the path length, say a delay, the constraint may no longer be relevant, and it needs to be updated manually. Dynamic constraints are those tied to a condition and not a date, such as "as late as possible," which – in some programs- delays the start of activity either to its absolute LS or the latest start without impacting the ES of successors.

Schedulers must minimize the use of constraints and must never use them as an alternative to logic. Having too many constraints in a schedule is often an indication of either poor scheduling or other problems, and they may interfere with functions such as schedule acceleration, risk analysis, or delay claim analysis.

Lags and leads:

A *lag* is defined as a minimum waiting period between the finish (or start) of an activity and the start (or finish) of its successor. The lag takes time, like an activity, but consumes no cost or resources.

A lead is a negative lag. Think of a lag as (after) and a lead as (before).

Lags are treated in CPM calculations just like real activities: added in the forward pass and subtracted in the backward pass. Leads are treated the same with their negative number.

Table 7.5 summarizes the difference among activities, events, and lags.

Table 7.5 Comparing activities, events, and lags.

	Type/Criteria	Activity	Event	Lag
1	Designation	☐	◇	A number over the relationship
2	Duration	>0	0	>0
3	Dates	Start *and* Finish	Start *or* Finish	Start and Finish (not calculated)
4	Cost	>0	0	0
5	Resources	>0	0	0
6	Status	Anywhere from 0 to 100%	0 *or* 100%	Anywhere from 0 to 100%
7	Has responsibility?	Yes	Maybe	Unlikely
8	Assigned a calendar?	Yes	No	Yes
9	Impact on project % complete?	Yes	No	No
10	Impact on progress pay	Yes	Maybe	No

Example 7.3 We are redoing the logic network of Example 7.2, but we added some lags as shown in the Table 7.6. You will notice that for some activities (F, H, L, and M), we split the immediately preceding activities (IPAs) to those with lags and those with no lags.

Activity	Duration	IPA	Lag
A	2	—	
B	7	A	
C	10	A	
D	5	A	
E	4	B	
F	4	B	
		C	4
G	3	C	

(Continued)

Example 7.3 (Continued)

Activity	Duration	IPA	Lag
H	6	C	
		D	3
I	3	D	
J	6	E, F, G	
K	8	F, G, H	
L	5	H	
		I	2
M	3	K	
		L	4

Solution:

The graphic solution is shown in Figure 7.16, and the numerical solution in summarized in Table 7.7.

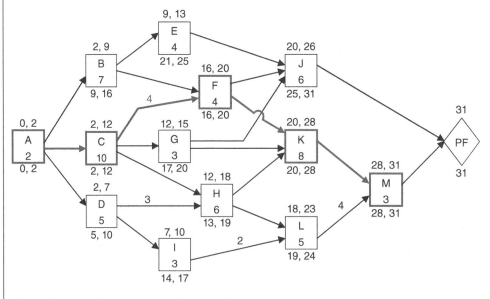

Figure 7.16 The final solution for Example 7.3.

Example 7.3 (Continued)

Table 7.7 Numerical summary of Example 7.3 and the solution.

Activity	Duration	ES	EF	LS	LF	TF	FF
A	2	0	2	0	2	0	0
B	7	2	9	9	16	7	0
C	10	2	12	2	12	0	0
D	5	2	7	5	10	3	0
E	4	9	13	21	25	12	7
F	4	16	20	16	20	0	0
G	3	12	15	17	20	5	5
H	6	12	18	13	19	1	0
I	3	7	10	14	17	7	6
J	6	20	26	25	31	5	5
K	8	20	28	20	28	0	0
L	5	18	23	19	24	1	1
M	3	28	31	28	31	0	0

Event times: When arrow networks were used, they showed event times as nodes in arrow networks represent an event (or events) of an activity's start or finish. In node networks, which we solely use nowadays, events are not shown conspicuously, but we can create nodes for important events and call them milestones: start milestones or finish milestones. This is much better than events in arrow networks because, in a network with hundreds of activities, we will have a handful of milestones only out of hundreds or thousands of events. This utilizes the "spotlight principle," allowing us to focus only on important events such as NTP, building dry-in, and Substantial Completion.

Near-critical activities: Near-critical activities are those with little float that may become critical. Think of a path that is only one or two days shorter than the critical path. If an activity on this path suffers a 2- or 3-day delay, this path will supersede the critical path as the longest path. The scheduler and project manager must pay attention to near-critical activities and paths. Near-critical activities take more importance if the risk involved in them is high.

But what defines the term "near-critical?" There is no consensus on a specific value defining "near-critical" and should not be one because our assessment differs from one situation to another. Generally, we may use the term that an activity is considered near-critical if its TF is less than 3–5 days. The scheduler can sort activities by TF or use a software "filter" to show only activities with TF < 5 days. The important issue is to keep these activities under the vision of the scheduler in case their float starts evaporating and they become critical. This process must be updated continuously since activities come in and out of the near-critical list.

Project float/nonwork days:

As indicated earlier in "Step #4: Create the Network and Perform CPM Calculations," it is recommended that the schedule contain a number of days as "time contingency" for events that likely happen and negatively impact the schedule but may not qualify the contractor for an extension of time. While we may agree on reserving such time contingency in the schedule, the question is how many days and where and how to implement this time in the schedule.

The author believes that activities, durations must be lean and should not be fluffed. The extra time (call it project float, management float, or contingency time) must belong to the project, not the individual activities, as demonstrated in Figure 7.17. Same as contingency money, this contingency time must be estimated and used carefully for events that likely will happen, but we do not know when, where, or how much/how severe. Contingency: time or money, must not be an excuse for poor estimating.

This project float must come in addition to known non-workdays such as holidays and shutdowns. It may be problematic for some people to have two finish dates for the same project, but there are options to resolve this issue such as:

– The earlier finish date can be given a name other than finish date, so it will not be confused with the contract finish date.
– The earlier finish date can be converted from a milestone to an activity with a duration equal to the project float. This activity can have any suitable name such as "preparation for completion" or the "Punch List" activity can absorb this project float.

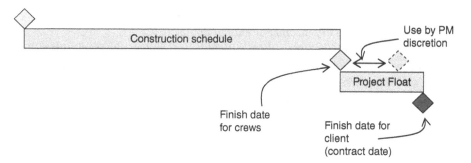

Figure 7.17 Project Float.

The important matter is that such practice is completely ethical and professional. The client must not object to it as long as the contractor is not falling behind without a legitimate reason and the project is completed on time.

Imposed Finish Date

In almost every project, there is a finish date set by the owner in the contract. The contractor's challenge is to reconcile this date with the calculated finish date. In the examples earlier, we started the backward pass in the CPM calculations, with the same finish date we just calculated in the forward pass. Now, we are going to start the backward pass with the imposed finish date. When doing so, we get into one of these two cases:

A. Imposed finish date ≥ (later than or equal) calculated finish date, or
B. Imposed finish date < (earlier than) calculated finish date.

In the first case, as shown in Figure 7.18, there is no problem, as the contract gives the contractor more time than their own estimate. The difference can be considered as project float. However, in the second case, as shown in Figure 7.19, the contractor has to either accelerate the schedule to meet the imposed finish date or request more time from the owner.

Example 7.4   Repeat Example 7.2 if the contract stipulates that the project must finish in 32 days.

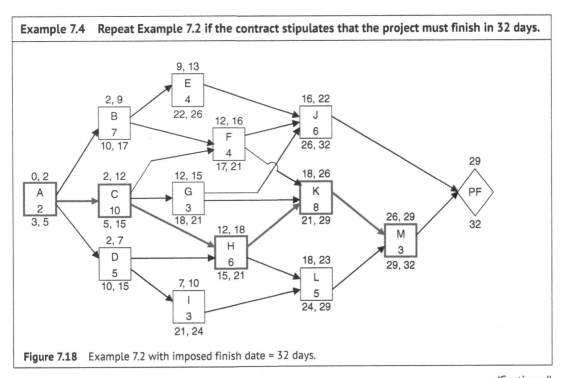

Figure 7.18   Example 7.2 with imposed finish date = 32 days.

(Continued)

Example 7.4   (Continued)

Solution:
We notice that every activity received 3 additional days of float. Thus, each of the activities on the critical path, A-C-H-K-M, has 3 days of float. Can they still be considered critical? If our definition of critical activities is being part of the longest path, then yes, they are critical. But, if our definition of critical is that any delay to these activities will cause a delay to the entire project, then no.

   As discussed earlier, it is recommended that the contractor keeps the calculated finish date (29 days) as the target date. The 3 days difference may be part of the project float.

Example 7.5   Repeat Example 7.2 if the contract stipulates that the project must finish in 25 days.

Solution:
In this network, Figure 7.16, we notice that several activities have their late dates earlier than their early dates, which result in negative float. This is a situation indicating that performing activities even on their early dates is not enough to meet the project's imposed finish date. The contractor has to either be allowed to extend the contract finish date (by 4, in our example) or accelerate the schedule to finish on the contract date (25, in this example).

   We also notice that all activities lost 4 days of float, which is the difference between the calculated and imposed finish date. Some activities now have negative float although they are not on the longest path, like F, G, and L. In this case, we can still call these activities critical, but call those on the longest path (A-C-H-K-M) the most critical. Other activities, B, D, E, I, and J, still have some positive float, but it is 4 days less than those calculated in Example 7.2.

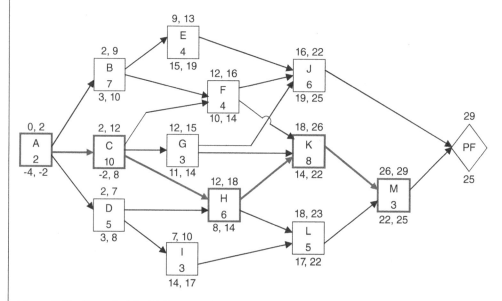

Figure 7.19   Example 7.2 with imposed finish date = 25 days.

We can tabulate the results in Table 7.8 below.

Table 7.8 Tabular Solution for Example 7.5.

Activity	Duration	IPA	ES	EF	LS	LF	TF	FF
A	2	—	0	2	−4	−2	−4	−4
B	7	A	2	9	3	10	1	0
C	10	A	2	12	−2	8	−4	−4
D	5	A	2	7	3	8	1	0
E	4	B	9	13	15	19	6	3
F	4	B, C	12	16	10	14	−2	−2
G	3	C	12	15	11	14	−1	−1
H	6	C, D	12	18	8	14	−4	−4
I	3	D	7	10	14	17	7	7
J	6	E, F, G	16	22	19	25	3	3
K	8	F, G, H	18	26	14	22	−4	−4
L	5	H, I	18	23	17	22	−1	−1
M	3	K, L	26	29	22	25	−4	−4

Negative float is a situation that occurs when an activity, even when performed on its early dates, fails to meet the project's imposed finish date or other constraints. Unlike positive float, negative float is a hypothetical, not real, "thing." It is like debt when counting someone's wealth: any negative balance implies that the person as broke! Schedules with negative float, especially those prepared before the start of the project, are not final. They indicate that the contractor's assumed durations and logic will not satisfy the imposed finish date, set by the contract. The contractor must either accelerate the schedule or request a later finish date from the owner, so in the end there must be no negative float.

For schedule updates, while the project is in-progress, the appearance of negative float is an indicator of a delay (also called slippage). It is an alarm that the project will finish behind schedule if work keeps going at the same pace. There could be many reasons for the delay, some are the contractor's fault and responsibility, while others are not. Again, the contractor must resolve the issue.

Note that while we have negative float, the calculation of free float, FF, changes slightly. It was previously calculated for an activity as the maximum amount of time we can delay this activity without delaying the ES of the successor(s), but since LS happens earlier than ES when imposed finish date < calculated finish date, we use LS for calculating FF. For example, activity F has two days of negative float. If we compare its EF date, 16, with the lesser of ES dates of the successors, 16 and 18, FF would be 0, but we must compare it with the lesser of LS dates of the successors, 19, and 14, so it is -2 days.

Precedence Diagrams

Precedence diagrams (networks) are node diagrams that allow four types of logical relationships:

1. Finish-to-Start, FS,
2. Start-to-Start, SS,
3. Finish-to-Finish, FF, and
4. Start-to-Finish, SF.

Precedence diagrams allow overlapping activities, mostly using combination SS+FF relationships between the same two activities. Because of this feature, the forward and backward passes become slightly more complicated: in every combination relationship situation, you get two sets of early dates and two sets of late dates. You need to reject one set and accept the other.

In precedence diagrams, there is more need for lags, especially with SS and FF relationships. In this chapter, we assume that activities are contiguous, meaning once we start the activity, work continue till its finish.

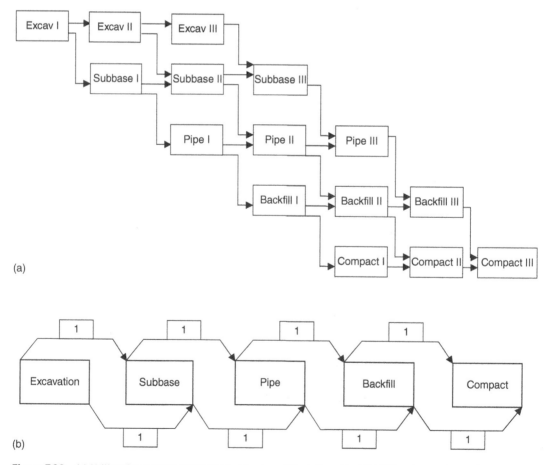

Figure 7.20 (a) Utility pipe project using the ladder-type FS relationship. (b) Utility pipe project using combination SS+FF relationships with 1-day lag on every relationship.

The SS+FF combination relationship allows sequencing overlapping activities without the need to break them, as we used to do in what was known as the ladder-type relationship, by splitting activities and then using only FS relationship. In Figure 7.20a, we show a 5-activity project, laying down a concrete utility pipe. We needed to break down every activity into three segments in order to allow an overlapping relationship among them, which resulted in increasing the number of activities to 15. The number of activities can be even more if we need to break each activity into more than three segments.

In Figure 7.20b, we maintained the number of activities but connected them with a combination SS+FF relationships with a 1-day lag on every relationship. Of course, the lag amount may vary. In this example, we showed a clear advantage to precedence diagrams over regular node diagrams that uses only FS relationships. But the argument has another side, with relationships such as SS, FF, and SS+FF combination, calculations and analysis may get a bit complicated. We will follow a balanced approach to make the schedule as simple and efficient as possible.

Precedence diagrams through examples:

Example 7.6

Table 7.9   Information for Example 7.6.

Activity	Duration	IPA	Rel. type	Lag
A	10	—	—	—
B	5	A	SS	—
C	3	B	SS	—

Solution:
What is interesting in the solution of Example 7.6 is the fact that the critical path is only activity A since its duration constitutes the duration of the "project." The grey-shaded bars for activities B and C represent their TF.

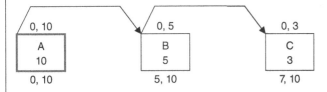

Figure 7.21a   Solution for Example 7.6.

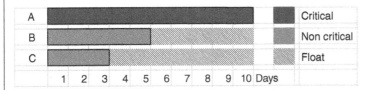

Figure 7.21b   Solution for Example 7.6 in a bar chart format.

Example 7.7 **We are redoing Example 7.6 after adding lags between the activities.**

Table 7.10 Information for Example 7.7.

Activity	Duration	IPA	Rel. type	Lag
A	10	—	—	—
B	5	A	SS	6
C	3	B	SS	1

Solution:

The critical path here included both activities A and B and the project took an extra day.

If the lag between activities A and B was 4 days or less, the critical path would be only activity A with project length = duration of activity A = 10 days.

On the flip side, if the lag between activities B and C was increased to 3 days, all three activities would be critical, and the project duration would be extended to 12 days. The graphic solution is shown in Figures 7.22 a through 7.22 f.

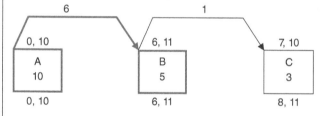

Figure 7.22a Solution for Example 7.7.

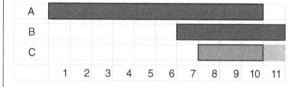

Figure 7.22b Solution for Example 7.7 in a bar chart format.

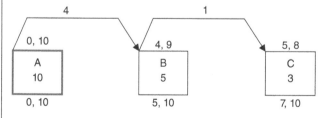

Figure 7.22c Solution for Example 7.7 with lag between activities A and B is only 4 days.

Figure 7.22d Bar chart solution for Example 7.7 with lag between activities A and B is only 4 days.

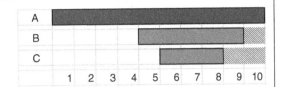

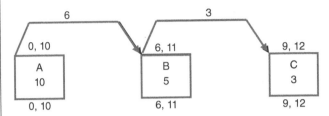

Figure 7.22e Solution for Example 7.7 with lag between activities B and C is 3 days.

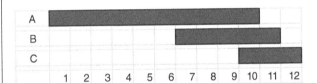

Figure 7.22f Bar chart solution for Example 7.7 with lag between activities B and C is 3 days.

Example 7.8 We are redoing Example 7.6 but replacing the SS relationships with FF relationships.

Table 7.11 Information for Example 7.8.

Activity	Duration	IPA	Rel. type	Lag
A	10	—	—	—
B	5	A	FF	—
C	3	B	FF	—

(Continued)

Example 7.8 (Continued)

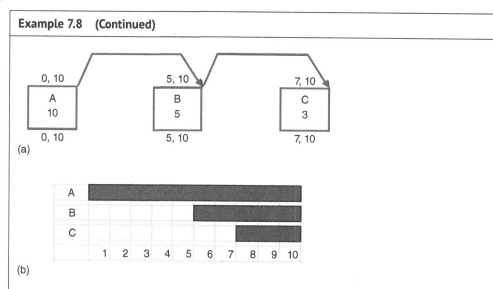

(a)

(b)

Figure 7.23 (a) Solution for Example 7.8. (b) Bar chart solution for Example 7.8.

Solution:
What is interesting here is that, unlike the case in Example 7.6, all three activities are critical. This is because of the FF relationship and that the first activity, A, has the longest duration. So, activities B and C must finish no earlier than day 10 (when activity A is finished), thus, they cannot start earlier than the date shown. The graphic solution is shown in Figures 7.23 a and b.

Example 7.9 We are redoing Example 7.8 but with lags between the activities.

Table 7.12 Information for Example 7.9.

Activity	Duration	IPA	Rel. type	Lag
A	10	—	—	—
B	5	A	FF	2
C	3	B	FF	1

Solution:

The graphic solution is shown in Figures 7.24 a and b.

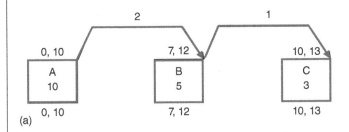

(a)

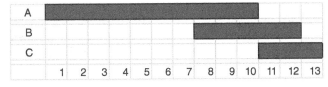

(b)

Figure 7.24 (a) Solution for Example 7.9. (b) Bar chart solution for Example 7.9.

Example 7.10 **Now, we are doing Example 7.6 with combination SS+FF relationships.**

Table 7.13 Information for Example 7.10.

Activity	Duration	IPA	Rel. type	Lag
A	10	—	—	—
B	5	A	SS+FF	—
C	3	B	SS+FF	—

Solution:

When we have combination SS+FF relationships, we will get two sets of early dates (one out of each relationship) and two sets of late dates: we accept one set of each and reject the other. For example, going from A to B: the SS relationship gives activity B the early dates (0, 5) while the FF relationship gives it the early dates (5, 10). Obviously, the first set is rejected because it violates the FF

(Continued)

Example 7.10   (Continued)

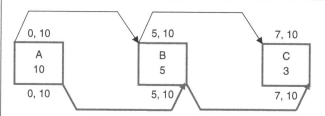

Figure 7.25   Solution for Example 7.10.

relationship. The same principle applies between B and C, producing (5, 8) and (7, 10). Again, we rejected the first set.

The solution, as shown in Figure 7.25, looks identical to that of Example 7.8, and it is indeed identical, but only because the FF relationship drives (dominates) in this case. Therefore, we labeled the FF relationships as critical while the SS relationships are not. Let us take other examples when different relationships drive.

Example 7.11   Redo Example 7.10 with some lags added. Note that we separated the relationships between the same two activities because one has a lag and the other does not.

Table 7.14   Information for Example 7.11.

Activity	Duration	IPA	Rel. type	Lag
A	10	—	—	—
B	6	A	SS	2
			FF	
C	5	B	SS	3
—	—	—	FF	—

Solution:
The graphic solution is shown in Figure 7.26, with driving relationships in thick lines.

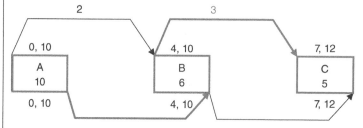

Figure 7.26   Solution for Example 7.11.

Example 7.12 **We made changes to the lags in Example 7.11.**

Table 7.15 Information for Example 7.12.

Activity	Duration	IPA	Rel. type	Lag
A	10	—	—	—
B	6	A	SS	5
—	—	—	FF	—
C	5	B	SS	3
—	—	—	FF	1

Solution:

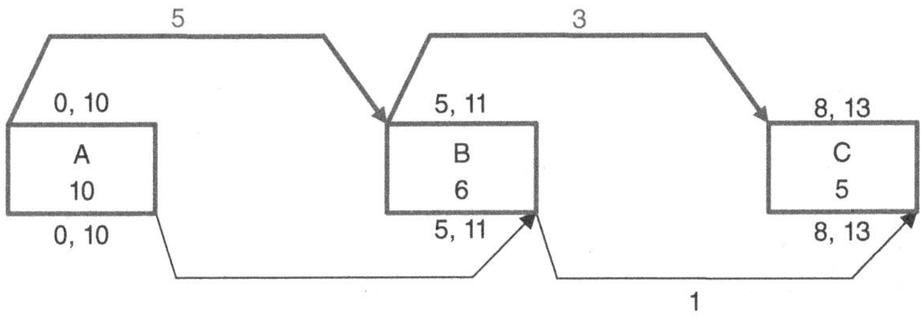

Figure 7.27 Solution for Example 7.12.

Types of Lags: Lags are three types:

1. *The wait lag:* The wait time is not part of either activities; the predecessor or successor, as shown in Figures 7.28 a and b.

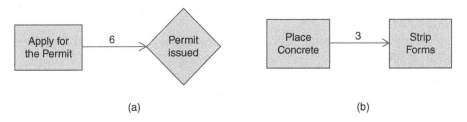

(a)   (b)

Figure 7.28 Two examples for the wait lag.

2. *The start lag* is when the successor depends on a portion of the predecessor, as shown in Figure 7.29. Once that portion is achieved, there is no more dependency between the two activities.

Figure 7.29 An example for the start lag.

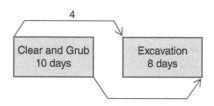

3. *The continuous lag* is when the predecessor and the successor are overlapping throughout their duration but there must be a buffer between them, as shown in Figure 7.30.

Figure 7.30 An example for the continuous lag.

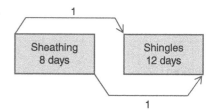

So, the lag in the three types mentioned above is: independent, dependent on a portion only of the predecessor, or dependent on the entire predecessor; respectively.

A lag as a percentage? The lag, at least the start and continuous types, represents the work portion that must keep the two activities apart. So, in Figure 7.29, the lag means the Excavation can start after 4 days of work on the Clear & Grub 4 have been completed, which is equivalent to 40% of it. So, can we show the relationship as a percentage, as shown in Figure 7.31?

Figure 7.31 A start lag as a percentage.

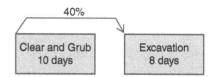

The concept in reality is more accurate than the number of days lag, but the CPM calculations convert it back to a number (by multiplying the percentage by the duration of the predecessor). In our example, the lag = 40% × 10 days = 4 days.

Different ways to show logic? There could be more than one way to show logic. Let us show the example in Figure 7.29 in different ways:

Figure 7.32 (a) Redoing the Example in Figure 7.29 with FS relationship and lead. (b) Redoing the Example in Figure 7.29 with FF relationship and lag.

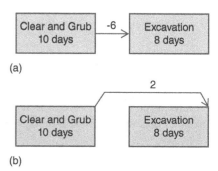

So, we can describe the three diagrams:

– In Figure 7.29, activity B can start at least 4 days after the start of activity A.
– In Figure 7.32a, activity B can start no more than 6 days before activity A finishes.
– In Figure 7.32b, activity B can finish at least 2 days after activity A finishes.

Simplifying Schedules

Simplifying the schedule starts with the terms of the contract and then includes steps during the creating of the schedule, the updating and control, reporting and documentation, and archiving. Here are a few hints for simplifying the schedule during the creating phase.

We can, and should, simplify schedules by using a few tips:

1. Break the work into homogeneous activities that are easily measured and controlled. Don't be concerned about a large number of activities.
2. Use FS relationship and avoid other relationships whenever possible.
3. Minimize the use of lags, and do not use leads.
4. Minimize the use of constraints.
5. Use activity codes along with software tools such as "Group and Sort" and "Filter."

An example of simplification is redoing the work of Figure 7.29 by splitting activity A (Clear and Grub). Note that we split this activity by the lag amount, resulting in two portions, one is critical (the predecessor of the excavation activity) and one is non critical (Figures 7.33a and 7.33b).

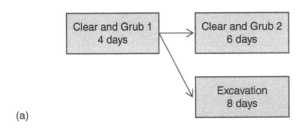

(a)

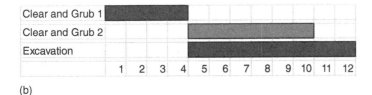

(b)

Figure 7.33 (a) Redoing Example 7.27 with only FS relationship and no lags. (b) The bar chart solution of Figure 7.31a.

Schedule updating

After construction starts, the schedule needs to be updated; reflecting actual performance and implementing any changes, to make sure it stays accurate. This topic will be discussed in detail in chapter 8, Project Controls.

Schedule Acceleration

Schedule acceleration is shortening the duration of the project schedule *without* reducing the project scope, till reaching the desired duration or the crash (absolutely minimum) duration, whichever comes first.

Schedule acceleration is also known as schedule compression or schedule crashing. It is covered under the topic time-cost trade-off. Schedule acceleration may be planned before construction starts or decided during the execution of the project (in this case, called recovery schedules).

The duration of a project is directly related to its cost and scope. Figure 7.34 shows the interaction of the three variables (cost, schedule, and quality[8]) with the scope and among each other.

– If we downsize the scope, the cost and duration decrease, and vice versa,
– If we accelerate the schedule, with the scope and specs unchanged, the cost will increase[9],
– If we change the specifications (specs), the cost and/or schedule may be impacted.

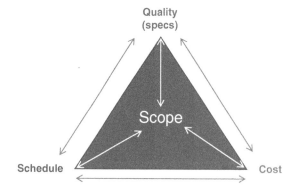

Figure 7.34 Interaction of, schedule, cost, quality, and scope.

There are several reasons why owners and/or contractors accelerate schedules:

1. The contractor may accelerate the project to make sure the schedule meets the contract's stipulated finish date. Sometimes, this is done at the bidding stage when the bidders are required to submit a CPM schedule.
2. To respond to market demand and maximize profit such as when there is high demand for projects. The contractor may accelerate the schedule, even when the contract/owner does not require so, but this will allow the contractor to start other projects.

---

8 Quality in the construction projects context is defined as conformance to specifications.
9 With some exceptions, as will be explained later.

3. In some public projects, the government agency accelerates the project for the convenience of the public.
4. Simply to reduce the cost. This is an interesting scenario when accelerating within certain guidelines and limits, it will result in decreasing the cost.

> Schedule acceleration must be studied by the contractor even if not required. In many cases, limited acceleration can save money.

5. The contractor may accelerate the project to avoid penalties (liquidated damages) for late finish and/ or get early finish bonus.

Methods for accelerating work in construction projects:

1. Invest more time and money in the planning phase.
2. Choose design-build contract.
3. Add contractual incentives and deterrents such as liquidated damages for late finish and/or bonus for early finish.
4. Revisit the schedule to find creative ways to optimize time and money.
5. Fast-track the project by overlapping the design and construction phases. This is likely to be coupled with design-build contracts.
6. Perform preconstruction services, value engineering, and constructability studies.
7. Utilize building information modeling (BIM) as a valuable tool for optimizing the construction process and detect potential clashes.
8. Improve communications among parties in both protocol and methods/technology.
9. Make certain crews work overtime as needed. Possibly add more crews and more shifts.
10. Acquire special materials, equipment, and technologies that can do the work faster and more efficiently. This includes using pre-engineered, pre-manufactured, modular, and/or preassembled units.
11. Offer incentives to crews and subs.
12. Improve project management methods and processes.

Keep in mind that some of the above methods require planning ahead of time. This underscores the importance of early and thorough planning. The Construction Industry Institute (CII) suggested more than 90 techniques for schedule compression. It further classified them by project phase or function[10]:

i. Ideas applicable to all phases of a project
ii. Engineering phase
iii. Contractual approach
iv. Scheduling
v. Materials management
vi. Construction work management
vii. Field labor management
viii. Start-up phase

---

10 Concepts and Methods of Schedule Compression, publication number RS6-7, 1988. Austin, TX: CII

The concept of schedule acceleration:

1. Since the critical path is the longest path in the network, and we need to shorten the duration of this network; therefore, we need to shorten the duration of the critical path.
2. We select the activity (or activities) on the critical path that is easiest and least expensive to shorten its duration.
3. As the critical path is getting shortened, at some duration, it will tie with the next-longest path. Then, we need to shorten the duration of both paths.
4. The more we shorten the duration of the project, the more paths tie as the longest (critical) paths in the network. This makes shortening the duration of the project more difficult and more expensive.
5. We stop when we either reach the desired duration or when the project is completely crashed.

Effect of Project Acceleration on Cost

The effect of project acceleration on direct and indirect cost is different, so we have to monitor direct, indirect, and total cost, as we accelerate the schedule.

1. Direct cost increases at an increasing rate, for two reasons. The more we accelerate, the more paths will tie for the longest path, so they all must be shortened. Also, as we accelerate more, we run out of easy/inexpensive methods and need to use more expensive methods. This is illustrated in Figure 7.35.
2. Indirect cost is usually constant through time. It represents "steady" expenses needed to run the project but not tied directly to any cost item in the project, such as office expenses and salaried staff. So, less project duration means less indirect costs. We can say indirect costs decrease at a constant rate[11], as illustrated in Figure 7.36.

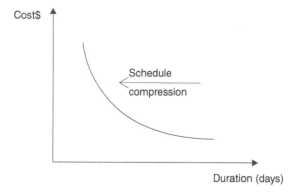

Figure 7.35   Effect of project acceleration on direct cost.

---

11 Aggressive acceleration may require more overhead per day or per week. Refer to Construction Project Scheduling and Control by Saleh Mubarak, 4th edition, 2019, Wiley.

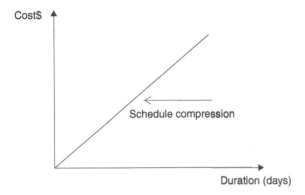

Figure 7.36 Effect of project acceleration on indirect cost.

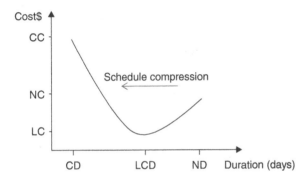

Figure 7.37 Effect of project acceleration on total cost.

3. Total cost: Assuming that the initial saving from the decrease in indirect cost is greater than the increase in direct cost, the total cost will initially decrease. However, as we keep accelerating, the direct cost increase will keep increasing so at some point, it will surpass the decrease in indirect cost, and the total cost – after hitting the minimum point – will start increasing at an increasing rate, as illustrated in Figure 7.37.

In Figure 7.37, we identify three points: normal, least-cost, and crash. They are defined as below:

– Normal duration (ND) is the amount of time required to finish the project under ordinary circumstances without any deliberate acceleration or deceleration. The term "normal" here may vary from one contract to another, and from one crew / situation to another.

   o Normal cost is the cost of a project that is performed within the normal duration.

- Least-Cost Duration, LCD, is the duration of an accelerated schedule to reach the point where the total cost of the project is minimum. If the duration of the schedule increases or decreases, the total cost will increase.
  - Least-Cost, LC, is the minimum cost of the project which is performed at its least-cost-duration.
- Crash Duration, CD, is the least possible duration for a construction project schedule. It is usually achieved by maximum schedule acceleration.
  - Crash Cost, CC, is the total cost of a construction project at its minimum duration, including the impact of crashing (maximum acceleration of) the schedule.

Comments on project schedule acceleration:

1. Project acceleration should be based on scientific and systematic principles. It is not random dumping of extra resources on the site. It is not a race to set records.
2. It should be done only to the extent needed.
3. When a contractor likes to accelerate several projects, he/she needs to prioritize them because they will compete for resources and attention.
4. There are physical and practical limits to project acceleration. This is a point that must be explained to owners who have unrealistic expectations and want the project finished in no time!
5. Unplanned or unorganized acceleration may result in adverse consequences in cost and time, defeating the purpose of acceleration.

Linear Scheduling Method (LSM)

Networks using the CPM are used for projects that consist of large numbers of activities with multiple paths from start to finish. Some construction projects consist of a few activities (usually with large quantities) that must be done in the same order or sequence, such as heavy construction projects (e.g., roads, earthwork, or utility piping). They may use another method called linear scheduling method, LSM[12], which is completely different. It was used in numerous large projects in the past, including the Empire State building in New York City, 1930–1931, and the English Channel Tunnel Project, 1988–1994.

LSM is a graphical scheduling method with linear and repetitive activities represented by lines with slopes proportional to their production rate. These lines are plotted in X–Y graphs where X (horizontal axis) represents time, while Y (vertical axis) represents work progress (location, distance, step, or floor), or the reverse. LSM focuses on continuous resource utilization.

Figure 7.38 shows how the LSM works with two activities: Activity B has a slower production rate (flatter slope), so its buffer with activity A gets larger as work advances. We must plan them in a way that does not violate the buffer requirement.

---

12 There are very similar, almost identical, methods to LSM, such as line of balance (LOB) method and vertical production method (VPM). Also, LSM method is known in some parts of Europe as time couplings method (TCM).

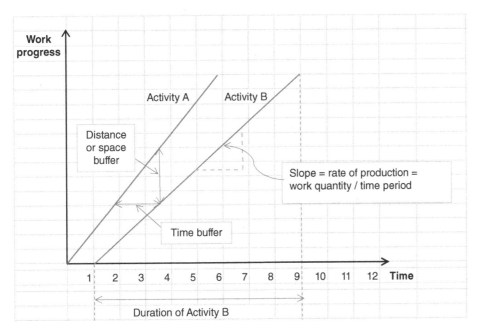

Figure 7.38 The concept of the LSM is demonstrated in a graph.

It is not unusual to use the *X*-axis to express work progress and the *Y*-axis to express time units, as shown in Figure 7.39. We then have to be careful in calculating the slope of the line representing the activity. For example, activity B in Figure 7.39, has a lower production rate than activity A but a higher rate than that for activity C.

When we have two or more activities, the production rate will differ from one to another. The horizontal distance between the two lines represents the float of the earlier activity. In the LSM, we call it the time buffer. The vertical distance represents the distance separating the two operations. We call it the distance buffer. When we define a time or distance buffer, we must not allow the successor to violate it, which means getting closer to the predecessor than the buffer. We will clarify this concept in the following example.

There are three simple steps necessary to build a schedule by using the LSM:

1. Determine the work activities. As mentioned previously, we expect only a few activities in LSM schedules.
2. Estimate activity production rates. Such estimation is similar to determining durations. We still estimate durations, but we are more concerned with production rates.
3. Develop an activity sequence, similar to determining logical relationships. Using CPM terms, relationships here are all start-to-start (with lags) with finish-to-finish (with lags too).

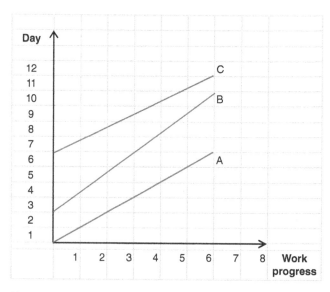

Figure 7.39 LSM graph example with *X*- and *Y*-axis reversed.

Example 7.13 We need to lay down 1,000 linear feet (LF) of an underground utility pipe. The basic activities and their production rates are:

A. Excavation (trench): 100 LF/day,
B. Prepare the subbase (gravel): 125 LF/day,
C. Laying the concrete pipe: 75 LF/day,
D. Backfilling: 200 LF/day, and
E. Compacting: 150 LF/day.

Draw the LSM diagram and calculate the project completion date. Keep a minimum of one day as a time buffer between any two consecutive activities.

Solution: Let us calculate the duration of each activity:

A. Duration = 1000/100 = 10 days
B. Duration = 1000/125 = 8 days
C. Duration = 1000/75 = 13.33 ≈ 14 days
D. Duration = 1000/200 = 5 days
E. Duration = 1000/150 = 6.67 ≈ 7 days

From the chart in Figure 7.40, we find that the project should finish in 22 days. Interestingly, if activities B and D can be slightly adjusted (slowed down), the project can finish in 20 days, two days earlier:

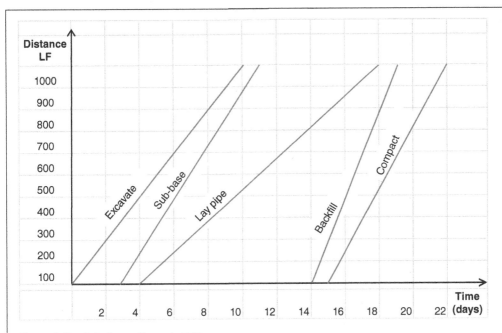

Figure 7.40 Solution to Example 7.13.

- Activity B: production rate 111 LF/day, duration = 9 instead of 8 days,
- Activity D: production rate 167 LF/day, duration = 6 instead of 5 days

This is demonstrated in Figure 7.41. The explanation is simple: when we adjust the production rate, up or down, to make the activities' work flow synchronized, we minimize wait time. For example, such production adjustment allows activity C, which takes the longest time, to start one day earlier. The same argument applies to activities D and E.

Keep in mind that production adjustment is not always possible if the activity depends on a major equipment that has certain speed. There might be other options available such as adjusting the production to other, preceding or succeeding, activities or doing the work in intervals by keeping the same production rate but taking breaks when the time buffer gets to the minimum (Figures 7.42 and 7.43).

(Continued)

Example 7.13 (Continued)

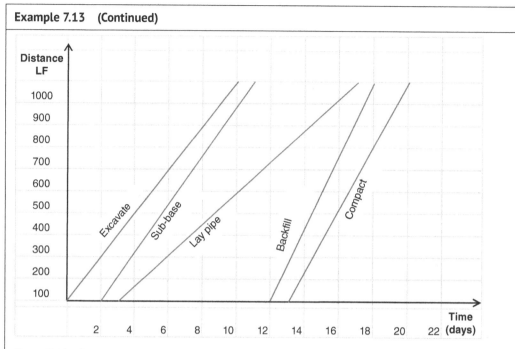

Figure 7.41 Solution to Example 7.13 with production adjustments to activities B and D.

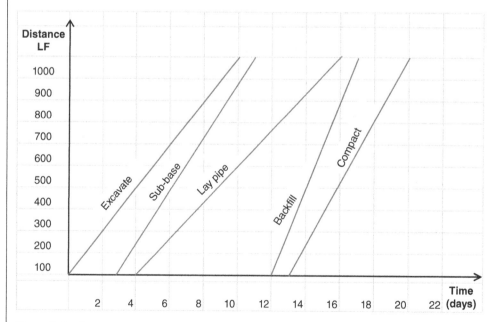

Figure 7.42 Solution to Example 7.13 with production rate to activity C adjusted from 75 to 83.3 LF/day.

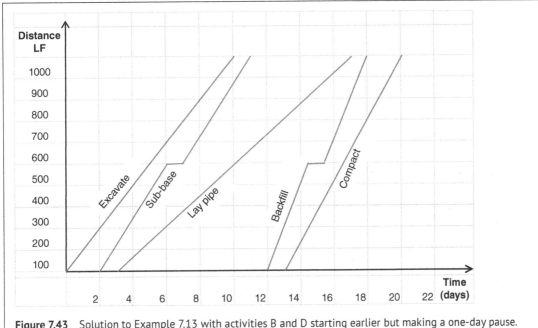

Figure 7.43 Solution to Example 7.13 with activities B and D starting earlier but making a one-day pause.

Final comment on LSM

LSM attempts to streamline operations and maximize resource utilization. Since there are usually a few activities in the LSM diagram, the scheduler will attempt to make adjustments for acceleration or better resource utilization. These adjustments include changes to activities' start dates, production rates, and/ or making stops (taking breaks) while progressing.

There are several computer software packages using LSM such as Trimble TILOS[13]. LSM has also been used in the form of diagrams that were drawn in computer-aided design (CAD) software, for example, for a major pipeline project. Any spreadsheet software, for example, Microsoft Excel as well as scientific computing environments, for example, MATLAB®, can be set up manually to perform calculations and draw linear schedule diagrams that can be updated during the project's execution.

Exercises

7.1 Define project scheduling, a schedule, and a project scheduler.

7.2 Define an activity, an event, and a milestone. Give an example of each.

7.3 Search online for a new product that improves project scheduling, particularly for the construction industry. It may be a concept, gadget, software, or something else.

---

13 https://constructionsoftware.trimble.com/products/tilos/

7.4 What is a bar chart? Why was it given the name Gantt charts?

7.5 What are the advantages and disadvantages of bar charts?

7.6 What is a logic network diagram? What are the types of networks?

7.7 In a node diagram, what does a node mean, and what does a line (arrow) mean?

7.8 Briefly describe the steps to create a project schedule.

7.9 What criteria do we use in breaking down the project into activities for creating a schedule?

7.10 Discuss the relationship between confidence in activity duration and the work breakdown structure.

7.11 You are the scheduler of a midsize construction project. You have the responsibility of breaking down the project into activities. The project manager likes to see a small number of activities, which will force you to combine certain activities. You know that such practice will strip the schedule from the ability to monitor individual activities and analyze production rates by crew. What would you do? How would you resolve the difference of opinion in a professional manner that does not compromise the quality of the schedule?

7.12 What are the methods used for calculating or estimating the duration of activities?

7.13 Why is accuracy important for calculating activities' durations?

7.14 What are the consequences of lack of accuracy in activities durations?

7.15 How do you assign logical relationships among activities? Do you sometimes have more than one way to show logic?

7.16 What are the 4 types of logical relationships in precedence diagrams? Which one is the most used?

7.17 In assigning logical relationships among activities, the scheduler must pay attention to external relationships. Give some examples.

7.18 What is resource loading and resource leveling?

7.19 What does implementing the schedule mean?

7.20 What is the critical path method? What benefits did it bring to project scheduling?

7.21 Define:
- **a.** The critical path
- **b.** Total float
- **c.** Early start date
- **d.** Early finish date
- **e.** Late start date
- **f.** Late finish date
- **g.** Free float
- **h.** Forward pass
- **i.** Backward pass

7.22 Mention some proper network drawing tips.

7.23 What is a constraint? How and why do we use constraints in schedules?

7.24 How do we assign and use work calendars in schedules? Can different activities have different calendars?

7.25 What is a lag? Compare lags with activities and events.

7.26 List the benefits of CPM scheduling in construction projects from the contractor's perspective.

7.27 Do all construction projects have the same need for CPM scheduling? Why or why not? Give examples.

7.28 What qualifications must a scheduler of a building project have? Can the same person be a scheduler for an industrial project? Why or why not?

7.29 Which is the longest path:

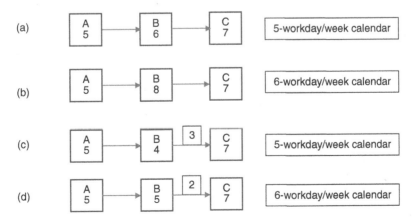

7.30 Draw the node network for the following project and perform the CPM calculations:

Activity	Duration	IPA	ES	EF	LS	LF	TF	FF
A	7	—						
B	10	—						
C	15	A						
D	8	A, B						
E	4	C, D						

7.31 Repeat exercise 30 with an imposed finish date of 24 days.

7.32 Draw the node network for the following project and perform the CPM calculations:

Activity	Duration	IPA	ES	EF	LS	LF	TF	FF
A	6	—						
B	2	—						
C	8	A						
D	5	A, B						
E	3	B						
F	6	D, E						
G	5	C, D						
H	2	F, G						

7.33 Repeat Exercise 32 with an imposed finish date of 18 days and then 25 days.

7.34 Draw the node network for the following project and perform the CPM calculations:

Activity	Duration	IPA	Lag	ES	EF	LS	LF	TF	FF
A	4	—	—						
B	9	—	—						
C	6	A	—						
D	4	A	—						
—	—	B	2						
E	4	B	—						
F	2	B, C	—						
G	6	D	—						
—	—	E	3						
H	3	F, G	—						

7.35 Repeat Exercise 34 with an imposed finish date of 21 days and then 30 days.

7.36 A change order was issued for the work in Example 7.2 that extends the duration of activity F from 4 to 8 days. Recalculate the finish date and check if this change order has any impact on the critical path and/or the project finish date.

7.37 If A, B, and C are three activities connected with SS relationships only with 2-day lag on each relationship. Their durations are 10, 5, and 3, respectively. What is the duration of the project?

7.38 If A, B, and C are three activities connected with FF relationships only with no lags. Their durations are 5, 10, and 3, respectively. What is the duration of the project?

7.39 If A, B, and C are three activities connected with FF relationships only with 2-day lag on each relationship. Their durations are 5, 10, and 3, respectively. What is the duration of the project?

7.40 If A, B, and C are three activities connected with FF relationships only with 2-day lag on each relationship. Their durations are 10, 5, and 3, respectively. What is the duration of the project?

7.41 If A, B, and C are three activities connected with SS and FF (combination) relationships with no lags. Their durations are 5, 10, and 3, respectively.
   a. What is the duration of the project?
   b. Which relationships are critical?

7.42 If A, B, and C are three activities connected with SS and FF (combination) relationships with 2-day lag on each relationship. Their durations are 5, 10, and 3, respectively. What is the duration of the project?

7.43 What do we mean when we describe an activity as near-critical? Are there well-defined criteria for such activities?

7.44 A schedule with a calculated finish date of 188 days. When we entered the imposed finish date, negative float appeared with a maximum absolute value of 6 days. What is duration of the project according to the imposed finish date?

7.45 The contractor may keep a number of days as project float (or management float), why? How does this concept work?

7.46 What are the types of lags? Give examples for each. Which ones can be eliminated by simplification of the activities and logic?

7.47 In resource leveling, what resources do we level? Why not all resources?

7.48 We can assign a cost amount (budget) for each activity, like $21,000), but there is a better way. What is it? What are its advantages over the assignment cost directly?

7.49 The forward pass does not involve the contract finish date, but the backward pass does. Explain this statement.

7.50 Generally, the GC does some of the work by its own forces and subcontracts other work. How can the GC and the subs do resource leveling?

7.51 What is an updated schedule?

7.52 What does the term "accelerating the project" mean?

7.53 Why would a contractor accelerate a project? Do you think studying the project cost-time trade-off is a good idea for the contractor even if the calculated finish date is within the contract limits?

7.54 An owner decides to build a 10-story office building that is expected to take 18 months. Later due to budget constraints, the owner decided to scale it down to 8 stories only, with a timeframe of 16 months. Is this reduction in duration considered project acceleration? Explain your answer?

7.55 A project manager may shorten the duration of a project using several methods. Mention six of these methods and discuss briefly.

7.56 Explain how accelerating a project schedule works.

7.57 Do an internet search or meet a project manager and discuss schedule accelerating methods. Pick one method and discuss it in detail.

7.58 In general, what is the effect of accelerating the schedule on the project's direct, indirect, and total cost?

7.59 What is a "recovery schedule"?

7.60 What type of projects is best for LSM application?

7.61 What are the steps for preparing an LSM schedule?

7.62 Define time buffer and distance buffer. Use words and graphs to explain your answer.

8

Project Controls

Introduction

The construction project's cost estimate and schedule are both predictions of future occurrences in terms of money and time. No prediction of events, years or even weeks in advance can be 100% accurate. Of course, the contractor needs to add contingency allowance to the cost estimate and contingency time to the schedule for the "unknown" factor. But what is more important, is to monitor actual performance against the predictions (cost and schedule baselines), detect and measure variances, and take corrective action whenever and wherever needed. This is what we call *project controls*. It is the other side of cost and time management, where we take our prediction to action to make sure the project stays within both baseline budget and schedule during execution.

The Baselines

A *baseline* is the originally approved plan for a project, including approved changes. It usually includes a baseline budget and baseline schedule. It is used as a benchmark for comparison with actual performance. It is the reference we use to make judgments at any point during construction:

- – If we are on, over, or under budget, and
- – If we are on, ahead of, or behind schedule.

Baseline budget is the project's original approved budget, including any approved changes.

Baseline schedule is a schedule prepared by the contractor before the start of the project – and usually approved by the owner – typically used for performance comparison. Therefore, the baseline schedule must be accurate and realistic. The baseline schedule is similar to the "As Planned" schedule. In fact, they are likely to be the same. The main, and perhaps only, difference between them is that the baseline schedule must be approved by the owner, while the "as planned" schedule may be for the contractor's own use.

Figure 8.1 summarizes the cycle of cost and time management in a traditional design–bid–build, DBB, contract delivery method. It starts with an approximate cost estimate and a summary schedule during

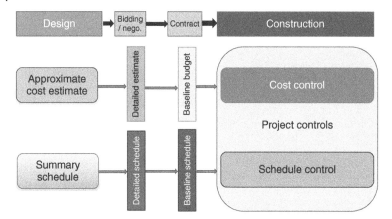

Figure 8.1 Cost and time management cycle.

the design phase. During the procurement phase (bidding or negotiating), the contractor candidate prepares, and usually submits, a detailed cost estimate and schedule. The detailed estimate and schedule may be tweaked and/or adjusted during the negotiation and contracting phase till they are finalized and approved by the owner, so they become baseline budget and baseline schedule. When the construction phase begins, actual performance is measured against these baselines.

Schedule Updating

When the construction phase starts, the schedule needs to be updated on a regular basis, reflecting work progress (actual) on the schedule, and comparing it to the baseline schedule. Most likely, this update includes actual cost and the comparison to the baseline budget. It is important to do so routinely and whenever necessary to detect and manage variances and implement changes. Schedule updating also provides the work amounts needed for progress payments from the owner to the general contractor.

An *updated schedule* is a revised schedule reflecting project information at a given *data date* regarding completed activities, in-progress activities, and changes in the dates, logic, cost, resources, constraints, calendars, and any other change at both activities and project levels.

Change orders are studied and implemented with schedule updates. Schedule updating provides vital information and guidance for the contractor's work plan and adjustments, whenever and wherever needed. Schedule updating procedures must be addressed in the contract agreement. Typically, the project schedule is updated periodically as the contract specifies. This can be weekly, biweekly, or others, but the important issue here is to make sure the schedule is updated on time, no matter what happens on the ground. The interval between updates must be balanced, not too short, so the project management team members will be overwhelmed with the process. But not too long, otherwise, it may be too late to implement any adjustment or corrective action. Also, it is possible that this period is shortened during "crunch time", such as when the project gets close to completion or when the schedule is accelerated.

While there is no specific standard process for updating schedules, the process must include the following steps:

a. Collection of work progress on all activities including amounts (or percentages) and start/finish dates. Depending on schedule details, the information may include cost and resource amounts (consumed and required), as of the end of business day of data date. This information is likely to be organized using a spreadsheet or other means that will be kept as a record for any possible review.

b. The scheduler, or the professional performing the update, enters all this information in the scheduling computer program and recalculates the schedule dates for incomplete activities as well as the entire project, with a focus on important milestones, the critical path, and the project completion date.

c. The project manager takes a look at the updated schedule and decides if it is acceptable, or some adjustments must be made. Based on that, the updated schedule is finalized.

d. The owner and all subcontractors get copies of the updated schedule. It is possible and likely that the scheduler customizes reports, so each party receives pertinent information.

There is a tendency among some contractors to relax the rules when things are going fine, and then they regret it when a claim dispute arises. The owner must insist on the periodic updates and tie the contractor's progress payments[1] to the updates. These updates represent valuable materials for potential analysis later and could be used as evidence to prove or disprove a claim.

The schedule update draws a timeline between "past"; what has been performed and reported, and "future"; what is remaining to complete the project. This timeline is called the *data date*, which is the date on which all progress on a project is reported. It is also called the as-of date, the *cut-off date*, and the *status date*. It is *not* the "current date" or the "time now," because the word *current* or *now* moves with time and is not tied to a specific time or date. So, the focus of every update should be the period between the previous data date and the current data date, as shown in Figure 8.2. No event prior to the current data date should be left without updating: it must be either marked as "Actual" or delayed beyond the current data date.

The update must include the information below:

1. Past information that includes what has happened since last update such as:
   a. Activities that have started, their actual start date, percent complete, remaining duration, and/or "Expected Finish Date", as well as the cost (or resource consumed) during this period.
   b. Ongoing activities (reported as started in previous update(s) but still in progress): new percent complete, remaining duration and/or "Expected Finish Date", and cost (or resource) consumed this period.
   c. Activities that are completed with actual completion date and cost (or resource) consumed this period and at completion.
      For started activities, completed or still in-progress, we calculate cost this period, cost to date, cost to complete (if incomplete), and cost at completion. We may also calculate cost variance.
2. Future: Any changes to the schedule or schedule-related items such as:
   a. Activities that have been added, along with their information (such as duration, logic, budget, calendar, resources, and constraints).

---

1 To be explained in detail later in the Cash Flow chapter.

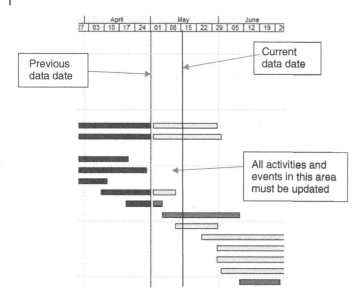

Figure 8.2 An in-progress bar chart showing the period to be updated.

   b. Any activities that have been deleted. The scheduler must be careful when deleting an activity, not to have interrupted logic or dangling activities.

   c. Activities that have changed in duration, logic, budget, calendar, resources, constraints, codes, or any other change.

   d. Any approved change to the "imposed finish date" for the entire project or a constraint date for certain milestones.

     Note that some future activities will have their dates changed as a result of the change in the predecessors' dates.

3. Any schedule-related, but not activity-specific changes such as:

   a. Cost and availability of resources.

   b. implementation of an approved extension of time,

   c. Change in calendar workdays.

   d. Change in responsibility.

4. Notes at both activities and project levels (explained later.) Also, the scheduler may need to look at the updated risk plan and see if it has an impact on the schedule.

   Making changes to events in previous update cycles must not be allowed unless there was an error to be corrected. In this case, the errors and corrections must be stated clearly.

   Some computer software programs allow users to mark future events (start or finish dates after data date) as "actual." This practice is not recommended and may have repercussions and misleading output data.

> Schedule updating is a must because things on the ground do not usually go as planned. The contractor needs to compare actual work progress to the planned and adjust work plan, if needed.

Changes to Future Events

Changes that happen to future activities and events (start/finish dates shifted earlier or later) happen either as a result of "pure progress" or as "intended changes." Let us look at Figure 8.3a: The contractor is planning these four activities in the sequence shown, with a total duration of 14 days. Drywall can start two days before the completion of Framing.

In Figure 8.3b, 4 days after starting the first activity, framing, the contractor faced some issues that extended the activity's duration from 7 to 9 days. This change will have a domino effect on the succeeding activities. So, even with no expected changes in the duration of the other three activities, the entire operation is expected to take 16 days, 2 days more than originally planned.

In Figure 8.3c, we are showing a different scenario, also 4 days after starting the first activity, framing. Work is going fine in the framing activity, with the expected finish date same as originally planned. However, due to another assignment, the painting crew cannot start the painting activity till day 12. This forced the contractor to put a constraint, delaying the start of painting activity from day 10 to day 12.

Repercussions of Wrong/Inaccurate Remaining Duration

When updating, many schedulers focus on what happened because it is the important component for progress payments. However, it is equally important to examine the remaining schedule and any other

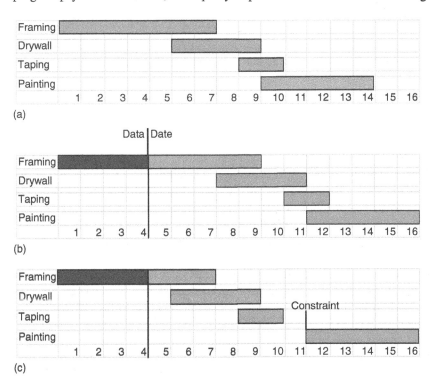

Figure 8.3 (a) Planned sequence for framing, drywall, taping, and painting, (b) operation framing, drywall, taping, and painting, after 4 days of start and a delay in finishing framing due to slower work progress, and (c) operation framing, drywall, taping, and painting, after 4 days of start with a planned delay for starting painting.

changes. These changes may include shifts in start / finish dates of future activities and events and changes in activities' duration. Such changes may have an impact on the project's milestones and/or completion date. They may also have an impact on work plans for crews, materials management, cash flow, and other aspects of the project management. Failure to see and acknowledge these changes may lead to work disruptions, idle crews, shortage of materials, chaos, and increased expenses. Let us consider this situation: an activity with original duration = 8 days:

1. After 4 days of work (actual duration, AD = 4), it was recorded in the schedule update as 50% complete. So, the software calculated the remaining duration as 4 days.
2. In reality, it was only about 37% complete. The remaining duration, if we assume production rate will remain the same, $= 4 * (100\%-37\%)/37\% = 6.81 \approx 7$ days.
3. This means the schedule, if not corrected, shows the completion of this activity 3 days earlier than it should be.
4. Now imagine the impact of this error on succeeding activities. If this activity is critical, it will also show an inaccurate project completion date.
5. Of course, it is possible that the drop in production rate in the first 4 days was due to unexpected circumstances that no longer exist in the remainder of the work, so production rate should be higher in the activity's remaining duration, and the contractor can estimate it accurately. It can be 4, 5, 6, or 7 days. It is also possible that the contractor adds more resources so the activity can still be finished on time. The important thing is that the remaining duration must be examined if actual production rate does not match the planned rate.

> When updating a schedule, we must pay attention to both "what happened" and "what is remaining to happen."

Tips on Schedule Updating

1. The schedule update process may differ slightly from one organization to another, but most organizations have schedule update templates, either paper or electronic, that the project manager or superintendent has to fill and send to the scheduler for updating the schedule in the computer system. The form must contain the information mentioned earlier. For simplicity, the form may exclude completed activities and activities to be performed far in the future (start date $\gg$ data date). The form must include a space for notes and signatures. The form may be customized further depending on the preference of the user organization.
2. One of the most important pieces of information in the schedule update is taking notes whenever needed. The notes can be at the activity or project level. Notes are needed particularly when things do not go as planned, to explain what happened and why. An example of a note at the activity level is: "Excavation was halted on day 3 because groundwater broke out. A pump was acquired and arrived on site on day 4. Excavation was resumed on day 6." An example of a note at the project level is: "A 10-day extension of time was approved by the owner. The new contract finish date is July 25, 2025."
3. The scheduler uses the provided information to update the schedule and produces preliminary report. After discussing the report with the project manager, the scheduler makes any needed adjustments

and produces the final report. The update report must be communicated to all parties (owner, subs, etc.) promptly. It will be more efficient if the report is customized, so each party receives a focused and concise report with relevant information. The contractor's own report must include a description of the situation, variances with their amounts, causes, and possible remedies and recommendations. For example, if the project slips by 5 days or the critical path changes, the report must include the cause for the slippage or change.

4. As mentioned earlier, the current schedule must be compared to the baseline schedule in order to determine the situation. But it is also a good idea to compare it to the previous update, especially when a remedy or adjustment was implemented recently. For example, the current update shows a 7-day negative float. This is bad news, no doubt, however, when looking at the previous update, we found out it had a 12-day negative float, which shows that we are on the right track to recovery.

   The scheduler, or the person responsible for updating the schedule on the computer, must do so in a way that can identify the source of any information or notes. It happens sometimes that a person in the project team provides the scheduler with information that constitutes a change in the schedule, such as a change in duration of an activity, the logic, or other. The scheduler must have the approval of the project manager for such a change. In fact, it is safer to have such requests for change sent in a written memo or email.

5. There are now new cell phone or tablet applications (apps) where the field supervisor creates "schedule daily reports," and the management staff can see the reports instantly. Although this may not constitute an official schedule update, it is a good idea for the management to get the updates on a daily basis and be alerted to important issues in between updates.

Who Should Do the Update?

In large construction companies and other organizations, there is usually a department or at least one professional dedicated to scheduling and project control matters. In smaller organizations, several responsibilities may be assigned to one person, such as the project manager or cost estimator. The important issue is that schedule creation and management must be handled by a professional who is knowledgeable in scheduling concepts and computer software. There is a potential problem when the person creating and managing schedules, implementing change orders, or making changes, is not knowledgeable enough in scheduling concepts and the software. It is easy to make a mistake without realizing it. Making an error in activities' dates or not realizing the impact of a change can be devastating when such a schedule goes to crews, subcontractors, and the owner. This is why in well-managed organizations, especially large ones, only professionals are given rights to edit, while others can view but not edit. In all cases, the project manager still has the responsibility of reviewing all scheduling reports.

Project Control

The definition of the term project control may differ slightly among professionals and regions. It is used in singular or plural tense (control / controls). Some professionals argue that document control is part of it while others, including the author, disagree. However, all professionals agree that the two major components of project control are time and money.

Project control is defined as the continuous process of monitoring work progress, comparing it to baseline schedule and baseline budget, detecting any variances (deviations from baselines), where and how much; analyzing the variances to find out the causes, and then taking corrective action whenever and wherever necessary to bring the project back on schedule and within budget. Project control is also called project tracking.

So, we can break down project control into four basic components:

1. Monitoring and measuring work progress,
2. Comparing actual progress to baseline budget and schedule,
3. Detecting variances: quantities and causes, and
4. Taking a corrective action wherever and whenever necessary.

While it is the obligation of the contractor to do all four components, the owner needs to do only the first three components. The owner's role is called project monitoring.

Project control is a cyclical iterative task: monitor and adjust work, one step at a time. It is like someone traveling through the desert to a destination 1000 miles away and needs to be at the destination at a certain time. Every 50 miles, he checks where he is now as compared to where he is supposed to be at this time. Also, he checks the time to make sure he is not late. If he finds himself off the track, then he needs to adjust his direction to align it with the target. If he is late, he needs to speed up to make sure he arrives on time.

> Both the owner and the contractor must be continuously informed of the project's progress and the schedule status.

Measuring Work Progress

Project control requires measuring work progress and calculating activities' percent complete. Percent complete is defined as an estimate, expressed as a percent of the amount of work that has been completed on an activity or a work breakdown structure component.[2] It shows how much work has been done in proportion to the total amount of work in that activity, work package, or project. Percent complete can be measured to elements from the smallest activity to a large program. In general, measuring percent complete to individual activities is important to determine:

1. Amount of work performed on this activity, and remaining duration, if the activity is still in progress. This information has an impact on succeeding activities.
2. Activity's cost (money or resources) during this period and to date, and then the cost to complete and cost at completion. This is necessary for progress payments and cash flow calculations.
3. Make any adjustments to the work plan (crews and materials).
4. Recalculate project completion date and important milestones.
5. Use these individual activities' percent complete to calculate/estimate the percent complete for a larger component (assembly, work package, or the entire project).

---

2 Project Management Body of Knowledge (PMBOK) by the Project Management Institute (PMI), 7th edition, 2021.

On the other hand, percent complete for the project may be used for:

1. Report to the client or the upper management (just to get an idea)
2. To calculate the overhead cost
3. Can be used for possible financial arrangements/adjustments or marketing
4. In projects (or subcontracts) that have one main type of work (excavation, concrete black wall, paving, etc.), the percent complete for the project becomes a true representative of the situation and may be used to authorize progress payments

The Confusing Part of Percent Complete

The tricky issue with percent complete, for an activity or larger component, is the fact that it may represent cost, time, effort (man-hours), physical/units, or others. Furthermore, it may represent actual or baseline to any of the variables mentioned. So, for the same situation and at the same time, you may have one percent complete = 25%, another = 40%, and a third = 50%, yet none of them is "wrong." It all depends on what does this percentage complete represent.

Methods to measure work progress and determine percent complete:

1. Units completed: A method of measuring percent complete for an activity that is made up of small / simple, similar, and repetitive units of work.

   Percent complete = Units completed/Total units.

   This method can be applied to the entire project if the project can be divided into similar types of work, such as sidewalk, road construction, or earthmoving projects.

   To simplify calculations, we assume the uniformity of units, which implies that all units require the same amount of effort to perform. This assumption may not be totally accurate, but it should be accurate to the point of ignoring the error. If we feel that some units require significantly different amounts of effort to perform, we can then separate them into different activities. For example, excavation for mixed areas with some sandy and other rocky soil. Another example is installing ceramic tile in both large areas with straight parallel walls as compared to small, crooked, narrow, or curved areas.

   The units in this method may be:

   a. Discrete / countable, such as concrete blocks, trusses, or electric or plumbing fixtures
   b. Measurables such as painting, installing carpet, concrete placement, or land area clearing. In industrial projects, we may use units such as gallons of water or megawatts of electrical power.

   Physical percent complete is another name for this method, although one can argue that in the physical percent complete, there is not necessarily a clear distinction of units, but we can measure the amount in units, such as the examples in part (b) earlier.

   This method can be applied to the entire project if the project can be divided into similar types of work, such as sidewalks, road constructions, or earthmoving projects.

2. Start / finish: This method works best for small activities or those with very short duration. The project manager can assign one of three stages:

   a. Has not started yet (0%)
   b. In progress (an arbitrary amount, say 40% or 50%)
   c. Completed (100%)

   When activities become so small, we can look at them as "almost events" with either 0% or 100% complete. Even though individual activities' percent complete may not be accurate, the combined percent complete should be fairly accurate.

3. Supervisor opinion: This is the most subjective method. It is used when no other method can suitably apply. For example, an "engine tune up" or "dewatering" operation. It relies totally on the judgment of the foreman or superintendent.

   The contractor may use this method to exaggerate the percent complete for early payment. The scheduler must be careful not to allow this method to be overused.

4. Cost ratio: This method measures cost consumed so far (cost-to-date) as a percentage of the total cost:
   o Cost ratio percent complete = Cost to date / total cost (Budget)

   It is used to check the cost consumption, and compare it to the baseline or other types of percent complete, such as physical or effort (man-hours). It also applies to those activities that are continuous and uniform throughout the entire or part of the duration of a project, such as supervision and other project-management type activities. It can also be applied to calculating overhead:
   o Overhead expenses are estimated to be $252,000 for a project with a duration of 18 months, at the rate of $14,000/month. After two months, actual overhead spending was $30,700. So, percent complete = $30,700 / 252,000 = 12.2%.

5. Man-hour ratio: This method is like the cost ratio method, keep in mind that man-hours (or labor-hours) is a measure of effort, not time:
   o Man-hour ratio percent complete = Man-hours to date / Total man-hours

6. Duration ratio: This method is suitable for tracking time, how much time passed as a percentage of the total duration.
   o Duration percent complete = Duration to date / total duration

7. The "work-unit" method: This method assigns weights to activities proportional to their duration. If we are using the day as a unit of time, we can call it the workday method. It is used when the schedule is not cost- or resource-loaded. For example, the 3-activity project shown in Figure 8.4, = 4/12 = 33% complete even though we are at the project's duration's midpoint.

   In Figure 8.4b, we have two bars for each activity: one for the baseline and one for the actual. The bar chart shows a one-day delay in the performance of activity A. So according to the work-unit method:
   o Baseline percent complete = 3 / 9 = 33.33%
   o Actual percent complete = 3 / 10 = 30%

8. Incremental milestones: This method is more suitable for large and complex or multistage activities. Each stage is given a "weight" that approximately equivalent to its percent share of the total effort in that activity. Then each stage is treated as "all or nothing."

9. Weighted or equivalent units: This method is used for large and complicated activities that are usually composed of several consecutive/overlapping sub-activities. We assign a weight to each sub-activity so that total weights = 100% and multiply each sub-activity's weight by the quantity of the total activity. This is the "equivalent weight" in units for each sub-activity.

There is a similarity between the "weighted or equivalent units" and "incremental milestone" methods. Both are used for relatively large and complex activities that are composed of several sub-activities. The main difference is that sub-activities in "incremental milestone" method must go in sequential order. In "weighted or equivalent units" method, sub-activities may be somewhat independent. However, it will be more efficient in both cases to simplify things by breaking the complex activity down into its sub-activities.

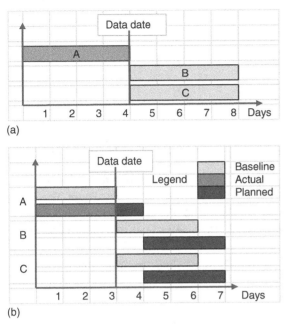

(a)

(b)

Figure 8.4 (a) Work-unit percent complete example, (b) work-unit percent complete example with baseline and actual bars.

All methods can be used to measure actual or baseline percent complete. Perhaps it is a good idea to do both to check the status in comparison to what it should be at this point. For example, in Figure 8.5, the baseline schedule shows that the project should be completed in 32 days. However, when we updated the schedule on day 19, we found out that the project started 3 days late and will take an additional day (duration at completion = 33 days, and the expected finish date is day 36).

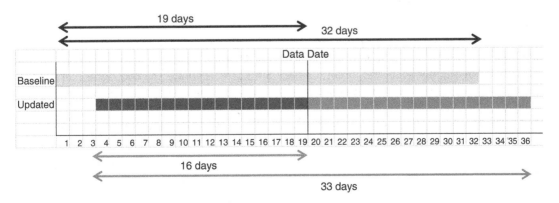

Figure 8.5 Duration percent complete: actual versus baseline.

Project duration percent complete:

- o Actual duration % complete = 16 / 33 = 48%
- o Baseline duration % complete = 19 / 32 = 59%

The following examples explain methods 1 to 7.

Example 8.1

Scenario A

A contractor signed a contract to build a masonry wall, 420 feet long and 8 feet high. The contractor will use 8-in. concrete masonry blocks (CMU) with horizontal and vertical reinforcement. The contractor will get paid $12/block. He will use a crew of five workers (3 bricklayers, paid $50/hour each, and 2 bricklayer helpers, paid $40/hour each). The crew can finish 340 ft^2 per day. The wall must be finished in 10 days.

   After 5 days, the contractor installed 1540 ft^2. Actual cost (including overhead and profit) was $19,220. What is the project status?

Solution, Scenario A
Total wall area = 420 * 8 = 3360 ft^2
Number of blocks = 3360 * 1.125 = 3780 (the block face measures 8″ by 16″ including the joint)

Baselines:
Budget (cost) baseline = 3780 * $12 = $45,360 (or $13.50/ft^2)
Duration baseline = 10 days
Man-hours baseline = 10 days * 5 workers * 8 hours/day = 400 man-hours

After 5 days:
Physical (units) percent complete = 1540 / 3360 = 45.83%
Cost percent complete = $19,220 / 45,360 = 42.37%
Duration percent complete = 5 / 10 = 50%
Man-hours percent complete = 200 / 400 = 50%

Analysis, Scenario A
The project is slightly behind schedule: We are at 50% of the duration but the physical percent complete is only 45.83%. The man-hours percent complete = 50%, which indicates that the daily production was slightly under the target = 1540 / 5 = 308 ft^2. Perhaps the contractor had a slow start due to mobilization and setting up the work or possibly the crew was unable to perform as planned.

   The project is slightly under budget as he consumed only 42.37% of the budget but completed 45.83% of the work.

Scenario B

We will make a slight change to the situation: 2 days after work started, the contractor decided to have his crew work 10 hours/day. This means the extra 2 hours are paid 1.5 times the regular rate.

This allowed him to finish 1950 ft^2 in 5 days. After that, the crew will go back to regular hours. Actual cost (AC) is reported = $26,850. Let us analyze the results:

Physical (units) percent complete = 1950 / 3360 = 58.04%
Cost percent complete (baseline)= $26,325^3 / 45,360 = 58.04%
Cost percent complete (actual) = $26,850 / 45,360 = 59.19%
Duration percent complete = 5 / 10 = 50%
Man-hours to date = 5 workers * (2 days * 8 hours/day + 3 days * 10 hours/day) = 230
Man-hours at completion = 430
Man-hours percent complete = 230 / 400 = 57.50%
The contractor can adjust the man-hours at completion by adding the overtime hours (3 days * 2 extra hours/day * 5 workers = 30 man-hours) to the total man-hours, then the percent complete (actual) = 230 / 430 = 53.49%.

Analysis:

Scenario B
So, we completed 58.04% of the project in 50% of the time and consumed 59.19% of the budget and 57.5% of the man-hours.
The project is ahead of schedule: We are at 50% of the duration and the physical percent complete = 58.04%.
The project is slightly over budget as it consumed 59.19% of the budget but completed 58.04% of the work. Perhaps this is attributed to overtime pay.
Actual man-hours percent complete = 57.50%. Daily production was over target = 1950 / 5 = 390, due to working overtime. But when we calculate production by hour (1950 / 46 = 42.39 ft^2), we find it almost same as the target production (340 / 8 = 42.5 ft^2).
Assuming regular hours for the rest of the project, the crew needs 1410 / 340 = 4.15 days, so the contractor will finish a few hours ahead of schedule.

When someone tells you "The project is 45% complete," make sure you know what is 45% complete: physical, units, time, cost, labor hours, or others.

Example 8.2

An excavation contractor signed a contract to excavate 14,000 CY of mixed soil at a unit price of $6/CY in 34 days. The total contract is $84,000.
   While the contract does not classify the soil types, the contractor believes that:

1. 8000 CY common earth/sand, excavated with production rate = 800 CY/day and cost = $3.25/CY^4
2. 3780 CY clay, excavated with production rate = 420 CY/day and cost = $6.24/CY
3. 2220 CY of rocky soil, excavated with production rate = 150 CY/day and cost = $15.50/CY

---

3 This number came from multiplying completed area, 1950 ft^2 by the contract unit price, $13.50/ft^2.
4 The unit cost shown for all types includes the contractor's overhead and profit.

The contractor has one crew (will switch equipment according to the type of soil). He intends to do the three types of excavation in series: common earth, clay, and then rock. The crew labor includes two workers (equipment operator and helper).

At the end of the project, the total man-hours = 2 workers * 8 hours/day * 34 days = 544 man-hours

Table 8.1 The breakdown of the soil quantities, unit price, and production.

Type	Quantity CY	Unit price	Total price	Production CY/day	Duration (Days)
Common earth	8000	$3.25	$26,000	800	10
Clay	3780	$6.24	$23,587	420	9
Rock	2220	$15.50	$34,410	148	15
Total	**14,000**	**$6.00**	**$83,997 ≈ $84,000**		**34**

After 8 days, the contractor received this report: 5900 CY of common earth has been excavated. Actual cost to date is $20,768. Excavation of clay and rock has not yet started.

What is the status of the project, and what is the percentage complete?

Solution:

Let us calculate the percent complete according to the methods mentioned earlier:

1. Units completed: Here, we treat all soil types as "units," regardless of classification:
   It is like we need to excavate 14,000 CY in 34 days,
   Actual percent complete = 5900 / 14,000 = 42.14%
   Baseline percent complete if soil classification is not considered = 8/34 = 23.53% this is based on production = 411.8 CY / day
2. Cost ratio:
   Baseline percent complete if soil classification is not considered = 23.53%
   Baseline percent complete using the contractor's soil classification and cost estimate for each type = 800 * 8 * $3.25 / $84,000 = 24.76%
   Actual percent complete = $20,768 / $84,000 = 24.72%
3. Man-hour ratio: Two laborers worked for 8 days, the total man-hours = 2 * 8 * 8 = 128
   Total man-hours budget = 2 * 8 * 34 = 544
   Baseline percent complete = 128 / 544 = 23.53%
   Actual percent complete = 128 / 544 = 23.53%
4. Duration ratio:
   Baseline percent complete = 8 / 34 = 23.53%
   Actual percent complete = 8 / 34 = 23.53% assuming the remaining duration is unchanged.
   Supervisor's opinion: The project manager believes the project is about 25% complete.

Project Status

The crew has been working regular hours for 8 days, so the duration and man-hour consumption is at 23.53% complete. Looking at the cost, it is a bit misleading because the baseline percent complete using the contractor's soil classification and cost estimate for each type = 24.76%, which is almost equals the actual percent complete, 24.62%. But the reality is that the contractor performed less than planned: he was supposed to excavate 800 CY of common earth per day, which means 6400 CY in 8 days, but he did only 5900. His "baseline" unit price is $3.25, which means his cost should be $3.25 * 5900 = $19,175 but it actually cost him $20,768, so he is slightly over budget. The same thing can be said of the man-hours because even though the man-hours consumption is going according to the baseline (16 man-hours/day) but he is getting less output than expected with the same amount of man-hours. In other words, in 8 days he consumed 128 man-hours and was expecting 6400 CY output but he only got 5900 CY.

Again, we canno't rush to judgment till all information is known and analyzed, but the current numbers show that the project is slightly behind schedule and over budget.

The analysis may be clarified further with the introduction of earned value management (EVM).

Reports are more than just numbers. In most cases, you need to explain the numbers in words.

Discussion on Measuring Work Progress and Percent Complete

As demonstrated earlier, calculating the percent complete for an activity, or a larger component, can be misleading if we do not know exactly what we are measuring and how. Therefore, it is important to:

1. Make sure activities are homogeneous in nature, have same production rates throughout their duration, and being performed by one crew. This is to avoid mixing production rates for two parts of the work, which may lead to erroneous results. In Example 8.2, a "cubic yard of soil" is not the same for all the excavation work. Production rate differs among the types of soil.
2. The person measuring work progress and percent complete, and the team reading it, must know exactly what each percent complete represents and how it is calculated.
3. The contractor may and should use several types of percent complete to measure different aspects of performance.
4. The contractor must be consistent and may not flip-flop percentages to make things look better than they really are. The owner must be careful in reading such numbers to make sure they are accurate and consistent.

Earned Value Management, EVM

Earned value management, also called earned value analysis, EVA, is an integrated cost-schedule approach to monitor and analyze the progress of a project. EVM, started in the 1960s as a method for integrated project cost and schedule control, was designed by the U.S. Air Force and named the Cost/Schedule Planning and Control System. In 1967, it became U.S. Department of Defense policy and was renamed Cost/Schedule Control Systems Criteria, or C/SCSC.

Earned value management concept works this way:

1. After work starts on the project, we report actual cost. We also calculate the "earned value" (EV) so far (actual quantity * contract unit price) and the "scheduled value" so far (scheduled quantity thus far * contract unit price).
2. Calculate schedule and cost variances so far.
3. Analyze causes of major variances and suggest remedies.
4. Extrapolate these variances to the end of the entire project.

Let us define some key terms:

- Actual Cost for Work Performed, ACWP: This is what it has cost the contractor so far. Also called Actual Cost (AC)
- Budgeted Cost for Work Performed, BCWP: This is what the contractor will get paid for work performed so far. Also called Earned Value (EV)
- Budgeted Cost for Work Scheduled, BCWS: This is what it would have cost the contractor so far if work proceeded on schedule and on budget. Also called Planned Value (PV)
- Cost Variance, CV, is the difference between the Earned Value of an activity and its Actual Cost:

$$CV = BCWP - ACWP$$

- Forecasted Cost Variance, FCV, is the expected Cost Variance of an activity or a project after its completion, assuming the continuation of productivity and job conditions without change.

$$FCV = CV \, / \, \%Complete$$

- Schedule Variance, SV, is the difference between work performed and work scheduled.

$$SV = BCWP - BCWS$$

But the equation above gives a monetary value for a time variable, so we need to convert it to time units by using the equivalency of daily planned budget:

$$Daily \, planned \, budget = Total \, budget \, / \, Total \, duration$$

$$SV(time) = SV(\$) \, / \, Daily \, planned \, budget$$

- Forecasted Schedule Variance, FSV, is the expected Schedule Variance (SV) of an activity or a project after its completion, assuming the continuation of productivity and job conditions without change.

$$FSV = SV(time) \, / \, \%Complete$$

Always strive for zero or positive variances.

- Cost Performance Index, CPI, is the ratio of the Earned Value of an activity to its Actual Cost.

$$CPI = BCWP \, / \, ACWP$$

- Schedule Performance Index, SPI, is the ratio of work performed to work scheduled.

$$SPI = BCWP \, / \, BCWS$$

Always strive for performance indexes ≥ 1.0.

Note that all numbers here should include overhead expenses and profit (OH&P.)

Both the variances, CV and SV, and the performance indexes, CPI and SPI, measure performance (cost and schedule) compared to the baseline. However, variances measure amounts in absolute amounts with units (dollars or days), while performance indexes compare the amount to the baseline in relative measure, expressed in a percentage. Let us consider two activities:

- Activity A: BCWP = $600, ACWP = $660. Then, CV = −$60, CPI = 90.9%
- Activity B: BCWP = $6000, ACWP = $6060. Then, CV = −$60, CPI = 99%

Now we can see that both activities have the same CV, but significantly different CPI. The $60 deficit is considered large for activity A but insignificant for activity B.

The following example explains how EVM works.

Example 8.3

A subcontractor agreed to install 9450 ft^2 of ceramic tile in 12 days with a unit cost of $5/ft^2. After 3 days, the subcontractors completed 2620 ft^2 at actual cost of $13,676. What is the status of the work activity?

Solution

Total Budget = 9450 ft^2 * $5/ft^2	=	$47,250
Daily planned production = 9450 ft^2/12 days	=	787.50 ft^2
Daily planned budget = 787.50 ft^2 * $5/ft^2	=	$3,937.50, or
Daily planned budget = $47,250/12 days	=	$3937.50
Percent complete (using units completed method) = 2620/9450	=	27.72%

So, we can consider the baseline budget = $47,250, and baseline schedule = 12 days.
After 3 days:

ACWP	=	$13,676 (reported)
BCWP = 2620 ft^2 * $5/ft^2	=	$13,100
BCWS = $3937.50/day * 3 days	=	$11,812.50, or
BCWS = 787.50 ft^2/day * $5/ft^2 * 3 days	=	$11,812.50
CV = BCWP − ACWP = $13,100−$13,676	=	-$576
SV ($) = BCWP − BCWS = = $13,100 − $11,812.50	=	$1287.50
SV (days) = $1287.50/$3937.50/day	=	0.33 day
FCV = −$576/27.72%	=	−$2078
FSV = 0.33 day/27.72%	=	1.18 days
CPI = BCWP/ACWP	=	0.96
SPI = BCWP/BCWS	=	1.11

The Project Status

The subcontractor is performing well. After 3 days (25% of the total task time), he completed 27.72% of the work, which makes him 0.33 days ahead of schedule. If his production stays at the same rate, he will finish the job a little more than one day earlier. However, the cost is slightly, $576, over budget. If the cost stays the same, he will finish the job with more than $2k over budget. The S-curve for the solution is shown in Figures 8.6 a and b.

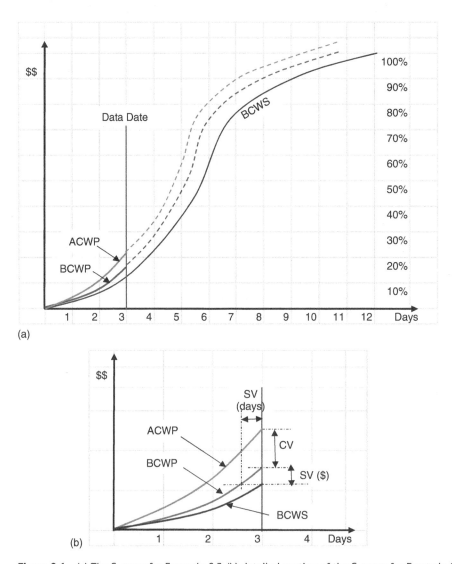

Figure 8.6 (a) The S-curve for Example 8.3, (b) detailed portion of the S-curve for Example 8.3.

The analysis above may not give a clear and complete picture of the situation. For example:

- Why was the production rate higher than planned? This can be attributed to one or more of several causes such as:
  a. The crew performed better than expected,
  b. The original estimate for production rate was too conservative, or
  c. The crew worked overtime, so the daily production was higher even though the production rate/hour was not.
- Why did the subcontractor spend more money than budgeted? Again, this can be attributed to one or more of several causes such as:
  a. The cost of materials and/or labor was more than expected,
  b. The crew worked overtime with pay scale of 150% for overtime hours, or
  c. There was an interruption in the work, so the crew got paid with no production during that time.

Any of the above, or even other possibilities may be the interpretation of our case. This underscores the importance of going beyond the numbers, exploring and explaining the root causes. We must not rush to judgment. Sometimes one indicator may be negative, but the contractor is aware and in acceptance of the situation in exchange for bigger gains or avoiding worse outcomes. An example of that is when a contractor pays more to accelerate the schedule to get bonus or avoid liquidated damages.

The concept of Earned Value is based on comparing actual performance to baselines, in both cost and time terms. The only problem here is that the analysis of both cost and time (schedule) is based on the cost, which always gives accurate results for cost, but not always accurate results for the schedule. For this reason, it is highly recommended that schedule information is obtained primarily from the CPM schedule and its calculations, not EVM. Earned value time indicators (SV and SPI) are calculated for all activities regardless of their criticality. So, it is possible that we get satisfying indicators ($SV \geq 0$, $SPI \geq 1.0$) because some noncritical activities are ahead of schedule while the critical path is behind schedule, which leads to wrong conclusion on the status of the project schedule.

Control Level

As mentioned in Chapter 7, the project breakdown system may go down up to six levels, with level 6 producing the smallest and most detailed components. Activities in the schedule are usually at level 5 but possibly anywhere between levels 4 and 6. Since there could be hundreds or thousands of activities, project controls may then apply efficiently at a mid-level such as level 3. Figure 8.7 shows a small sample of breakdown, based on the CSI MasterFormat. When indicators are satisfactory, there is no need to dig deeper, but when they are not showing satisfactory results, then the project manager has to drill down to a more detailed level till the source of the problem is identified.

Dealing with Variances

When doing the project (schedule and cost) updates by comparing actuals to baselines, there will always be variances, positive or negative. Analyzing and managing these variances on a constant basis are

Title	Level
03 00 00 Concrete	1
03 10 00 Concrete Forming and Accessories	2
03 11 00 Concrete Forming	3
03 11 13 Structural Cast-in-Place Concrete Forming	4
03 11 13.13 Concrete Slip Forming	5
03 11 13.16 Concrete Shoring	5
03 11 13.19 Falsework	5
03 11 16 Architectural Cast-in Place Concrete Forming	4
03 11 16.13 Concrete Form Liners	5
03 11 19 Insulating Concrete Forming	4
03 11 23 Permanent Stair Forming	4

Figure 8.7 Work breakdown structure, partially taken from the CSI MasterFormat.

important. Neither the baselines nor the performance on the ground is perfect, so the variances can reflect issues with either or both.

When a negative variance – cost or schedule – is detected, the project manager needs to know:

a. Where did it happen? (Affected items),
b. The amount in absolute and relative terms,
c. Why? and
d. What should be done about it?

Keep in mind that cost indicators are usually independent. This means, a negative cost variance for an activity may not necessarily impact succeeding or other activities. A scheduling variance for an activity, on the other hand, will likely impact succeeding activities.

Small negative variances may be tolerated and absorbed by contingencies or compensated by positive variance in other activities, unless they keep getting worse. However, large negative variances raise red flags and require attention.

Positive variances are always welcomed, but a large positive variance may be an indication of an over-estimate in the baseline.

In all cases, these variances are used for both effective management of the project, and to improve the contractor's database of production rates and cost for future estimates. Therefore, keeping notes is important, especially when numbers are not "normal."

Exercises

8.1 Define schedule and budget baselines.

8.2 Define schedule updating. What is an updated schedule?

8.3 Define project control and project monitoring. What roles do the contractor and the owner play in these processes?

8.4 Discuss the relationship between project schedule updating and project control.

8.5 Why is project control needed for construction projects?

8.6 What is the data date? Why is the name "current date" not suitable?

8.7 What kind of information is needed for updating schedules?

8.8 How frequent should schedule updating be performed?

8.9 Why is schedule updating important for progress payments?

8.10 What are the steps involved in updating a schedule?

8.11 Schedule updating reports on the progress achieved since the last update, but what happened may and does have an impact on what will happen. Explain this statement, with a focus on re-examining remaining durations and re-calculating future dates.

8.12 After a schedule update, you found out that the critical path has −3 days float. What does this mean? You are supposed to write the update report to the project manager, but you need to find out the situation. In simplified language, mention the steps you take for writing an accurate report.

8.13 When the project falls behind, there could be several reasons why this happened. Mention all the possible causes.

8.14 You are the project control manager of a major construction company that has many projects going on, led by project managers with varying knowledge of scheduling software. The CEO wants each project manager to update their own projects. Do you agree with that decision? Why or why not? Justify your answer.

8.15 What are the four components of project control?

8.16 Define percent complete for an activity.

8.17 Compare between percent complete for an activity and percent complete for a project. Which one is more important for project control?

8.18 What is the most important thing about percent complete when reporting on project progress?

8.19 Is it possible to see different percent complete estimates for the same activity at the same time in different reports? Explain your answer.

8.20 You have a schedule on Oracle Primavera P6 that is neither cost- nor resource-loaded. The program can still give you the percent complete for activities and the project. What method do you think it uses?

8.21 Is it a good idea to compare the current schedule to both the baseline and last update? Why?

8.22 You are the project manager of a hotel project that is planned to be completed by the end of October 2025. On November 29, 2024, 3 months after starting the project, you received the project control report, which has good news (the project is slightly under budget) and bad news (the project is 5 days behind schedule). Upon investigation, you found out that the issue was caused by a delay in the concrete mix delivery, and it does not qualify for an extension of time. Since it is relatively early, you have quite a few options for schedule recovery. List your options and indicate which option(s) is the likely one.

8.23 Will you change your likely option in Exercise No. 22, if the contract stipulates heavy liquidated damages for delay? Discuss the impact of the importance of finishing on schedule on your decision.

8.24 A roofer is working to cover a 2500 ft^2 roof. he/here needs to first install rafters, sheathing, felt, and finally shingles. he/here assumes the previous tasks make up 35%, 20%, 15%, and 30% of the total roofing activity. At a certain point, he/here got the following information on the subtasks:

Subtask	Unit	Total quantity	Installed quantity
Rafters	EA	150	102
Sheathing	ft^2	2720	1200
Felt	SQ	27.2	10
Shingles	SQ	27.2	5

Calculate the estimated percent complete for the roofing activity.

8.25 The earned value, EV, analysis is about comparing what happened to what should have happened. Explain this statement regarding both budget and schedule.

8.26 A mason contracted with a general contractor to build an exterior wall out of 8″ CMU. The wall is 164′ long and 8′ high. The mason must finish the wall in 6 days and gets paid $8 per block. At the end of day two, the mason has installed 440 blocks. His actual cost (including his overhead and profit) was $3344. Analyze the situation regarding both budget and schedule.

8.27 A flooring subcontractor is planning to install three types of flooring in an existing building:

Type	Quantity	Unit price	Total price
Carpet	2600 ft^2	$3.45/ft^2	$8970
Vinyl tile	1200 ft^2	$4.13/ft^2	$4956
Ceramic tile	3330 ft^2	$5.94/ft^2	$19,780
Totals	**7130 ft^2**		**$33,706**

The subcontractor was given 10 days to finish all work. Assume he/here has crews to start all three jobs concurrently. Four days after the start of the work, he/here found out:

Type	Quantity installed	Actual cost
Carpet	1500 ft^2	$4725
Vinyl tile	800 ft^2	$3483
Ceramic tile	750 ft^2	$4366

Perform earned value (EV) analysis on the task above:

A Analyze each task (carpet, vinyl tile, and ceramic tile) individually; find out their status regarding budget and schedule.
B Analyze the entire flooring activity as one unit; find out their status regarding budget and schedule.
C How do you put together the answers to parts A and B above?
D Based on performance so far, do you believe the subcontractor will finish the job on time? Within budget? Justify your answer.
E Draw the S-curve for the entire task.

8.28 How reliable is the Earned Value method for analyzing cost and time? How can you use it effectively?

8.29 What does the "control level" mean? How would you decide the control level for a schedule with over 2500 activities?

8.30 ABC Tiles contracted to install ceramic tile in a 4-floor office building, with a typical floor area of 4250 square feet. The contract stipulates that the work must finish in 34 days at a cost of $5/ft^2.

A What are the project's cost and schedule baselines?
B Assuming linearity in work, what are the daily planned production and budget?
C After 5 days, the crew finished 4100 ft^2 at a cost of $15,220. What is the status of the project?
D Would you change your mind on the status of the project, part c, if you learned that the crew worked only on rooms (open spaces) and have not started any of the bathrooms and kitchens? What does this tell you regarding project controls?

9

Risk Management

Introduction

Construction projects are inherently risky endeavors. The project is exposed to many risks that can impact its success. There are no two projects that are identical[1], and all projects are subject to many uncertain conditions such as weather, craftsmen's skill level, deliveries, accidents, labor strikes, fires, rejection of shop drawings, and even cyberattacks. An unexpected price spike in a major item in a fixed-price contract or a change in material availability for basic construction item can be a major risk to the contractor. Managing these risks is a part of managing the project with a main objective to minimize negative impacts of risks on project objectives.

What is a Risk?

A project risk is an uncertain event or condition that, if it occurs, has a positive or negative effect on at least one project objective, such as time, cost, scope, or quality[2]. Project risks are both threats and opportunities for the expected outcomes of a project, and hence its success. Threats tend to reduce the success of meeting the project goals, and opportunities tend to increase the success of the project.

A risk may have one or more causes and, if it occurs, one or more impacts. This means that any event that can be qualified as a risk (and included here), has to be uncertain. So, there are at least two likelihoods for its impact and/or consequences. Of course, risks can have an array of likelihoods. Also, although the combination of consequences may be under one title; for example, a hurricane or pandemic, the likelihood and severity of impact may vary and is unknown in advance. Everything that has ever gone wrong on a project is a potential risk on the next project. Another example of risk is a labor strike. A strike is known in nature, but not in certainty, timing (start/finish, and length), or outcome.

---

1 A reminder that the "project" in this context is the process of creating the final product, namely the construction process, not the final product itself. This is why projects are unique, even if the final product is identical.
2 PMI PMBOK, 7th edition, 2021.

Project Management in the Construction Industry: From Concept to Completion, First Edition. Saleh Mubarak.
© 2024 John Wiley & Sons, Inc. Published 2024 by John Wiley & Sons, Inc.
Companion Website: www.wiley.com/go/nextgencpm

It is interesting to note that risk is a major factor in a contractor's estimated desired profit. We can modify the saying "no pain, no gain" to no risk, no profit! In fact, this point is explained in Chapter 6: Project Budgeting and Cost Management.

Risk in this context does not include safety issues that do not have a direct impact on project objectives. Safety is important, of course, but it is addressed in another chapter under the title (Health, Safety, and the Environment). However, it is possible for the same incident or event to impact both risk and safety. For example, an accident onsite is a safety matter but it may be a risk as well if impacting the project's objectives (cost/schedule). For this reason, we will use the term "risk," as defined above, as an event that primarily has the potential to impact project objectives, and the term "hazard" to mean an event or situation that has the potential to cause death, injury, or property damage. Again, the two terms are not mutually exclusive. It is possible that a risk poses a hazard and vice versa.

Risk Management

Risk management is the process of identifying, assessing, prioritizing, and continuously monitoring risks and implementing strategies to avoid, transfer, mitigate, or accept them. Risk management is a continuous process designed to examine uncertain events that may occur during project execution and to implement actions dealing with those uncertainties to achieve project objectives, reducing the negative impact, and maximizing the positive impact of risks in the project. This chapter will provide an overview of the principles of risk management in construction projects, including the main elements and tools used to manage risk effectively.

The Risk Management Plan

The general contractor must prepare a risk management plan as part of the project management plan that describes how risk management activities will be structured and performed. The key benefit of this process is that it ensures that the degree, type, and visibility of risk management are proportionate to both risks and the importance of the project to the organization and other stakeholders[3].

Who has control? An important question in the risk management plan activities.

The risk management plan is the output of inputs plus tools and techniques. The inputs include mainly the project charter, the project management plan, and project documents. The tools include data collection and analysis, meetings (brainstorming), and expert judgment.

The earlier we anticipate and prepare, the better we can manage!

---

3 PMI-PMBOK, 7th ed., 2021.

The Risk Register

The first step in risk management is to read your contract and know your contractual responsibility in the performance of the project, including the project's specifications and drawings. The next most important step is to identify the possible risks. This step is accomplished by creating the risk register, which is a checklist of potential risks developed during the risk identification phase of risk management. The risk register is usually prepared during a brainstorming meeting(s) with the major project stakeholders, often called a risk workshop. Previous risk registers must be utilized, most likely through consolidation in a "master risk register." This list is a logical place to collect lessons learned on a corporate level from many project experiences. It is continuously updated and customized. The update adds any new risks that were not considered before, such as pandemics that became more visible after the COVID-19 pandemic in 2020. Customization takes into consideration the specific risks that may occur on the project. For example, an earthquake is a risk in California, while a hurricane is a risk in Florida. A labor strike is more likely when there are signs of labor discontent or disagreement with the management.

There are register templates that include elements such as risk identification ID, description, categories, probability, priority, response, and ownership.

In addition to the risk register, there are other methods for risk analysis such as the decision tree analysis, Monte Carlo simulation, SWOT (Strengths, Weaknesses, Opportunities, and Threats) analysis. The risk manager may use more than one method.

Performing the Qualitative and Quantitative Risk Analysis

Qualitative risk assessment is based on a person's perception or judgment, so it tends to be more subjective. It focuses on identifying risks to measure both the likelihood of a specific risk event occurring during the project life cycle and the impact it will have on both the budget and the overall schedule should it occur. We use terms like very low, low, medium, high, and very high to capture the nature of each risk. Together, the likelihood and the impact, determine the "exposure," which may be recorded in a risk assessment matrix (or any other form of an intuitive graphical report) to move to the next step in the risk management plan.

The risk may have a continuous domain of values, such as the wind speed: it can take any value between 40 mph and 150+ mph (if we do not consider wind with speed below 40 mph as a risk). The impact, being a function of speed, can also be a continuous variable. For the purpose of analysis, it is easier to convert the continuous variable to an ordered categorical variable, with categories 1, 2, 3, and so on. Thus, each wind speed category will have a defined impact. Both Figures 9.1 and 9.2 demonstrate the risk based on the impact (severity or consequences) and probability (likelihood), in matrix and chart formats.

Quantitative risk analysis is based on verified and specific data, so it is more objective. It comes after, and based on the result of, the qualitative analysis. Quantitative risk analysis is the process of calculating risk based on data gathered with a goal to further specify the impact of the realized risk on the associated cost or scheduled activity. This is achieved by using a variety of methods for data collection and analysis such as observations, surveys, and statistical data. Data from past projects can be a valuable source, provided the data is compatible with location, time of the year, and other project-specific conditions.

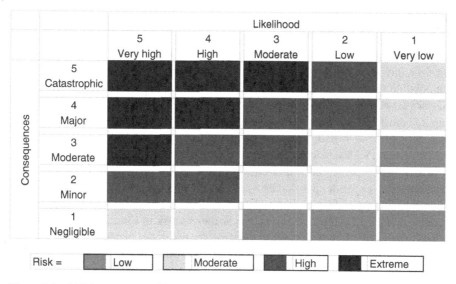

Figure 9.1 Risk impact/probability matrix.

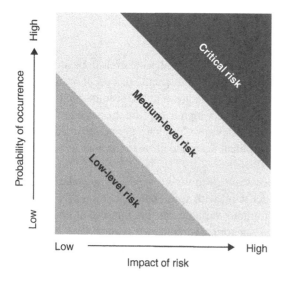

Figure 9.2 Risk impact/probability chart.

For example, a contractor may observe the weather conditions in the previous years and calculate the "severe weather" events with their probability and impact on past projects. The contractor can also evaluate deliveries, categorized by country of origin, carrier, or other, and their likelihood for timely arrival. The contractor may study the correlation between jobsite accidents and extended overtime work hours, with accidents classified based on their type, causes, and consequences (on project objectives).

Now, that we have completed the risk identification and analysis, we need to complete the risk management plan. This includes creating risk handling strategies, or contingency plans to alleviate the impact of each project risk we have previously analyzed. The threats can be addressed in one of these ways:

a. Avoid the risk by eliminating the risk. For example, buying a local item instead of the original item that was supposed to be shipped internationally. Thus, eliminating the likelihood of a delivery delay.
b. Transfer the risk; for example, the general contractor hired a company to secure the site against possible theft or vandalism. Sometimes, when the risk is outside the scope of the project or when the proposed response would exceed the project manager's authority, the risk is *escalated* to the party or level that has authority over the risk or the response.
c. Mitigate the risk by reducing the probability of the risks happening, the impact, or both. For example, the contractor is facing the risk of late delivery for a critical item. He may work on expediting the delivery and/or study the CPM[4] schedule to examine the impact of the item's late delivery and create a recovery plan.
d. Accept the risk by acknowledging it and choosing not to actively avoid, transfer, or mitigate it. The contractor must continue to monitor the situation and be ready for the consequences with little or no action. The contractor estimates that the cost of taking a mitigating action may be higher than the impact itself.

Keep in mind that these actions are not necessarily mutually exclusive, they can be combined. For example, if there is a risk of strong winds and heavy rain, the contractor may:

a. Avoid: Delay the delivery of some items to the site and/or move other items away from the site
b. Transfer: Require every subcontractor to protect their own materials, equipment, and installations.
c. Mitigate: Provide protection to those items (materials, equipment, and structures) that will remain on the site, and
d. Accept: Leave items that do not get impacted by the risk (e.g., a stack of concrete blocks).

The author had experience when working for an architectural firm that also acted as project management consultant (PMC) on a historical restoration project. The owner hired the subcontractors directly, while the PMC had the responsibility of supervising the subs and coordinating among them, with no general contractor. The two-story building had been abandoned for years and it was empty. All doors and windows were taken down and laid on the floor for repair or replacement. The typical size of these doors and windows was larger than average since the ceiling height of the building was 20 feet.

A hurricane was looming in the direction of the project site as a potential direct hit. The contractor had two options:

1. Mitigate the risk by boarding the building, which required closing all (over 160) openings with plywood sheets. This required extensive effort and cost.

---

4 Critical Path Method.

2. Accept the risk since the building was empty, except for:

   a. Light equipment and tools that were taken away to a safe place, and
   b. A shipment of new doors was placed inside the building away from windows, covered with heavy tarp, and topped with concrete blocks. The ground floor was three steps above the site level, so the chance of water flooding the building floor was almost nil.

The project manager chose the second option, accepting the risk. If the building was occupied with many valuable items that were difficult to transfer, the project manager may have chosen option 1 to mitigate the risk.

> Whenever the term "risk" is mentioned, people think of threats... but we need to also think of a risk as a possible opportunity.

Risk may be an option for the contractor, such as taking on a risky project or using a new technology, or non-optional such as the risk of a fire, earthquake, hurricane, or labor strike. The contractor's options in risk management include avoiding the risk if it is optional. In this case, the contractor will weigh the benefit-to-cost ratio of the risk. But for a compulsory (non-optional) risk, the contractor must manage the risk in the most efficient way.

Risk Monitoring and Control

The risk plan must be updated, both periodically and when needed. This is close to, but not exactly the same as, project controls (cost and time) when we compare the "as built" to the "as planned." The risk management plan update starts with the register:

- Risk events that have expired or have been eliminated are closed. This includes events that either happened or will not happen. For example, a hurricane passed already, with or without an impact on the project. Another example is a labor strike threat, but the parties agreed on a new labor agreement, so the risk is no longer there. A third example is a critical shipment with a possibility of late delivery, but already arrived.
- Risks with more information newly available, are updated in terms of likelihood and consequences/impact. This includes all the risks that remain, but we now know more about them.
- New risks may arise, so they must be added to the register and the plan.

After the register is updated, the actions in the plan must be reviewed and updated, as needed, as illustrated in Figure 9.3.

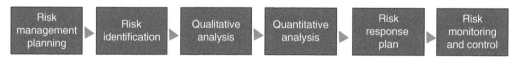

Figure 9.3 The steps in the risk management plan.

Risk Shifting in Contracts

Risk shifting in construction contracts is a common practice if the terms of this shift are favorable to both parties. Some owners like to shift the risk to contractors during negotiation. Sometimes, they do so without realizing the price they pay. Generally, the more the risk is shifted to the contractor, the higher the cost and, sometimes, the longer the duration time of the project will be. However, this may be the right choice in some cases provided the owner understands the benefits and cost of such a decision. For example, the use of clauses stating that geotechnical reports and information are provided to bidders for information only, and the owner is not responsible for any usage or interpretation of the geotechnical information.

In fact, the choice of the contract type is basically a matter of risk allocation. Fixed-price contracts tend to allocate all or most of the risk to the contractor, while cost-plus-fee contracts shift the risk to the owner. In both cases, the party that takes the risk can modify the terms to shift some of the risk back to the other party. For example:

- In fixed-price contracts, the general contractor can add a clause that any price escalation of certain item(s), usually the big-ticket items such as lumber, concrete, or steel, over an annual rate of 5%, will be charged to the owner.
- In cost-plus-fee contracts, the owner can shift back some of the risk by requiring a guaranteed maximum price (GMP) from the contractor and / or making the fee sliding (goes down after certain threshold.)

In the end, it is a game of risk management that must be planned well for the best interest of all parties. The Construction Management Association of America, CMAA, has this recommendation for risk management:

> *"Following the timely identification and assessment of risks, a rational approach to risk allocation can proceed based upon the following general principles:*
>
> - *Risk should be assigned to the party who can best control it.*
> - *Risk should be assigned to the party who can bear the risk at the lowest cost.*
> - *Risk should be assigned to the Owner when no other party can control the risk or bear the loss.*
> - *Assumption of risk by the other parties to the construction process results in increases in cost (visible or hidden) to the Owner[5]."*

The last paragraph is important, as happens in many situations, one party (usually the owner) may shift a risk to the other party (usually the contractor), seemingly at no cost. In reality and most likely, the other party will estimate the cost of taking responsibility for this risk and add it to the total price tag. It is like negotiating with a used car dealer for a "free" 3- or 6-month warranty, and the dealer accepts. The customer may or may not realize that he/she could have negotiated for a lower price in lieu of this warranty. Again, we emphasize that such shift may not be a bad idea, but the party that shifts the risk must know and accept the price of such shifting.

---

5 The Construction Management Association of America, CMAA.

The Expected Value Theory

Expected Value is one of the approaches for estimating the effects of any given risks, and can be used, in combination with a risk-based approach to set up a contingency budget or time. Statisticians combine the likelihood of different outcomes of an event with their corresponding potential consequences in what is known as the expected value:

$$E(X) = \sum_i x_i \cdot p(x_i)$$

where:

$E(X)$ is the "expected value" of an event, X, such as an accident, a hurricane, or shop drawing rejection
i is the domain of all possible outcomes, such as catastrophic, severe, significant, mild, light, negligible
$p(x_i)$ is the probability of occurrence of outcome i
x_i is the value of consequence of outcome i

For example, if a hurricane is expected to hit with these probabilities and outcomes:

Hurricane wind speed	General impact	Probability	Cost ($)	Delay (days)
Under 70 mph	Negligible	25%	$0	0
70–89 mph	Low	40%	$40,000	3
90–109 mph	Moderate	25%	$150,000	5
≥ 110 mph	Severe	10%	$800,000	12

The expected value of the hurricane on the cost = 25%*$0+40%*$40,000+25%*$150,000+10%* 800,000 = $133,500.

The expected value of the hurricane on the schedule = 25%*0+40%*3+25%*5+10%*12 = 3.65≈4 days

This does not mean that the hurricane will likely cause a financial loss of $133,500 and a delay of 4 days. These numbers are the result of statistical calculations of all possible outcomes multiplied by their impact, and they may be used to set up contingency budgets and time. The project manager can take a conservative approach and increase them at his/her discretion. The numbers are also used to evaluate the possible responses to the risk: avoid, transfer, mitigate, or accept.

Another important issue in this context is the third factor[6], which is resiliency and absorbency of the party bearing the risk. For this matter, we must consider the worst-case scenario: If it happens, can I (or my organization) take it and absorb it? To demonstrate this point, we take an example of investing in the stock market for two people. The investment is $1 million; however, one investor is worth $100 million, while the other has only $1 million. If the worst-case scenario means the total loss of the investment, with a 10% likelihood, can both investors afford it? The likelihood and consequences are the same for both investors, but the resiliency and absorbency are not the same. The first investor can take that loss without major impact, but the other may be devastated. Another example of a construction project with

---

6 After the risk likelihood and consequences.

significant risk. A strong company may be willing to take the risk, even with the possibility of the worst-case scenario, but another company cannot afford taking this risk.

This resiliency and absorbency can be defined in this context as the strength, financial and other, of the individual or organization, to withstand the impact of the worst-case scenario and stay functional without being impaired or significantly damaged.

Resiliency and absorbency may be the deciding factor in whether to buy an insurance policy. The person (or organization) must examine all possible scenarios and outcomes of any potential risk. If they can afford taking it while staying "alive," then they probably do not need to get insurance; otherwise, they do[7].

Expect the Expected!

One of the main characteristics of risks being considered in this chapter is uncertainty, which means that the likelihood of the event is > 0 but < 100%. In some cases, the risk event is almost certain. You wonder then if you can classify it as a risk or a certain event. But, in addition to the question of "how likely", we also try to know more about it like where, when, how, how much, etc. Take for example the weather in the summer in central Florida. Thunderstorms and heavy rain are expected in a significant portion of July and August days although they are not known exactly in timing, severity, and duration. When a contractor plans the project during this period, he must take this information into account. From a claim perspective, if the contractor had 12 "rain days" in July, and he requests extension of time for this reason, the owner may grant him only what is more than the "normal severe weather days" for that location during that month. So, if there is an average of 8 days of severe weather in July (based on data from the past 20 years, for example), the contractor may get only 4 days extension. The risk here is divided between the owner and the contractor: What is expected (the 8 days) is the responsibility of the contractor but he may be granted an extension of time for the exceptional or unexpected (4 days). Of course, this argument covers time only. The contractor will be responsible for the cost and damage resulting from severe weather in all cases.

Risk management is about managing probable risks, which include those with likelihood >0 but <100%. So, when we say, "expect the expected", we mean events that are certain or almost certain, but the extent and severity are not exactly known. For example, if you live in a major city and commute daily to your office, you must expect heavy traffic when you leave home to the office in the morning and must be prepared for it. When you plan a project that will last several months, and there is a history of severe weather in that area and during that period or other conditions that tend to reduce productivity, then you must expect these events and be prepared for them. When you acquire materials from a vendor with a history of delays, you must expect the delay.

Some scientists suggest that global warning and climate change are making certain natural phenomena such as rain, floods, storms, tornados, typhoons, hurricanes, and even earthquakes more frequent and stronger[8]. This has an impact on: Design codes, materials (and possibly methods), project risk, and project safety. With this in mind, the contractor may keep this in mind when creating the risk plan, expecting somewhat worse weather conditions and natural disasters!

Expect the expected and make plans for the unexpected!

---

7 This is likely the case when an organization says we are "self-insured". It means we can take the hit!

8 https://www.edf.org/climate/how-climate-change-makes-hurricanes-more-destructive?utm_source=google&utm_campaign=edf_none_upd_dmt&utm_medium=cpc&utm_id=1606920135&gclid=Cj0KCQjwuLShBhC_ARIsAFod4fKrX Qn667deYNGS7LV7-2mvkj8odt7CDDrFgPDnbIUpH9AaXwUImREaAuKvEALw_wcB&gclsrc=aw.ds

The Concept of Contingency

Contingency, as a sum of money allocated aside for unexpected expenses, was explained in Chapter 6, Project Budgeting and Cost Management. The concept also applies to time, and it was discussed in Chapter 7, Time Management. Both cost and time contingency are correlated with risk when estimating their amounts. The higher the risk, the more contingency the contractor must allocate.

The contingency, both money and time, must neither be used as an excuse for poor estimating or performance nor get mismanaged. It also must not be mixed with the budget or duration of individual activities, but rather stay allocated to the project as a whole and be under the control of the project manager and based on his/her discretion.

If the contingency amount was estimated delicately and professionally, it is expected to be depleted during the execution of the project, if things occurred "normally," which means as expected, not better or worse. If things occurred better than expected, some of the contingency will remain unused by the end of the project. Conversely, if things occurred worse than expected, not only the contingency money and/ or time will be used, but also the project manager likely to have used extra funds (meaning completing the project over budget and likely with less profit) and/or needed more time (meaning either finished behind schedule or had to expedite it.) The contingency is a buffer that must be estimated with balanced attitude.

Risk at the Corporate Level

Every company or organization must consider risk not only at the project level, but at the organization level as well, considering all possible scenarios combined. We know that the probability of two independent events happening at the same time is the product of the two probabilities: $P(A \text{ and } B) = P(A) \times P(B)$.

So, if the probability of financial loss in project P_i, is Pr_i so the probability of financial loss for n projects combined is $Pr_N = Pr_1 \times Pr_2 \times Pr_3 \times \ldots Pr_n$, which means having all projects in loss at the same time. This will likely be a very small likelihood.

However, the above-mentioned equation is valid only if the risks in the projects are independent. If the risks in two or more projects are dependent, then the probability of the risks happening in the projects at the same time depends on the correlation among them, but it will likely be higher than the case of independent project risks.

Take for example weather risk: If the contractor is considering the risk of extreme weather in two projects in the same geographic area (with same weather pattern), then there is up to 100% correlation between this risk for the two projects, meaning that severe weather will either happen or will not happen to both projects. If the two projects are not in the same area but close enough to have a correlation, say one project in South Florida and the other in the Bahamas, the risks are correlated with coefficient >0 but <1.

Most major construction companies diversify their projects in terms of risk, so they do not face the worst-case scenario in multiple projects at the same time. The important matter for the leadership of the organization is to make sure that all realistic scenarios are considered, and that the outcomes are within the organization's capacity to absorb and sustain. Of course, there are risks that may not be totally predictable such as the COVID-19 pandemic and the spike in some materials prices that came later with it, especially lumber. What made things even worse was the demand increase for housing and other types of construction. This is why contractors and other business owners must be both proactive and reactive.

There is no strategy that is 100% guaranteed or "fool-proof." The essence of business is risk, but there is a major difference between uncalculated/random and calculated and managed risks.

Conclusion

Effective risk management is critical to the success of construction projects. By implementing effective risk management practices, construction professionals can ensure that projects are completed on time, within budget, and to the satisfaction of all stakeholders. Furthermore, effective risk management can help to minimize project risks, increase customer satisfaction, and enhance the reputation of the organization.

Exercises

9.1 Define a project risk. Give an example with both positive and negative potential (opportunities and threats).

9.2 Define risk management and risk management plan.

9.3 What is the risk register? How do we create it? Discuss both preparing a project risk register and the company's policy regarding this matter.

9.4 An international construction company wants to create a risk register master list that must be used by all projects. How would you approve this list? Keep in mind that risks differ by type of project and region.

9.5 Briefly explain the qualitative and quantitative risk analysis with a practical example.

9.6 Investigate a project risk that you can apply "Avoid, transfer, mitigate, or accept" to it. Briefly discuss how each action is taken.

9.7 You managed a project in New York City for a major owner's corporation. Now, you will be responsible for managing a similar project in Los Angeles, CA. You can use the risk management plan, including the risk register, from the previous project. How would you utilize it for the new project?

9.8 Discuss risk monitoring and control during the project's progress. What changes and what remains unchanged?

9.9 Most people, by nature, like to avoid risks by shifting them to others. What is risk shifting? Does it cost anything to shift risk? Discuss in detail.

9.10 Who should take responsibility for the risk in each of the following scenarios:
  a. The construction site and operations during the construction process.
  b. The owner and contractor can accept certain risk, but the contractor can do it at a lower cost.
  c. This is a project with creative and unusual design with lots of unknown.

9.11 Compare the cost risk and time risk in any typical construction project, including similarities, differences, means of identifying the risks, and responses to each.

9.12 There is a risk that may affect the cost but has no impact on time. The worst-case scenario is a loss of $100,000. The contractor can buy an insurance policy that protects against losses from this risk. Under what circumstances you would recommend buying this policy?

9.13 Talk to a project manager with a local construction project and ask how risk management is handled. If no formal risk management plan is used, discuss how the project is protected against typical project risks. Discuss the benefits of formal risk management.

9.14 Mention five different risks that may have to be considered in a typical:
  a. Infrastructure construction projects consisting of bridges and/or roadways.
  b. Mid-rise project.

9.15 You are an infrastructure contractor and plan to start a project in the downtown of a major city. You are contemplating the renting of an expensive underground micro-tunneling machine. It will drastically reduce the disturbance to above-the-ground traffic. Now, you are comparing the two options: Using the micro-tunneling machine or open trench excavation. Discuss the options along with their risks, and the factors that will make you lean either way.

9.16 What would a contractor do for schedule risk planning if he plans to work on a road project in an area during a time known for adverse weather? Make any assumptions as needed.

9.17 From past experience, you believe that the chances for approval of a subcontractor's shop drawings are:
  a. 50% from first submission
  b. 80% from second submission
  c. 100% from third submission

  If estimated durations for shop drawing activities are: 45 days for first submission, 14 days for first review, 14 days for each re-submission, and 7 days for each review (after the first). What is the duration you assign for the entire cycle of the above process, using the Expected Value equation?

9.18 What is the duration you assign for the entire cycle in the previous exercise, based on the "worst-case scenario"? Do you believe that such duration is realistic? Why or why not?

9.19 You are the project manager of a condominium project near the beach in Melbourne, Florida. The building's concrete skeleton is complete, and you already started the finishes on the lower floors. You are informed that a major hurricane is approaching and will likely hit the area in about 6 days. Predictions vary in the path and strength of the hurricane with the optimistic scenario (25–30%) that it will totally avoid the project area and the pessimistic scenario (15–20%) that it will hit the project area with winds reaching 150 MPH. How would you react to this risk?

9.20 Your company decided to take on a project to build a research center on an island in the Pacific Ocean. You can travel there by flying to another larger island about 300 miles away, and then take a "taxi flight" on a small aircraft to the destination island. This project can be an opportunity for other projects. Create a risk management plan for this project.

9.21 An investor is thinking about buying an old 10-story building that has been abandoned for many years. Its condition has deteriorated and needs unknown repairs and restoration. His plan is to turn it into a modern office building. You are a general contractor in the same area of this building. The investor approaches you to take on the project, which can be a great opportunity but also carries many risks. Itemize the potential risks, starting with the type of contract, and how you would address them.

9.22 In risk management, what does "Expect the Expected" mean?

9.23 How does the concept of contingency work in risk management? Address both money and time contingency.

9.24 A project that has several risks and several construction companies are interested in it. You are a risk expert giving advice to clients. You believe that company A can and should accept this risk while company B must refrain from it. Why? Mention some factors that influenced your decision.

10

Managing the Contractor's Cash Flow

Introduction

According to several studies and surveys, failing to efficiently manage cash flow is the number one reason for the failure of construction companies. In fact, problems with cash flow often do not arise from just lack of money, but also from not having it available when needed.

Managing cash flow is arguably one of the most important tasks for the survival and profitability of the organization. Its importance comes from two facts: first, necessity and availability for money – and in particular cash or liquid money – as the bloodline of the business that without its operations will cease. Second, it is the often-overlooked side of the business since the usual focus is the technical side. As mentioned in other chapters, any businessperson may be skilled and even brilliant in their technical field, but this does not necessarily include the ability to manage the business. Thus, having qualified professionals for this task is of utmost importance.

What is Cash Flow?

Cash flow is the total amount of money being transferred *into* and *out of* a business, especially as affecting liquidity[1]. Basically, it focuses on the transfer of money, cash, and equivalent, with their timing, from one party's point of view. It includes financial commitments, expectations, and liabilities. Keep in mind that we are using the word "cash" in loose/broad terms. It includes any money that can be used, transferred, exchanged, or immediately utilized, no matter what form it is in. It does include fixed assets.

Cash flow is part of the financial management of the organization, which includes planning, organizing, directing, and controlling the financial activities such as procurement and exchange of funds of the organization. It is applying general management principles to financial resources for accomplishing the goals of the organization within its limits and constraints.

---

1 Oxford dictionary online.

Project Management in the Construction Industry: From Concept to Completion, First Edition. Saleh Mubarak.
© 2024 John Wiley & Sons, Inc. Published 2024 by John Wiley & Sons, Inc.
Companion Website: www.wiley.com/go/nextgencpm

Figure 10.1 (a) Cash flow diagram for a bank loan from the borrower's perspective, (b) cash flow diagram for a bank loan from the lender's perspective.

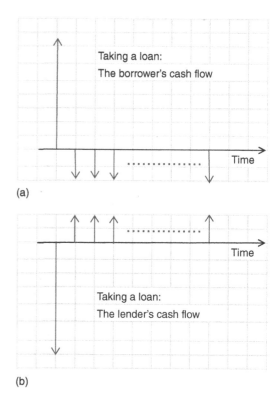

(a)

(b)

Cash flow diagrams for the same project or transaction are drawn from different perspectives. For example, a simple bank loan to an individual is drawn from the borrower's perspective as shown in Figure 10.1a, and from the lender's perspective as shown in Figure 10.1b.

The two most important benefits of cash flow diagrams for the construction company are:

A. Calculating the cost of (or interest on) borrowing money so it can be added as an expense, and
B. Calculating the maximum expected debt (highest negative cash flow, in absolute value) so the contractor can be prepared for it.

> Failing to manage the company's cash flow is one of the main reasons for the failure of construction companies.

Contract terms that may impact the contractor's cash flow

The contract agreement between the owner and the contractor contains relevant items that may impact the contractor's cash flow such as:

1. Advance payment, if any: This is a practice in some countries, where the owner makes a payment of 10–20% of the value of the contract, and then recovers it by deducting the same percentage from every progress payment. This payment helps the contractor in expenses related to hiring laborers from another country, transport and process them, and take care of their living expenses.

2. Mobilization payment: This is a lump-sum amount that the owner pays the contractor for expenses related to mobilization to the project site. It helps the contractor with the initial expenses that are not part of the progress payments.

3. Retainage (also called retention) is a portion of the eligible progress payment that is held by the owner until the contractor fulfills his/her contractual obligations. The contract usually specifies the amount, usually 10% of the total payment, and the conditions of the retainage. In large projects, it is customary to either reduce the retainage percentage or stop retaining any money by the owner after the project reaches a certain completion stage, such as 50%. Retainage negatively impacts the contractor's cash flow.

4. Liquidated damages (LDs) and/or bonus: A liquidated damages clause is a stipulation in a contract of a monetary amount that must be paid by the contractor if the contractor fails to satisfactorily complete the project by the contract finish date. LDs, are usually assessed per day of delay and they may increase after a certain number of days (for example, $1000 per day for the first week and $1500 per day for the second week). On the other hand, the bonus clause is a stipulation in a contract of a monetary amount that must be paid to the contractor if the contractor satisfactorily completes the project earlier than the contract finish date. The contract may have either LDs, bonus, or both, with and without a cap on the maximum amount on each.

5. Purchased but not installed materials: This is an important issue for the contractor. Many contracts do not allow the contractor to get reimbursed for materials purchased but not installed. In this case, the contractor must weigh the benefits versus the drawbacks of buying and storing materials that will be used later.

6. Progress payment request cycle and terms, such as terms and conditions for the payment request to be eligible, the time limit given to the owner to review and approve the request, and other related terms. The process may vary depending on the type of contract, for example, if it is lump sum, unit price, or cost-plus-fee contract. The contract may carry other terms that affect the payment process and the amount of payment.

7. Schedule of values[2] that assigns unit price for each cost item in the project. This is necessary for assessing the value of the work performed every period and used for progress payments.

8. Type of percent complete to be used for payment requests: As was explained earlier, there are many types of percent complete that can be used for measuring work performance. The contract must specify the types used in the progress payment request, perhaps with the mathematical equation for its calculation.

9. Change (Variation) orders process: This includes the steps and stipulations for issuing, reviewing, and approving change orders, initiated by the owner or contractor.

10. Final payment and fiscal close out: The terms here are important for the final settlement. After the project's physical completion, there will still be other financial and legal steps to take, including settling all disputes and invoices. The contractor will be impatiently waiting for the final payment, which usually makes up a good percentage of the project sum.

Using popular contract templates makes it easy for both sides to go through these items, although it can and does happen that the contracting parties change some of these terms.

---

2 Schedule of values was explained in chapter 6, Project budgeting and cost management.

Consequences of Cash Flow Problems

There are many reasons for the failure of construction companies, but ineffective financial management comes at the top[3]. Having available liquid money is essential for sustaining the operations. Lack of cash flow will result in negative consequences such as:

1. Additional expenses such as interest charges and penalties. In some cases, lack of cash delays acquisition of materials and/or equipment, which forces the contractor later to expedite the shipping and delivery, adding more expenses.
2. Less options when paying others, because when a company defaults on its payments, this may lead to a reduction in its credit rating, which – in turn – may result in failure to buy with credit and/or pay higher interest rates for borrowing money.
3. Tarnished reputation to the clients and vendors, and low morale for employees.
4. The possibility of having to limit/restrict operations because of lack of financial resources.
5. The possibility of selling needed assets to pay bills.
6. The possibility of bankruptcy, restructuring or final.

Managing Cash Flow at the Corporate Level

Typically, cash flow for construction companies works this way:

A. Main office expenses continue throughout the year, but not necessarily at a constant level or even consistent way. The amount and frequency differ. They can be:
   a. Continuous such as salaries and utilities,
   b. Periodical such as taxes, maintenance, and external legal fees, and
   c. Accidental/incidental such as a repair, donation, or a party or bonus to celebrate an occasion[4]
   Expenses in parts a and b are predictable in timing and roughly in amount. Expenses in part c are mostly unpredictable; however, some of them are optional (improvement to the office, party), while others are required such as repair to the office air-conditioning system.

B. Project expenses go along with each project itself but start earlier. In a typical situation, the project cash flow from the contractor's perspective starts negative (in debt) till about the project's mid-point, as will be explained later in this chapter. This forces the contractor to either borrow money or use their own funds, which is also an expense (lost opportunity.) After the successful completion of the project, including the fiscal closeout, the company is expected to make a profit, which adds to its positive cash flow.

The organization also has to manage all projects' risks in its financial planning. Risk management must be done at both the individual project and the corporate level. Risk and expected profit are often, but not always, correlated.

---

3 https://www.constructconnect.com/blog/6-reasons-why-construction-companies-fail,
https://www.clockshark.com/blog/construction-company-failure,
https://projul.com/blog/8-reasons-why-construction-companies-fail/
4 If this bonus or celebration belongs to a specific project, it may come from its own budget.

When a company considers taking on a project with relatively high risk, it must take into consideration the risk factors at the project level (this was covered in the risk chapter) but also the risk level at other projects, and thus the risk at the organization level as compared to the strength and resiliency of the company. A well-established strong company may be able to take on a risky project that a new company or a company with a weak financial situation cannot.

So, balancing the company's overall cash flow is very important as projects pump in positive cash flow periodically to offset the required expenses in both the main office and some projects that are still in negative cash flow. The main office must keep a contingency budget (completely different and independent from individual projects' contingency allowances) for unexpected expenses. Monitoring and controlling the overhead expenses, including this contingency allowance, is the duty and responsibility of the leadership of the company.

There are many reasons for cash flow problems with construction companies such as:

1. Lack of proper planning/anticipation of financial commitments and expectations.
2. Lack of or poor risk management for planning and preparing for the expected unknown.
3. Lack of a contingency plan and funds.
4. Contractor's slow and/or unorganized processing, billing, and collection of invoices from clients.
5. Failure to read the terms of the contract.
6. Failure to consider and/or manage own "other expenses"; mainly overhead.
7. Investing a high percentage of capital while leaving little liquidity. It is like "I have the money... but just cannot get it now!"
8. Disputes, mostly with the owner.
9. Issues beyond the control of the contractor.

Time Value of the Money

One of the basic concepts in finance is the time value of the money, which implies that a dollar today is not the same as a dollar tomorrow. So, when the contractor spends money for the construction operations and then gets reimbursed by the owner in a few weeks, there is an added expense here: the time value of the money. This is equivalent to the cost of borrowing that money or the cost of the "lost opportunity"; that is, if the contractor invested that money somewhere else.

In addition to the lag between spending money and getting reimbursed, the contractor's cash flow often includes some holding on the money for retainage, disputes, or other, which adds to the interest cost.

The equation we need here is the compound interest:

$$F = P(1+i)^n$$

where:
   F is the future value
   P is the present value
   i is the interest rate per period
   n is the number of interest periods

> Planning for the project financing is a major task. The contractor has to anticipate the payments, in and out, and their timing, so there will be no shortage of cash. Also, the cost of interest on borrowed money has to be added to the project expenses.

Note that the terms future and present are relative. So, if today we like to calculate the average inflation rate between the years 2010 and 2020, the index in 2010 will be the "present," while the index in 2020 will be the "future," even though they both are past. Also note, that i and n correspond to the same time period, regardless of the interest nominal rate or the APR[5] that banks show in their advertisements and contracts. Thus, if the interest accrues monthly, i in the equation must reflect the monthly interest rate, and n = the number of months.

Progress Payments

Typically, construction projects cost considerable amounts of money and last for a long time. Unlike financial transactions in other contracts, where the client pays the entire amount at the time of the transaction or within a short period of time, the money exchange between the owner of a construction project and the contractor happens in installments during the construction phase and based on measured work performance. These installments are called progress payments. Typically, the general contractor (GC) reports all work performed by own crews and subcontractors, to the owner, in an invoice called progress payment request.

Progress payments are periodic (usually monthly) payments made by the owner to the GC (and then by the GC to the subcontractors) upon the approval of the payment request. Owners usually retain a portion of the progress payment, typically 5–10%, until the successful completion of the project. They may also make other deductions for a variety of reasons such as:

- Recovery of advance payment
- Recovery of reimbursement of the cost of materials purchased previously but not installed
- Disputed amount till the dispute is resolved
- Penalties, if any

Figure 10.2 summarizes the cycle.

The progress payment cycle works this way:

a. Two or three days before the submission of the request, the GC receives all payment requests from all subs who performed work this past cycle. The GC rolls them with its own invoices into one payment request and submits it to the owner on the new progress payment cycle day.

b. The owner reviews the request within the time allowed in the contract, usually 2–4 weeks. After that, the owner may approve the request as is or request clarification, or even dispute certain items. The owner's dispute can be based on a unit price or quantity (or percent complete) of an item or items, the calculations, other, or a combination of these factors.

---

5 Annual percentage rate.

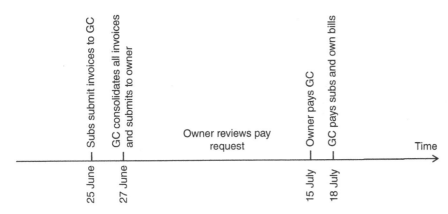

Figure 10.2 The progress payment cycle.

c. The owner pays the GC the approved amount after making any deductions such as explained earlier. Upon receiving the owner's payment, the GC pays for his/her own subs.

d. The owner usually requires lien waivers (also called lien releases) from the GC and every party that received money.

e. After the completion of the project, the GC makes a request for final payment. This final payment includes reimbursement for work performed during the last period (with no deductions), in addition to any previously held amount (mainly retainage). If no dispute arises, the owner makes this final payment, which represents the final financial settlement between them.

The cash flow diagram is summarized in Figure 10.3.

f. In some countries, the owners keep half of the retainage, 5%, for the period of the basic warranty (usually one year or 400 days). This is done as a preemptive measure in case there is a problem with

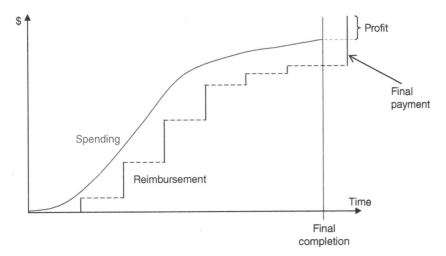

Figure 10.3 Typical cash flow diagram for a construction project from the contractor's perspective.

the project that is covered by the warranty, and the GC fails or refuses to repair. The owner then can find a professional to do the repair and deducts the repair cost from the retained fund.

Requirements for Preparing Cash Flow:

1. A detailed (itemized) cost estimate that shows the cost of every activity. The cost of the activity may not be linearly distributed over the duration of the activity. This distribution is decided by the contractor and could be subject to the approval of the owner.
2. A detailed CPM schedule showing when every activity gets performed. The schedule must be cost- or resource-loaded. This way, the schedule will show the spending by how much and when.
3. The overhead must be included. The contractor may allocate it to (distribute it over) individual activities or load it to a hammock activity that spans the entire project.

Example 10.1   **A general contractor made an agreement to build a small office building. His/her estimated cost, not including the cost of borrowing money is $2,481,000, and the duration to take 9 months. He/she anticipates his/her spending to be as shown in Table 10.1.**

The owner's reimbursement payments lag by a month and the owner retains 10% till the final and satisfactory completion of the project. Borrowing money costs the contractor 1% per month. Draw the cash flow diagrams and calculate the cost of borrowing money and maximum debt.

   Solution: Looking at Table 10.2, we can answer both questions:

- The cost of borrowing money is $36,158, which is not a negligible amount. In fact, it is almost 1.5% of the cost of the project, or about 30% of the 5% anticipated profit allowance.
- The maximum debt is $506,800 in the 6th month of the project.

Table 10.1   Anticipated spending in the 9-month project, per month.

Month	Spending
1	$175,000
2	$272,000
3	$311,000
4	$327,000
5	$343,000
6	$364,000
7	$298,000
8	$228,000
9	$163,000
Total	$2,481,000

Table 10.2   Solution to Example 10.1.

Month	Spending	Reimbursement	Net cash flow	Cumulative cash flow	Future value[a]
1	($175,000.00)	$0	($175,000)	($175,000)	($191,394.92)
2	($272,000.00)	$157,500	($114,500)	($289,500)	($123,987.09)
3	($311,000.00)	$244,800	($66,200)	($355,700)	($70,975.36)
4	($327,000.00)	$279,900	($47,100)	($402,800)	($49,997.60)
5	($343,000.00)	$294,300	($48,700)	($451,500)	($51,184.19)
6	($364,000.00)	$308,700	($55,300)	**($506,800)**	($57,545.40)
7	($298,000.00)	$327,600	$29,600	($477,200)	$30,496.91
8	($228,000.00)	$268,200	$40,200	($437,000)	$41,008.02
9	($163,000.00)	$205,200	$42,200	($394,800)	$42,622.00
10		$394,800	$394,800	$0	$394,800.00
Total	($2,481,000)	$2,481,000	$0	$0	($36,157.64)

[a] Future Value is calculated using single payment formula.

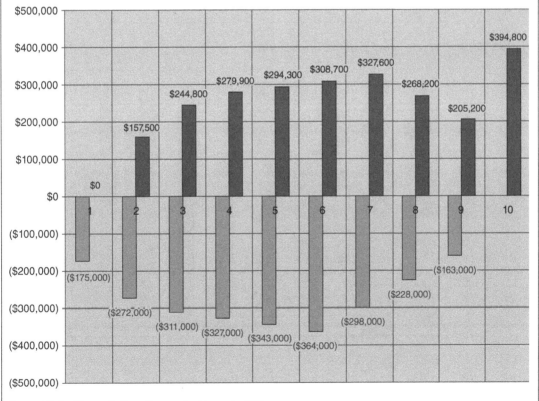

Figure 10.4   The cash flow diagram for Example 10.1.

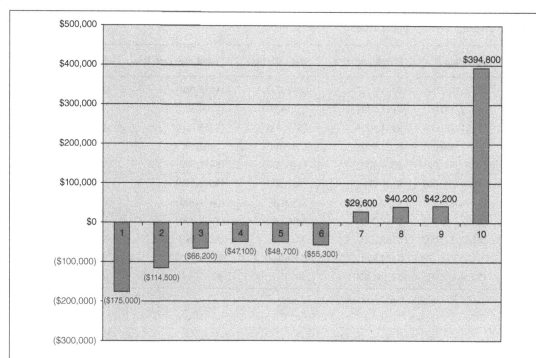

Figure 10.5 The net cash flow diagram for Example 10.1.

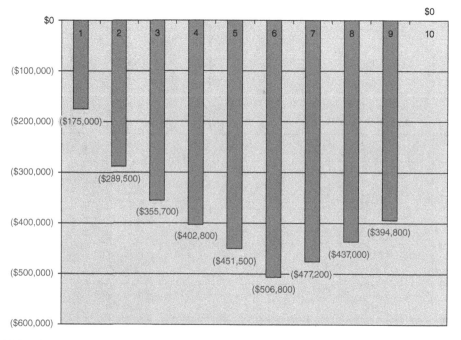

Figure 10.6 The cumulative cash flow diagram for Example 10.1.

To highlight the importance of time value for the money, let us **repeat** the same solution with one change: Make the owner's reimbursement payments lag by two months. The answers change significantly:

- The cost of borrowing money is $62,258, which is about 2.5% of the cost of the project, or half of the 5% anticipated profit allowance.
- The maximum debt is $815,500, also in the 6th month of the project.

The results are shown in Table 10.3 (Figures 10.3–10.6).

The cash flow diagram from the owner's perspective is shown in Figure 10.7. It starts with progress payments to the contractor (negative cash flow) and then the takeover of the project when completed (large positive "cash flow"), and finally making the final payment (negative cash flow).

The project takeover after completion is considered positive cash flow because while the project is under construction, it is "owned" by the GC, in the sense that the contractor is responsible and liable for all its matters. The handover by the GC and the takeover by the owner is legally an important step, where the owner becomes in control, with the responsibility and liability for the project. All warranties usually start from that date.

Table 10.3 Solution to Example 10.1 with assumption of 2-month payment lag.

Month	Spending	Reimbursement	Net cash flow	Cumulative Cash flow	Future value[a]
1	($175,000.00)	$0	($175,000)	($175,000)	($193,308.87)
2	($272,000.00)	$0	($272,000)	($447,000)	($297,482.39)
3	($311,000.00)	$157,500	($153,500)	($600,500)	($166,218.50)
4	($327,000.00)	$244,800	($82,200)	($682,700)	($88,129.53)
5	($343,000.00)	$279,900	($63,100)	($745,800)	($66,981.92)
6	($364,000.00)	$294,300	($69,700)	**($815,500)**	($73,255.40)
7	($298,000.00)	$308,700	$10,700	($804,800)	$11,134.46
8	($228,000.00)	$327,600	$99,600	($705,200)	$102,617.98
9	($163,000.00)	$268,200	$105,200	($600,000)	$107,314.52
10		$205,200	$205,200	($394,800)	$207,252.00
11		$394,800	$394,800	$0	$394,800.00
Total	($2,481,000)	$2,481,000	$0	$0	($62,257.66)

[a] Future Value is calculated using single payment formula.

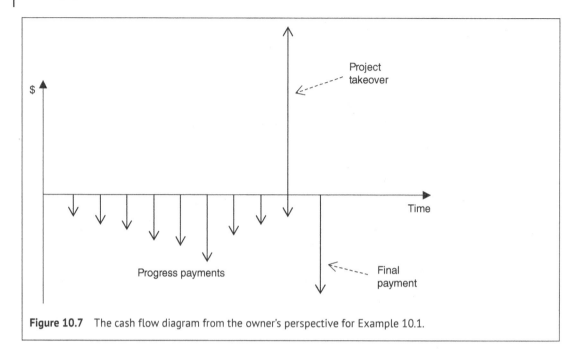

Figure 10.7   The cash flow diagram from the owner's perspective for Example 10.1.

Cash Flow Issues with the Project Schedule

1. When the contractor levels the resources, activities – both critical and noncritical – are held in their time position. Shifting them even within their available float will disturb resource leveling. In this case, the cash flow diagram must be calculated after the resource leveling.
2. The cash flow diagram must be re-calculated after each schedule update. Things on the ground rarely, if ever, go as planned. Therefore, changes can and do happen.
3. In case the contractor needs to accelerate the schedule, these issues must be considered:
   a. Progress payments will likely be higher for two reasons. First, because we do more work per unit of time, so the progress payments will reflect more work quantities than before acceleration. Second, because acceleration will likely cost additional money.
   b. Everything will be at "fast pace," including reviews and approvals.
4. In some rare cases, the owner decides to decelerate the project due to cash flow problems (from the owner's side). It is true that such deceleration will reduce progress payments because less work is performed per unit of time. However, the total overhead expenses may increase because overhead is continuous throughout the duration of the project: longer project duration means more overhead cost. Also, interest amount may increase.
5. The importance of accuracy and good recordkeeping cannot be overemphasized. The contractor may need to respond to the owner's inquiry about a number in the payment request and must always be ready to explain and prove any number in the request. The record may also be recalled later in the future in case a dispute arises.
6. If the contract does not require the owner to reimburse the contractor for materials purchased but not installed, the contractor needs to decide if the materials will be purchased early and stored till the time of installation, or they will be ordered and delivered just before their installation.

Buying the materials early may get the contractor a volume discount. For example, he needs blocks for the entire building, so he buys the entire needed quantity. Doing so also gives the contractor peace of mind for having the materials when needed. The drawback is having to pay for them and not being reimbursed for some time (frozen capital). They may also pose a challenge for storage if the site is tight or insecure enough. The balance between the "Just-in Time" and "Inventory Buffer" theories was discussed in Chapter 6.

7. Companies that do business internationally may have to do cash flow with different currencies. It is possible that the contractor deals with two different currencies in the same project. This may be a challenge as exchange rates usually fluctuate.

Cash flow and Activities' Float

When the contractor prepares the CPM schedule, all noncritical activities will have different values of float time, which means a range of dates that the contractor can move the activity within, without delaying the entire project. We discussed float in Chapter 7, Time Management, but we note here that the choice of dates for these activities will impact the project's cash flow. Figure 10.8 shows the project cash flow in three different scenarios, representing going with early, planned, and late dates. Although the total cost by the end of the project is the same for all three scenarios, we notice that the spending at any data date is different among the scenarios, depending on the performance time of the activities. Performing activities as early as possible results in front-end loading of the cash flow curve (convex). While it is the opposite when we perform activities as late as possible (concave).

The contractor's choice of activities' dates within their float amounts depends on several factors such as resource utilization and leveling, risk management, materials management, and work plan. Thus, it is likely that the contractor will not go for either full scenario, as soon as possible or as late as possible, but somewhere in between, where some but not all activities are delayed within their available float time. The contractor must take all factors into account when creating and submitting the schedule to the owner for review and approval. The choice of activity dates impacts the cash flow for both the contract and the owner. So, it is not a good idea, and may not be acceptable, to submit a schedule to the owner and then make significant shifts in activity dates, even if this does not delay the entire project. The owner will likely have their own cash flow prediction, and may not be ready for progress payment request that is significantly higher than predicted.

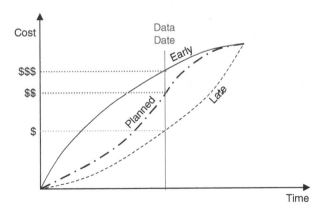

Figure 10.8 Cash flow variation with choice of noncritical activity dates.

Financial Planning in Volatile Times

One of the characteristics of construction projects is that they take time to finish but the parties' commitment is usually done earlier. It may happen that the commitment is made when the economy is booming and the market is high, but shortly after, the economy slows down, and the demand drops. It is indeed a dilemma for everyone: the developer, the contractor, and even the owner. If they do not commit, they may miss an opportunity and the prices may increase. If they commit, there may be a recession or a slower economy. Sometimes a developer starts a large project, residential or commercial. The project may take a few years to complete. It may consist of one indivisible scope such as a high-rise building or multiphases/multiunits such as a residential community. Even in the case of multiphases/multiunits, the developer may have to build the infrastructure for the entire scope first. This may include rezoning and permitting, underground utilities, roads, and other services.

The smart businessperson would plan for several scenarios in case the economy turns up or down despite the fact that there is risk in every plan. The risk and cost of the plan have to be weighed against the potential profit. If the economy starts showing signs of slowing down, the developer may decelerate, pause, or scale down the project. However, a construction contract is not a simple switch that can be turned off instantly. It will take time for a gradual change in the scope or pace of the work.

The cash flow during these times will be a crucial matter. Wise management of business, including the financial aspects, can be the difference between those companies that survive tough times and those who do not.

The Contractor's Options During Slow Economic Times

When the economy slows down, construction follows suit, and contractors are faced with tough decisions to make. Here are some options for the contractor, with the emphasis that they are not mutually exclusive. In fact, many can and should be combined:

1. Operate as lean as possible: reducing overhead to a minimum. Options include, but are not limited to, downsizing the office and car fleet. Some contractors are so casual with overhead expenses to the point of taking them for granted or simply not having enough time to examine them. It is a good idea for the contractor to review these expenses and classify them into three categories: those that cannot be eliminated or reduced, those that cannot be eliminated but can be reduced, and those that can be eliminated.
2. Lend/rent resources to other contractors. This includes mainly equipment but possibly employees and even office space. The options may also include selling equipment that is not needed or negotiating to get out of long-term leases.
3. Accept lower profit margins to increase the chance of getting projects.
4. Put on hold or cancel any unnecessary plans that require budget, especially large ones.
5. Find a niche that can give the contractor an advantage. This can be a specialty such as renovation or a new technology that is in demand.
6. Be a visionary and anticipate changes in the market (supply and demand), economic conditions, technology, and other matter. Make plans so you can "hit the ground running!"
7. Spend any extra time, for you and your employees, on continuing education.
8. Seek new opportunities and new markets, perhaps in another country.
9. Renegotiate contracts with owners, subs, and vendors for better terms and conditions. Many debtors prefer leniency over delinquency.

10. Check your billing and see if there are any uncollected bills.
11. Always invest in relationships and good reputation, built on trust and quality.
12. Involve your employees as partners. Be honest and realistically optimistic.
13. When applying any austerity measures, start with yourself before applying them to employees. This will motivate the employees and make them look at you as a leader, not just a boss. Cost-cutting is a must in some cases, but it can be a double-edged sword. You do not want to cut an operation that can generate more money than its cost.
14. Try not to take more loans than you can afford. You may solve an urgent problem but create bigger problems in the near future.

How Much Cash On-hand Must be Kept Available?

Although much of the operations and expenses the contractor needs depend on funds to be received later from clients, construction companies, like any other business, need to have immediate access to liquid and available funds. There is no magic formula that calculates how much these funds must be, as an amount or event percentage of operations. However, a good businessperson would have a "diversified plan," such as:

A. A minimum balance in the bank
B. A preapproved loan
C. Assets that can be liquidated in relatively short time
D. Certain creditors that can extend their deadline

These options are not necessarily in order, and they are not mutually exclusive. The contractor must always keep an eye on the total debt and liabilities, and make sure they are within the company's financial limitations. The important thing for the contractor is to be ready for such occurrences, and not to allow a delayed payment to have a negative domino effect on their business. Using credit cards with partial payments, that is, less than the full balance, is discouraged because of the amount of interest and the way it is calculated for most credit cards. Smart businesspeople do not get lured by credit cards or loan offers that look enticing in the beginning but turn around later to be vicious and oppressive.

Loans Between Attraction and Reality

Many financial institutions make it easy and attractive to take loans, even when you do not need them. Merchants also make it easy and attractive to "buy now and pay later," which encourages people and companies to borrow and/or buy merchandise based on credit. Many people take loans beyond their financial capability and end up in troubled situations that lead them to severe austerity measures or even bankruptcy. Reading the fine print, long-term planning, and discipline are very important in these situations. Here are some points people have to keep in mind when taking a loan:

1. Many loans have easy and attractive payment schemes for a limited period that can be a year or so. After that, not only do tougher conditions apply, but also if the borrower cannot fulfill the loan contract stipulations, many of the attractive incentives will be taken back retroactively. For example, using a traditional credit card, you will not pay any interest on purchases if you pay your entire balance by the due date. Assume you buy merchandise on April 11 for $1000. Your credit card cycle closed on April 26. You received your statement on April 29, informing you of your balance, and giving you a payment deadline of May 25. Now, if you pay the full balance by the due date, you pay no interest at all. If you make partial payment, no matter how much, the interest will accrue starting the date

of purchase, April 11, retroactively. If you pay the full amount but you made it late, the same thing will apply to you in addition to penalty fees.

2. The effective interest rate is often higher than the declared nominal one, depending on the accrual of the interest. For example, if the annual percent rate (APR) is 18.9% and the interest is accrued daily, then the effective rate is 20.8%[6].

3. Many merchants give incentives for full early payment. Some merchants add to the invoice a term like *2/10 NET 30*, which means a 2% discount if the bill is paid fully within 10 days of the delivery date, or the full amount is due in 30 days. After 30 days, interest and/or a penalty may apply. Customers who default on their payments may not be able to even buy on credit. Instead, they must pay the full amount at the time of purchase.

4. Many of the companies providing services to the contractor, such as insurance and cleaning, give 5–10% discount if the fees are paid annually in advance. Doing so saves money and administrative work.

The above points underscore the importance of good financial control for the contractor, which can be summarized in:

1. Good planning with estimate of the cash flow, possibly with "plan B" in case things do not work out the way the contractor plans.

2. Reading any financial commitment contract carefully is important. Sometimes, the help of a contract professional may be needed.

3. Practicing good control of spending, especially overhead expenses. The company has to prioritize its spending, differentiating between urgent versus nonurgent, and important versus unimportant.

4. Keeping a sufficient amount of liquid money that can be used for urgent cases. This can be in form of "organization's contingency," or it can be through assuring a minimum financial cushion always.

5. Continuously keeping up with commitments, expectations, and liabilities.

Exercises

10.1 What is the cash flow for a company in general, and construction company in particular.

10.2 What does management of cash flow include?

10.3 Do an internet search for reasons for the failure of construction companies. List the top five reasons. How does management of cash flow rank among them?

10.4 Does the cash flow diagram for the lender differ from the borrower's diagram? Explain.

10.5 What are the two most important benefits of cash flow diagrams for a construction company?

10.6 Mention five contract terms that may influence the contractor's cash flow.

---

6 Using the equation, Effective rate $= (1 + i/n)^n - 1$

10.7 Mismanagement of cash flow has negative consequences. Mention five of them.

10.8 Briefly discuss cash flow management at both the project and corporate levels.

10.9 What are the common reasons for cash flow problems with construction companies?

10.10 What are progress payments? Mention the cycle steps for these payments.

10.11 What are some of the usual or possible deductions off progress payments? Which ones will be paid later and when?

10.12 What type of information is needed to prepare a cash flow diagram?

10.13 A general contractor contracted to build a house. His/her estimated cost, not including the cost of borrowing money is $843,500 and the duration is 6 months. He/she anticipates his/her spending to be as shown in the table below.

Month	Spending
1	$85,000
2	$122,000
3	$171,000
4	$255,000
5	$143,000
6	$67,500
Total	$843,500

The owner's reimbursement payments lag by a month and the owner retains 10% till the final and satisfactory completion of the project. Borrowing money costs the contractor 1% per month. Draw the cash flow diagrams and calculate the cost of borrowing money and maximum debt.

10.14 A general contractor contracted to build an apartment building. His estimated cost, not including the cost of borrowing money is $3,898,000 and the duration is 12 months. He anticipates his spending to be as shown in the table below.

Month	Spending
1	$225,000
2	$272,000
3	$371,000
4	$407,000
5	$443,000
6	$464,000

Month	Spending
7	$498,000
8	$385,000
9	$293,000
10	$212,000
11	$165,000
12	$163,000
Total	$3,898,000

The owner's reimbursement payments lag by a month and the owner retains 10% till the final and satisfactory completion of the project. Borrowing money costs the contractor 0.75% per month. Draw the cash flow diagrams and calculate the cost of borrowing money and maximum debt.

10.15 Repeat Exercise 14 if the monthly interest rate is 1%.

10.16 Repeat Exercise 14 if owner's reimbursement payments lag by 2 months.

10.17 Repeat Exercise 14 if the monthly interest rate is 1% and owner's reimbursement payments lag by 2 months.

10.18 Repeat Exercise 14 if the owner keeps half of the retainage for one full year after the completion of the project.

10.19 Examining the results of Exercises 14 through 18, draw a conclusion regarding the importance of preparation of cash flow diagrams.

11

Project Quality Management

Introduction

Construction projects are complex and multidisciplinary, involving numerous stakeholders and a range of activities. Quality management in construction makes the third side in the Cost-Schedule-Quality "Golden Triangle," shown in Figure 11.1. It represents the compliance with, and fulfillment of, the specifications stipulated in the contract documents. Quality management for the contractor is a process that starts with planning, continues through execution, and goes beyond project completion. This chapter will provide an overview of the principles of quality management in construction projects, including the main elements and tools used to achieve quality objectives.

What is Quality?

Linguistically, quality of something, goods or services, has been defined using different terms but all definitions revolve around two traits: having high degree of excellence, such as performance or durability, and achieving high degree of satisfaction to the client or receiver.

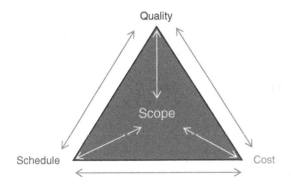

Figure 11.1 The "Golden Triangle" in construction projects.

Project Management in the Construction Industry: From Concept to Completion, First Edition. Saleh Mubarak.
© 2024 John Wiley & Sons, Inc. Published 2024 by John Wiley & Sons, Inc.
Companion Website: www.wiley.com/go/nextgencpm

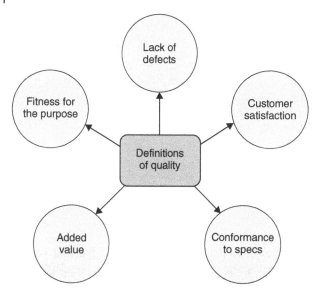

Figure 11.2 Definitions of quality.

In general, quality is regarded as the degree of excellence of something, but measuring this degree requires a reference, especially in contract agreements. This is why quality in construction contracts is defined as conformance to contract requirements, not a general degree or measurement of goodness. Figure 11.2 combines all definitions of quality.

For example, an agency wants to build temporary housing for refugees. It is required, according to the contract agreement, that these housing units withstand normal use (wear and tear) and normal natural environmental elements for five years. However, some of these units started showing cracks and defects after six or seven years of use. Or in a different scenario, some units started showing defects within the first five years, but because they were subjected to heavier use than what was specified in the original design. Do we consider either case as a quality defect? Of course not, since the units conformed to requirements for the period and type of use specified in the contract.

Quality, or specifications to be precise, includes materials and workmanship. Defining, measuring, and inspecting the workmanship is more difficult and subjective than doing so for materials.

Another quality-related definition is accuracy and precision. Accuracy is an assessment of correctness, while precision is an assessment of exactness. Accuracy measures how close results are to the true or known value. Precision, on the other hand, measures how close results are to one another. They are both useful ways to track and report on project results.

> In construction projects, quality means conformance to specifications.

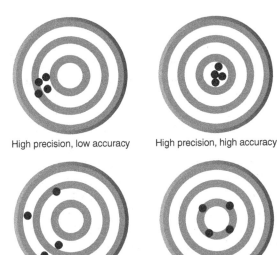

High precision, low accuracy　　High precision, high accuracy

Low precision, low accuracy　　Low precision, high accuracy

Precision versus accuracy.

Total Quality Management (TQM)

Total Quality Management (TQM) is a management framework based on the belief that an organization can build long-term success by having all its members, from low-level workers to its highest-ranking executives, focus on quality improvement and, thus, providing goods and/or services that attain customer satisfaction.

- It is organization-wide, not project-specific!
- It is an organizational culture!

The Concept of Total Quality Management

TQM is a concept that was developed for instilling the culture of quality across the organization. Thus, quality will move up from inspecting the final product, to controlling the entire process, and finally becoming a culture for the organization, including all levels of workers, and including all work processes, even those not directly related to products, such as the administration. Quality must come not only in the product itself, but also in the services that come before and after.

It is believed that once quality becomes a culture in the organization, products and services come naturally carrying the traits of quality. Such culture even though may be successful in attaining quality products and services, it also requires continuous improvement, which is part of the TQM culture. In other words, quality is not a fixed level that you achieve, but it is a continuous quest for an "illusive perfection." It is like a championship in football or basketball, it takes a collective and comprehensive effort to go to the top, but it also takes a continuous improvement to stay up there.

History of TQM

Although the use of the terms "quality," "quality control," (QC) and "quality management" is old,[1] it is believed that the concept of TQM started in Japan in the early 1950s, as several Japanese industry leaders were inspired by presentations made by American professor Edwards Deming (1900–1993) for the Union of Japanese Scientists and Engineers (JUSE). Deming's message to Japan's chief executives was that improving quality would reduce expenses while increasing productivity and market share.[2] This led to a quality revolution in Japan, perhaps also motivated by national responsibility in the aftermath of World War II for Japan's economy and infrastructure. In the late 1970s and early 1980s, Western countries faced stiff competition from Japan's products that enjoyed both high-quality and competitive prices. They adopted similar quality standards with the motto "If Japan Can... Why Can't We?[3]" Firms began reexamining the techniques of QC invented over the previous 50 years and how these techniques had been so successfully employed by the Japanese. In 1984, the United States Navy was seeking recommendations from experts as to how to apply their approaches to improve the Navy's operational effectiveness. The Navy branded the effort as "Total Quality Management, TQM" in 1985 and its study spread throughout the US Federal Government, triggering several follow-up and spin-out studies and actions. The private sector also made a quest for improving products and services, following TQM principles to remain competitive in both government and private contracts.

The key concepts in the TQM effort undertaken by the Navy in the 1980s include[4]:

1. Quality is defined by customers' requirements.
2. Top management has direct responsibility for quality improvement.
3. Increased quality comes from systematic analysis and improvement of work processes.
4. Quality improvement is a continuous effort and conducted throughout the organization.

TQM enjoyed widespread attention during the late 1980s and early 1990s, and still popular in certain industries, such as construction. Other industries, such as manufacturing, adopted other standards/concepts such as ISO 9000,[5] Lean manufacturing, and Six Sigma, which also aim at improving performance and products, and ultimately becoming more competitive and profitable. Figure 11.3 summarizes the concept of TQM.

> "Always treat your employees exactly as you want them to treat your best customers." Stephen Covey
> "Clients do not come first. Employees come first. If you take care of your employees, they will take care of the clients." Richard Branson

---

1 SAE International (formerly the Society of Automotive Engineers), with major objective in standardization and quality management, was established in 1905. The American Society for Quality (ASQ), formerly the American Society for Quality Control (ASQC), was established in 1946. In addition to other organizations in other parts of the world under different names but with a focus on quality control, quality assurance, or quality management.

2 See W. E. Deming's 14 Points For Total Quality Management, https://asq.org/quality-resources/total-quality-management/deming-points.

3 https://en.wikipedia.org/wiki/If_Japan_Can..._Why_Can%27t_We%3F.

4 Houston, Archester (December 1988), A Total Quality Management Process Improvement Model (PDF), San Diego, California: Navy Personnel Research and Development Center, pp. vii–viii, OCLC 21243646, AD-A202 154, archived (PDF) from the original on October 21, 2013, retrieved 2013-10-20.

5 The International Organization for Standardization (ISO) was established in 1947 but its first standards, ISO 9000, was first published in 1987.

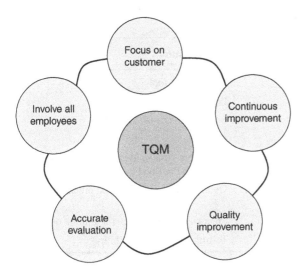

Figure 11.3 The concept of TQM.

Elements of Quality Management in Construction

Quality management plan: A component of the project or program management plan that describes how applicable policies, procedures, and guidelines will be implemented to achieve the quality objectives. It includes the following elements:

1. Quality planning: This involves defining the quality objectives and requirements for a project, as well as developing a plan for how they will be met.
2. Quality control: This refers to the systematic and continuous monitoring of project processes and outputs to ensure that they meet the established quality requirements.
3. Quality assurance: This involves implementing quality control procedures, as well as reviewing and verifying the effectiveness of these procedures.
4. Quality improvement: This involves continuous monitoring and improvement of project processes and outputs to enhance project quality over time.

Quality Assurance versus Quality Assurance

The terms "quality assurance" (QA) and "quality control" (QC) are often used to refer to ways of ensuring the quality of a service or product. The terms have different meanings. The PMI defines QA as "The managerial processes that determine the organization, design, objectives, and resources, and that provide the project team, client, and shareholders with performance standards and feedback on the project's performance.[6]" It also defines QC as "The technical processes that examine, analyze, and report the project's progress and conformance with performance requirements.[7]"

6 https://www.pmi.org/learning/library/quality-management-9107#:~:text=Quality%20Assurance%20%2D%20The%20managerial%20processes,feedback%20on%20the%20project's%20performance.
7 ibid

Table 11.1 Comparing quality assurance to quality control.

Quality Assurance	Quality Control
A managing tool	A corrective tool
Process oriented	Product oriented
Looks at the entire production process	Looks at the finished product
Proactive	Reactive
Prevention: aims to prevent defects with a focus on the process used to make the product	Detection: aims to identify (and correct) defects in the finished product
Uses a quality system to assure adherence to specifications. The system will be continuously assessed for its adequacy	Uses testing for the finished product
Practiced by the contractor (producer) only	Practiced by the contractor and the owner
Everyone's responsibility	Testing team's responsibility

Both QA and QC are powerful techniques, and both must be performed to ensure that the deliverables meet your customers' quality requirements. Quality assurance minimized the likelihood of an error or defect in the product, while quality control tests to verify such an outcome. This is why quality programs usually are entitled QA/QC. Table 11.1 shows a comparison summary between QA and QC.

Who Sets Quality Standards?

Every construction contractor's organization has its own set of standards, but when it works on a specific project, it is also bound by the specifications mentioned in the contract agreement of this project. The contractor usually stays within both standards (the higher of the two): company's standards and the project contract. Having good standards for the organization pays off and gives it a good reputation. However, company's own standards may differ from one project to another, depending on the type and level of finish for that project.

In construction projects, mostly in the residential building industry, specifications are set in two different situations: custom or standard projects. In custom projects, the design, including the specifications, is performed according to the owner's demands and needs. The contractor has to build the project according to the specifications prescribed in the design documents. On the other hand, standard projects are those when the contractor builds multiple units according to standard design (repeated frequently and known as cookie cutter or tract homes) with specifications set by the contractor, developer, or investor. The customer cannot make changes, especially architectural, structural, and other major ones, but may have limited choices in the finishes, such as paint, flooring, or plumbing fixtures. These choices may come as an "upgrade package" that usually includes certain upgrades, bundled by the contractor.

If we make an analogy with the auto industry, the first type is similar to placing an order with an auto manufacturer, for a car with specifications according to the client's demands. The second type, which is the common one, is when the auto manufacturer creates several car models, each with certain specifications that buyers cannot make any major change. The manufacturer may provide several lines of the same car model, with different features to suit different tastes and budgets of different buyers. Keep in mind that even with the full-customized product (auto or construction project), the client's specifications are still subject to restrictions such as safety requirements, government regulations, and technical limitation of the manufacturer or contractor.

It is also interesting that in many industries, such as auto and residential building construction, some corporations create two or more companies, under the same ownership: each company caters to different customers. Usually one has standard (average) products, while the other has upper-scale products. This is a marketing strategy as the luxury brand name catches the attention and interest of certain segments of customers, but it also has to do with quality as each line has its own specifications, staff, and customers.

When clients set the specifications, they must use professional guidance in aligning what they like or wish to have, what they need, and what they can afford. Many options and upgrades sound enticing but their cost may add up to a level that is beyond the owner's budget. In general, the contractor must help the owner maximize the value of the project by selecting options and upgrades with highest value to the owner.

Setting the specifications is a challenge for the owner, especially when budget is tight. It needs careful exchange with the designer and the contractor. The designer has the professional duty of "translating" the owner's needs and demands into specifications. The contractor must also advise the owner in making any change that will add value to the project and service to the owner.

The contractor can set higher/more stringent quality standards than those in the contract but cannot go lower! Here is the list of specifications requirements, starting from the "bottom of the pyramid," but intersecting in many cases:

1. Legal requirements by law and local codes.
2. Defined by the contract agreement.
3. Individual item's warranty by manufacturer or vendor.
4. Contractor's organization standards.
5. Industry professional standards.
6. Customer's expectations.

Safety requirements in the specifications must never be a matter of negotiation or compromise.

The contractor is legally bound by the law and the contract agreement. The contractor would like to make the customer always satisfied, but the customer's expectations can be subjective and even illusive. This is why quality must be defined in more objective terms in the contract specifications. For satisfying the customer, the contractor needs to prioritize:

1. Fulfilling contract and other legal obligations: This is a must.
2. Going above and beyond the agreement: Sometimes, the customer may request something that is outside the contract terms or the basic warranty. The contractor is not obliged to respond to such request, but may choose to do it as a courtesy, if the work amount or value is reasonable, and also if this is in the best interest of a continued relationship with this customer.

3. Showing that you care even if you cannot respond to customer's demand: If the contractor cannot respond to a customer request, at least he/she can show courtesy by guiding the customer on how to get the work done. The contractor can refer the customer to a professional who can do the work or guide the customer to how can the issue be resolved.

4. Keeping an eye on your business and profit, especially for the long term. The contractor must have one eye on current business while the other eye is on the future. Construction companies are businesses, not charities. Focusing on current business only and insisting on maximizing profit in every project and transaction will make the contractor "penny-wise pound-foolish". For example, a dispute with a client on a repair for a minor quality fault like a leaking faucet or imperfect painting, may ruin future business. On the other hand, focusing on the future and long-term plans only, may get the contractor in trouble if the cost of immediate commitments escalate beyond control.

Kaizen (Continuous Improvement)

Kaizen, also known as continuous improvement, is a long-term approach to work, that systematically seeks to achieve small, incremental changes in processes in order to improve efficiency and quality.

Continuous improvement stems from the belief that perfection is illusive; you can get closer and closer, but you can never hit it! This means that there is no ceiling for improvement!

Kaizen can be applied to any kind of work, but it is perhaps best known for being used in lean manufacturing and lean programming. If a work environment practices kaizen, continuous improvement is the responsibility of every worker, not just a selected few.

Why Continuous Improvement?

1. Because perfection is illusive. There is no ceiling for improvement. The minor improvement you make, may be the advantage that gives you an edge over your competitors. Improvement has many sides such as reducing cost, increasing production (with the same input level), improving the performance or durability of the product, improving customer satisfaction, reducing accidents, reducing reworks, and others. These sides are not mutually exclusive, in fact, several of them may come together.

2. Because humans tend to relax the rules with time, especially when the management itself starts weakening the enforcement of the rules. This is why observations and comparisons to established standards are strongly recommended. So, the improvement can be either by re-aligning production with the standards (baselines) or by "raising the bar" and finding ways to challenge the standards.

3. Because of competition: If you are not making any improvement, some competitors are doing it and they will have an advantage over you. Again, this advantage can be in quality, cost efficiency, innovation, safety, or other.

4. Because of human turnover. This is yet another factor, as people in leadership positions usually leave their footprint in their work environment, even with the company standardizing all processes. When a new leader takes over their position, there will be a need to realign and redirect the work and team members. This is a chance for the new leadership to introduce improvement.

5. Because of technological advancement. This is arguably one of the most important factors as we face a rapidly changing world of technology. The change is not limited to devices but includes everything from tools and equipment, communication methods and devices, computers and software, materials,

and even concepts. Technology is advancing at a rapid pace that made "doing business as usual" a potential cause for failure in business.

6. Because laws, regulations, and standards change. Laws and regulations change from time to time in response to new requirements and developments due to local or global changes.
7. Because what worked best for one situation may not work best for other situations. This point probably combines all the points above. The industry is constantly looking for challenges, innovations, and standing out.

Lean Construction

The concept of "lean production" originated in the manufacturing industry. It is a systematic method for waste minimization within a manufacturing system without sacrificing productivity. Lean also considers waste created through overburden and waste created through unevenness in workloads.

The Toyota Production System, and later on the concept of Lean, was developed around eliminating the three types of deviations that show inefficient allocation of resources. The three types are Muda (waste), Mura (unevenness), and Muri (overburden)[8] are shown in Figure 11.4.

The concept was adopted in the construction industry as "lean construction," which is a combination of operational research and practical development in design and construction with an adoption of lean manufacturing principles and practices to the end-to-end design and construction process. Unlike manufacturing, construction is a project-based production process with no two projects that are identical.

Lean construction is a project delivery process that uses Lean methods of maximizing stakeholder value while reducing waste by emphasizing collaboration between teams on a project. The goal of Lean construction is to increase productivity, profits, and innovation in the industry.[9]

Project Management That Solves Problems

Lean delivery aims to solve these problems within the design and construction industry. By using a Lean operating system, designers and builders are able to deliver projects on time and within their budgets, minimize waste and rework, improve safety, and achieve higher levels of customer satisfaction.

The Lean Construction Institute (LCI) believes that Lean can be implemented into any project business plan through focusing on the six tenets of LCI:

1. Respect for people
2. Removal of waste
3. Focus on process and flow
4. Generation of value
5. Continuous improvement
6. Optimize the whole

Lean construction is about elimination of overloading, uneven loads, and waste.

---

8 https://theleanway.net/muda-mura-muri
9 https://leanconstruction.org/lean-topics/lean-construction/.

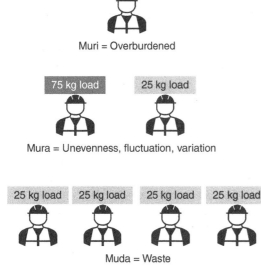

Muri = Overburdened

Mura = Unevenness, fluctuation, variation

Muda = Waste

No Muri, Mura, or Muda

Figure 11.4 Relationship between Muda, Mura, and Muri. The objective is to load 4 bags of cement, 25 Kg each (total 100 Kg). The normal carrying capacity for a worker is 50 Kg.

What is the Last Planner System?

The Last Planner System (LPS) is a system for project production that promotes the creation of a predictable workflow among various parties so that it achieves reliable results. The LPS® allows potential hurdles to be found and addressed before they slow down the flow.

The LPS focuses on generating and maintaining flow in work processes by promoting conversations between construction team members so they can identify problems before those issues interrupt the flow of work. This is done through pull planning.

Main Steps in Introducing a Quality Assurance (QA) System

As explained earlier, quality assurance is a system that includes the entire process in order to assure the required quality specifications in the final product. Such QA system must:

1. Establish awareness of quality
2. Develop quality system

3. Introduce the system
4. Evaluation of the system

Any system has to be well-designed, well-introduced and taught to employees, and well-enforced.

> Quality assurance is a comprehensive process, not just inspection of the final product. In fact, by carrying out efficient QA system, we are minimizing the chance of any defects in the final product.

The preparation of the project quality plan includes:

1. Policy and company profile
2. Organization and responsibilities
3. Procedures
4. Method statements and work instructions
5. Inspection and test plans

Inspection and Test Plans

An essential feature of QA is the collection of data. Inspection plans are lists of checkpoints for specific work items. The inspection plan is a table, typically listing:

1. *Work item*, e.g., "concrete slab pour"
2. *Who is doing the inspection*, e.g., "site engineer"
3. *According to what*, e.g., "specifications/drawing XX"
4. *Frequency of the inspection*, e.g., "every pour"
5. *Criteria for acceptance*, e.g., $\pm 5\,$mm

Quality Tools for Measure and Analysis

There are seven basic tools used by quality project management professionals for planning, monitoring, and controlling issues related to quality. It is also known in the industry as 7 QC Tools, are used within the context of the PDCA[10] Cycle to solve quality-related problems[11]:

1. Cause-and-effect diagrams
2. Flowcharts
3. Check sheets
4. Pareto diagrams
5. Histograms
6. Control charts
7. Scatter diagrams

---

10 PDCA stands for Plan-Do-Check-Act.
11 https://www.projectmanagement.com/contentPages/wiki.cfm?ID=460921&thisPageURL=/wikis/460921/The-seven-basic-quality-tools#_=_.

Check sheets, also called control sheets, are simple tally sheets that are used as a checklist to gather data. According to the American Society for Quality, a check sheet is a structured and prepared form for collecting and analyzing data. This is a generic data collection and analysis tool that can be adapted for a wide variety of purposes and is considered one of the seven basic quality tools[12]. There are two types of check sheets often used:

1. Defective item check sheets listing number, category, location, and cause of defect items. Many types used in factories, but not used in construction.
2. Inspections check sheets to make certain that work has been carried out correctly. List of items that are checked and approved by inspection person (e.g., site engineer).

In construction projects, two types of check sheets are often seen:

1. During construction check sheets: These are filled out on-site.
2. After substantial completion: "Snag lists" or "punch list." These are lists of minor outstanding items created when facility is handed over to the client. When all snags are rectified, the facility can be handed over.

Nonconformance: A nonconformance report (NCR) is a document to manage works that have been performed not in accordance with specifications. ISO 9000 requires documented procedures for handling non-conforming product, and the NCR is the typical instrument in construction.

Quality Records

Evidence documents that show how well a quality requirement is being met or how well the QA system is performing. For construction, we typically include:

1. Filled-out check sheets
2. Daily diary
3. Concrete test records
4. Closed-out non-conformance reports
5. Rectified snag list

Review of Shop Drawings

Shop drawings as drawings, diagrams, schedules, and other data specially prepared by a distributor, supplier, manufacturer, subcontractor, or contractor to show some part of the work. Submittals, which include shop drawings and other items and administrative documents, are how the contractor communicates what it intends to construct or what it or its subcontractors have designed and planned to construct. This is important especially when the subcontractor or fabricator modifies the original design. Shop drawings are typically required for prefabricated components. The review and approval process for shop drawings formalizes the method for a contractor to demonstrate how it will accomplish these

---

12 https://asq.org/quality-resources/check-sheet.

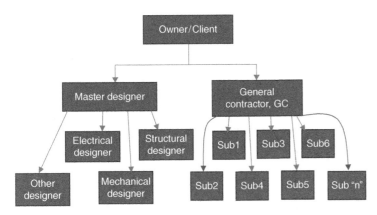

Figure 11.5 The traditional arrangement of contractual responsibilities.

design obligations. The submittal process also allows the design professional to review the design drawings and other documents and make sure they comply with the design intent.

The big question is: Whose responsibility is it to review shop drawings? The answer is: The master designer,[13] but it also depends on the contractual agreement. In the traditional arrangement, as shown in Figure 11.5, the owner has two contracts: one with the master designer and one with the general contractor (GC). Any submittal by the GC or its subs, fabricators, and vendors, has to go to the master designer through the owner. This communication protocol may be elaborated or modified in the contract, but the responsibility will still lie in the master designer's lap. The master designer then usually forwards it to the specialty designer who originally designed that component.

If it is a design–build contract, shop drawings review will be an "internal matter" within the design–build organization since both the designer and builder are under one entity, which will be responsible for the review.

To illustrate the importance of shop drawings review, we take a quick look at the Hyatt Regency, Kansas City, Missouri, Walkway Collapse case on July 17, 1981. The disaster happened when two suspended walkways collapsed during a dance festival, leaving 114 people dead and more than 200 injured. The original design called for one cable stretching from the ceiling (sixth-floor level) to the bottom of the second-floor skywalk, as shown in Figure 11.7a. The fabricator proposed an alternate design with a splice in the mid-level, which is the bottom of the fourth-floor skywalk, as shown in Figures 11.6 and 11.7b. The splice, which was approved without sufficient analysis, was inadequate. The skywalks performed normally for just over a year (the hotel was opened on July 1, 1980), but as mentioned earlier, structures are tested under extreme conditions. This time, there was a music festival on the ground floor, and the audience was on both skywalks, in addition to the ground floor. Not only there was a heavier than usual load on the skywalks, but the dynamic load resulting from the dance of the people exacerbated the impact. This loading, which was within the original design criteria, exposed the inadequacy of the splice joint, and led to its collapse.

---

13 The term master designer refers to the entity responsible for the overall design of the project. In most cases, the architect is the master designer for vertical construction projects, while the civil engineer is the master designer for horizontal construction projects.

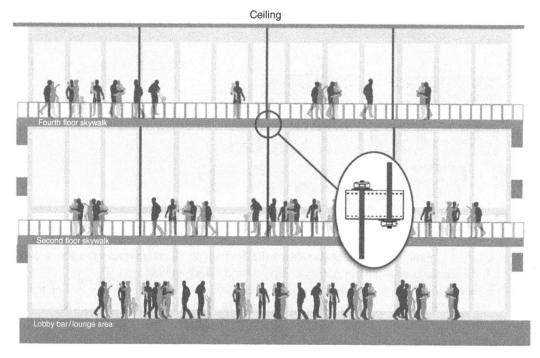

Figure 11.6   The Walkways of the Hyatt Regency, Kansas City, MO, before the collapse.

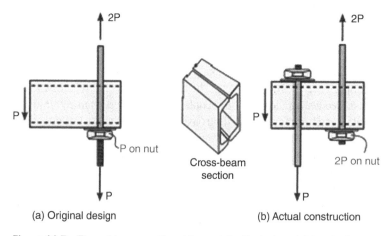

Figure 11.7   The cable connection: (a) as originally designed, (b) as built.

The consequences were dire for many people. The design company lost its engineering licenses in Missouri, Kansas, and Texas, and lost its membership with the American Society of Civil Engineers.[14]

---

14 For more details, refer to Hyatt Regency walkway collapse – Wikipedia.

Cost of Quality

1. Cost of conformance, which is basically the cost of the company's quality efforts, including both appraisal and prevention costs. Cost of conformance includes:
   1. Inspection of direct hire and subcontractor work
   2. Inspection at vendor source of supply
   3. Inspection of shipments
   4. Review of shop drawings
   5. Training costs
   6. Facilitator costs
   7. Salaries of quality staff
   8. Meetings of the steering committee and quality improvement teams
   9. Administration of the quality management program

2. Cost of nonconformance, which combines internal and external failures. Contractors pay a significant price for poor quality resulting from accidents, waste, rework, inefficiencies, poor subcontractor performance, and poor communication. These costs are estimated to be between 5% and 30% of the construction cost of a facility, not to mention the schedule delays. In addition, there are intangible "hidden" costs such as lost sales due to low customer loyalty.

The total cost of quality is the addition to the proactive portion (conformance and appraisal) and the reactive portion (repair, replacement, and rework). Figure 11.8 shows the curve of both types of cost and

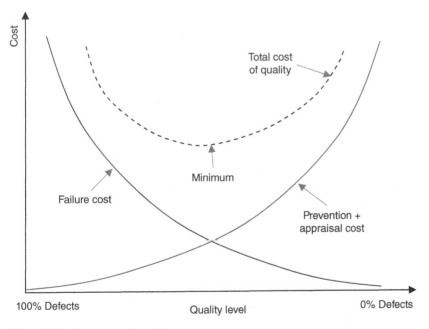

Figure 11.8 Total cost of quality.

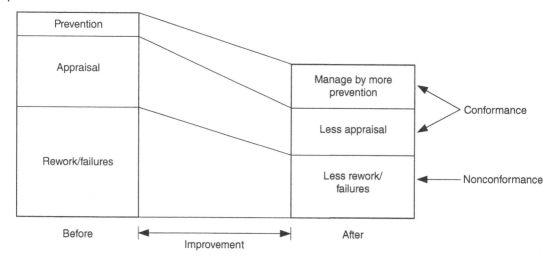

Figure 11.9 Total cost of quality before and after improvement.

the curve representing the total cost. It shows the point where the total is minimum. Figure 11.9 shows the total cost of quality, divided by category, before and after the improvement. While the cost of prevention may slightly increase, the costs of both appraisal and failure/rework will decrease.

Is Quality Expensive?

The short answer is no. In fact, the concept of TQM was originally built on the assumption that improving quality would reduce expenses. We can think of the cost of not having adequate quality. This will likely include the cost of rework in terms of money, time, reputation, and morale. Some business owners, including contractors, think that by cutting corners and compromising quality, they can save money. They take a gamble that their product will not fail, but the odds are against them. Sooner or later, the product will fail, and this may cost them more than what they saved earlier. This is what the author calls the "Unbuckled Seatbelt Paradigm" syndrome. This is when people do not do things right and hope that "things will be okay." In one of the author's observations of a newly built upper-scale shopping mall, it looked so nice and fancy; however, when heavy rain fell, the building had water leaks everywhere. Some people said, "the leaks happened because of the heavy rain." In fact, the leaks happened as a result of poor quality. Thanks to the heavy rain that helped expose it! Now, it will cost them many times what would have cost them originally if they did it right.

In another case study the author witnessed a large 2-story house was built with 2-ft floor trusses carrying the second floor. The trusses were similar to the one shown in Figure 11.10a but had plywood on one side as shown in Figure 11.10b for added strength.

The plywood panel on one of these trusses had a small "bump," less than an inch, at the top side of the truss. This bump happened to be under the master bathroom on the second floor. It is surprising that none of the crews who worked on the bathroom (framing, floor tile, plumbing, and shower door) observed or reported the issue. Also, none of the supervisors and inspectors noticed this issue. The

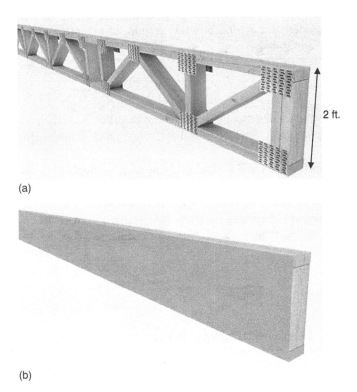

(a)

2 ft.

(b)

Figure 11.10 (a) Floor truss without the sideboard, (b): floor truss with the sideboard.

owner later noticed that the shower door was not plumb. He contacted the GC, who sent the subcontractor who installed the door to check the issue. The sub said the door was installed right but the floor was not level. It took the GC several investigative visits and finally, they had to tear down the entire bathroom and its floor to find that protruding piece of plywood. It took only seconds to cut it out and make the truss level, but it took a lot of work and expense to redo the bathroom and correct this quality defect. In addition to the thousands of dollars in losses to the GC, there was tremendous inconvenience and frustration to the owner. This is an example of a minor quality defect that resulted in high cost. The emphasis here is on the fact that all of these negative consequences could have been prevented by applying simple quality management processes.

Many of the quality defects, like the case just mentioned, start very insignificant but escalate and cause tremendous damage and cost, not to mention the negative impact on morale, relationships, and loss of use for part or all of the project. Take an example that happens frequently: a minor defect in the roof of a house that allowed rainwater to leak. This leak was not detected until it caused damage to the roof and ceiling elements and perhaps more.

> Heavy rain, wind, loading, or other extreme conditions, within design criteria, must not be thought of as causes for failure or faults. They are rather the real test for the quality of products and services.

> If you compromise on quality requirements, whether materials or workmanship, for monetary gain, you are likely to pay several times that gain later on!

Construction industry facts about quality[15]:

- It is estimated only 10–15% of GC have a formal written quality management program.
- It is estimated only 5–10% of trade subcontractors have a formal written quality management program.
- Federal government work generally requires that GCs and all trade subcontractors comply with the project-specific QC requirements and a full-time QC manager.
- The Construction Industry Institute (CII) has reported that the average contractor makes 10 errors out of 100 items – the best contractor made 1 error out of 200 and the worst made 19 errors out of 100.
- Customers who implement a robust quality management program can reduce their error rate significantly.

Quality Improvement

A research study by the CII has key findings and implementation tools. Here are some of the findings of the CII study[16]:

1. Demonstrated Management Commitment: If management is committed to rework reduction by putting time and resources into an effective system, then rework will decrease.
2. Staffing for Quality: Providing committed and professional staff for training, inspection, and other quality functions.
3. Quality Pre-Project and Pre-Task Planning: Planning, at the project and activity levels, is the key to minimizing mishaps and maximizing productivity. It must be conducted before mobilization, including constructability analysis, and also project-specific quality and safety plans.
4. Quality Education: Increasing training on quality issues, identifying quality rework problem areas, increasing full-time quality staff, and having field personnel analyze pre-task quality efforts all contribute to less rework.
5. Worker Participation and Involvement: The involvement of workers at all levels is needed to resolve the continuing problem of rework.
6. Subcontract Quality Management: It is recommended, if not required by the contract, that subcontractors comply with the general/prime contractor safety and quality plans. Some projects require that the subcontractor submit their own safety plan for the project. Quality planning should follow a similar pattern. Prequalification of subcontractors by the GC for safety and quality is also recommended.

---

15 CII "Making Zero Rework a Reality"– November 2005, https://www.construction-institute.org/resources/knowledgebase/best-practices/quality-management/topics/rt-203/pubs/rs203-1.
16 Do It Right the First Time (Best Practice), RT-203 Topic Summary, CII - Topic-Summary-Details (construction-institute.org).

7. Quality Rework Tracking: Tracking and investigation of rework or defects is very helpful for the improvement of both safety and quality. The contractor must investigate any issues, defects, rework, quality errors and near-errors, and accidents and near-misses. The contractor must take steps to avoid repeating the incident.

8. Drug and Alcohol Testing Program: The contractor must make sure that the workplace is free from any substance such as alcohol and drugs. Moreover, the contractor must use all means to observe and remove any onsite worker who might be under the influence.

9. Contract Type: Lump-sum contracts assign most of the risk to the GC, so it was observed that such contracts have less rework than reimbursable contracts. However, the choice of the contract type is subject to other factors.

10. Zero Field Rework Self-Assessment Opportunity Checklist: This checklist can assist in identifying areas for improvement to further strengthen a site construction quality process on the journey to zero field rework.

What is interesting is the link between quality and safety management. As we all know, accidents cause human and property losses. The lack of following safety procedures, regarding the worker, machine, or environment, may lead to both safety and quality issues.

Quality Management and Project Success

Quality management is critical to the success of construction projects. By implementing effective quality management practices, construction professionals can ensure that projects are completed on time, within budget, and to the satisfaction of all stakeholders. Furthermore, effective quality management can help minimize project risks, increase customer satisfaction, and enhance the reputation of the organization.

Exercises

11.1 Define quality in the general meaning and then specifically in construction projects.

11.2 What is the "golden triangle," and how does quality interact with the other components?

11.3 What is the concept of TQM? How does it work?

11.4 What is the quality management plan in construction, and what are its elements?

11.5 Define quality assurance (QA) and quality control (QC). Mention the role of both the owner and the contractor in a typical QA/QC program in construction.

11.6 Meet with a construction project manager. Ask them to tell you about a quality issue they had in this or previous project.

a. If that experience was negative, what did the project manager learn from it? If they had the same issue today, what would they do differently?

b. If that experience was positive, what made it positive? What would have happened if they had chosen an alternative action?

11.7 Who sets quality standards for a construction project? Discuss the roles of the owner, designer, and contractor in this effort.

11.8 What is Kaizen? Why do we need continuous improvement?

11.9 What is the concept of lean production? How do we help make work more efficient with no overloading, uneven loads, or waste?

11.10 What are the six tenets of the Lean Construction Institute (LCI)?

11.11 Briefly describe the inspection plan, both format and content.

11.12 What are check sheets? Mention their types.

11.13 What do quality records typically include in a construction project?

11.14 Discuss the importance of reviewing shop drawings, particularly when the matter relates to safety.

11.15 Research a case where failure to properly review shop drawings resulted in either an accident, defects, or other negative consequences. (Your case must be different from the one mentioned in the chapter.)

11.16 Talk to a contractor about lack of quality in a construction project and its consequences. Summarize the findings.

11.17 What are the items included in the cost of quality, both conformance and nonconformance.

11.18 You are a general manager for a construction company. Your company suffers from the high cost of rework and repair. You are making an improvement plan. Which expenses do you expect to reduce: prevention, appraisal, and/or rework and repair? Explain your answer.

11.19 How do you respond to the CEO or manager of a construction company who refuses to approve a quality improvement plan because it is "too expensive"?

11.20 We discussed the cost of quality: conformance and nonconformance. Do you believe there is a relationship between quality and time (schedule)? Explain your answer.

12

Project Health, Safety, and the Environment (HSE) Management

Introduction

The construction industry comes with many hazards such as falls from heights, getting hit or crushed by moving equipment, injuries from sharp tools, falling objects, chemicals, explosives, electrocutions, or fires. Some injuries and health issues may result from exposure to excessive noise, dust, vibration, heat, repetitive motion, chemicals, heavy loads, and other health hazards. This includes everyone on the project site, including workers, management, and even visitors. These hazards are exacerbated by the fact that construction work is performed mostly outdoors, and subject to all weather conditions. Every year over 1000 construction workers die in work-related accidents,[1] not to mention many more injured, some with permanent disabilities.

Construction operations may also have a negative impact on the environment by spilling or dumping chemicals and other hazardous or nonbiodegradable waste, which may impact the land, vegetation, habitat, and water sources. Taking care of the job site's safety and environment is a legal, ethical, and professional obligation. Not only that but violations of safety and environmental regulations are likely to result in financial losses to the contractor, both directly and indirectly.

Differentiating between health, safety, and the environment (HSE) management and risk management is important: HSE management focuses on protection of people, property, and the environment in and around the project site. Risk management, on the other hand, focuses on managing events that may impact project objectives, mainly cost and schedule. There could be an overlap between the two areas of project management. A hurricane, fire, or earthquake, for example, may pose risk to people and property, and it may impact the project objectives. Such an event must be covered by both management areas but with different perspectives.

Definitions

HSE is an acronym for Health, Safety, and the Environment. The three aspects have been combined in most construction projects under one management. Sometimes, the acronym is written as HSSE, with

---

1 The number of construction fatalities in the past three decades peeked at 1,297 in 2006, started declining to 781 in 2011, and then started going up again to 1,034 in 2020 and 1,015 in 2021.

Project Management in the Construction Industry: From Concept to Completion, First Edition. Saleh Mubarak.
© 2024 John Wiley & Sons, Inc. Published 2024 by John Wiley & Sons, Inc.
Companion Website: www.wiley.com/go/nextgencpm

the second "S" stands for security. The objective of the HSE/HSSE management in a construction job site is to create a safe and healthy environment for everyone and everything.

OSHA is an acronym for Occupational Safety and Health Administration, as part of the executive branch of the government of the United States of America. It was enacted by the US Congress in 1971. OSHA has helped transform America's workplaces in ways that have significantly reduced workplace fatalities, injuries, and illnesses.[2] Over the past five decades, OSHA has developed more governance and issued more regulations. Also, the US Congress has given OSHA more powers by significantly raising the fine limits and giving OSHA officers more powers to protect the safety of workers and others on the job site. Willful and major OSHA violations may lead to criminal charges against the employer (contractor, in our context), which can lead to harsher penalties than a simple fine.[3,4,5] OSHA's jurisdiction covers all workplaces of all industries.

The creation of OSHA was a major step in the right direction for the protection of workers' safety and health in the workplace. OSHA's regulations combined with Workers' Compensation insurance, provide strong incentives and deterrents for employers to provide safe workplace and prevent death, injuries, and work-related illnesses to workers. Workers also have the obligation to follow safety measures, and the right to refuse work if deemed unsafe.

EPA stands for the Environmental Protection Agency, which is another governing body, both at the federal and state levels. EPA was created at the federal level on December 2, 1970, to protect human health and the environment.[6] Every state (in the United States of America) has its own department, agency, or commission to protect the environment, but subject to federal regulations. Every construction project is subject to both federal and state EPA rules and regulations, in addition to any other local regulations.

Safety

Construction site safety is an aspect of construction project management, concerned with protecting site workers and others, on- and offsite,[7] from death, injury, disease, or other health-related risks. Protection takes several forms, as will be detailed later, mostly preventative but also reactive, sometimes simultaneously. For example:

A. The contractor has to take all measures to prevent a worker working on scaffolding or other heights from falling.
B. If, however, the worker falls down, he/she must be protected by fall-protection gear/safety harness, so he/she does not get hurt.

---

2 OSHA at 50 | Occupational Safety and Health Administration.
3 Resurgence in Criminal Prosecutions for OSHA Workplace Safety Violations - Spencer Fane LLP.
4 Federal Employer Rights and Responsibilities Following an OSHA Inspection-1996 | Occupational Safety and Health Administration.
5 Historical Notes | Occupational Safety and Health Administration (osha.gov).
6 The Origins of EPA | US EPA.
7 Other than workers, safety protection must cover all those onsite, including employees, visitors, and even those offsite who may be impacted by the construction work by incidents such as flying and falling objects, noise, dust, hazardous fumes, or others.

C. If, however, the worker does get injured, the contractor must react promptly, both with on-site emergency first-aid and by getting prompt medical assistance.

We mentioned safety in Chapter 2, Architecture, Engineering, and Construction, in the context of design safety, which applies to the project (structure) during its service life. However, safety in this chapter applies to the work during the construction phase. It includes all the practices and conditions that may affect the safety of people and property, onsite and even offsite if caused or impacted by the construction activities.

> Safety in construction projects is a comprehensive process, both proactive and reactive. The contractor must make sure everything onsite, including materials, equipment, structures, and practices, is safe and compliant with legal safety requirements. Also, the contractor must be ready for any safety incident that may happen.

OSHA's role and powers have increased since its inception with several amendments.[8,9] For example, OSHA issued Hazard Communication Standard (HCS) in 1983 (updated later), giving workers the right to know which chemicals they may be exposed to, and requiring employers to provide workers with medical and exposure records. These standards came on the heels of a landmark decision by the U.S. Supreme Court, affirming that workers have the right to refuse unsafe tasks.[10] In the 1990s, OSHA issued the Process Safety Management standard, which provided new and stronger protections for workers from falls, bloodborne pathogens, toxic substances, and working in confined spaces, longshoring and marine terminals, and laboratories. In 2020, OSHA launched a historic response to protect workers during the COVID-19 pandemic. In July, 2023, legislation (S. 2504/H.R. 4897) was introduced, but not yet approved, directing OSHA to issue an enforceable heat standard for workers exposed to high temperatures.

Safety violations currently are classified by OSHA into four tiers[11]:

1. WILLFUL: A willful violation is defined as a violation in which the employer either knowingly failed to comply with a legal requirement (purposeful disregard) or acted with plain indifference to employee safety.
2. SERIOUS: A serious violation exists when the workplace hazard could cause an accident or illness that would most likely result in death or serious physical harm unless the employer did not know or could not have known of the violation.
3. REPEATED: A Federal agency may be cited for a repeated violation if the agency has been cited previously for the same or a substantially similar condition and, for a serious violation, OSHA's regionwide (see last page) inspection history for the agency lists a previous OSHA Notice issued within the past

---

8 Historical Notes | Occupational Safety and Health Administration (osha.gov).
9 OSHA at 50 | Occupational Safety and Health Administration.
10 Whirlpool Corp. v. Marshall :: 445 U.S. 1 (1980) :: Justia US Supreme Court Center.
11 Federal Employer Rights and Responsibilities Following an OSHA Inspection-1996 | Occupational Safety and Health Administration.

five years, or, for an other-than-serious violation, the establishment being inspected received a previous OSHA Notice issued within the past five years.

4. OTHER-THAN-SERIOUS: A violation that has a direct relationship to job safety and health, but is not serious in nature, is classified as "other-than-serious."

Recordkeeping: There are certain forms required by OSHA such as[12]:

- OSHA Form 300 is the log of work-related recordable injuries and illnesses.
- OSHA Form 300A is the summary of work-related injuries and illnesses.
- OSHA Form 301 is the injury and illness incident report.

Health

The "health" aspect in HSE deals mostly with any action or situation that may impact health, in direct or indirect way, on a short- or long-term basis. It may be considered the safety side that is "other than accidents." This includes factors that, if persist, will cause an illness or health risk to the workers such as work under extreme conditions (excessive heat or cold, exposure to dust or toxic vapors and gases, high noise, repetitive vibration, or radiation), carcinogen materials, fatigue, work with unergonomic positions, and handling heavy loads.

The Environment

The environment on and around the job site must be protected against negative action or impact such as pollution (to land, air, water resources, or others), unlawful removal of vegetation, endangering wildlife, harming historical sites, unlawful dumping of waste materials, unlawful use of or tampering with raw materials such as soil or water, unlawful burning, unlawful changing the status of wetland, or subversion in any other way. The construction operations must be sustainable, which means leaving the environment in and around the job site after the completion of the project, same as or better than it was before the operations.

Project HSE Management

An HSE management program is an integrated approach where health, safety, and environment are effectively managed to reduce hazards in the workplace. The objective of a safety management program is to provide a structured management approach to control hazards. The term *control* has been used throughout project management in different contexts. Controlling risks or hazards in the HSE context means taking all the necessary steps, proactively and reactively, to eliminate, minimize, or otherwise manage operations, objects, situations, actions, and events that may cause harm to people, property, and the environment, as a direct or indirect result of the construction operations.

---

12 Recordkeeping - Recordkeeping Forms | Occupational Safety and Health Administration (osha.gov).

The key elements of a successful HSE management system include:

a. Safety plan and inspection checklists
b. Risk[13] assessment and monitoring
c. Reporting and documentation
d. Training and induction

An HSE plan is a document that includes:

1. An establishment of the HSE management system
2. Implementation of the HSE policy, and
3. Achievement of the HSE objectives effectively

It is similar somewhat to cost and schedule management: We create the plan (cost or time) before the start of the project. We then implement and execute the plan with continuous monitoring and comparison between the actual and the plan. We detect and analyze variations and make any necessary adjustments.

Safety is Everyone's Responsibility

Unlike other project management areas, safety is not subject to joggling, compromise, or even economic valuation. It is unacceptable in emergency situations to delay action or even wait for supervisor's orders. It is unacceptable to blame someone else when you can prevent or eliminate a danger, even when this someone has direct responsibility. Safety always takes precedence over any other activity or consideration. Of course, with responsibility usually proportional to authority, those in charge of safety management are responsible first, but we need to emphasize that this does not mean others have no responsibility. In general, these are the key principles of health and safety:

- Employer is responsible for maintaining a safe and healthy workplace.
- Employees should be involved in developing policies and programs.
- There should be no sanctions for health and safety-related activities.
- Employer should implement the best and most effective practices/policies to protect workers from hazards.
- Accidents are preventable!

Common Causes of Construction Accidents[14,15]:

Construction accidents happen for many reasons. Some of the top causes of injuries on construction sites include the following:

1. *Slips, trips, and falls*: Slips and falls are common on construction sites. They often cause injuries, including broken bones and sprains. Some falls are fatal, especially when falling from great heights. Falls can happen from scaffolding, ladders, window openings, and roofs or elevator shafts. Tripping is always possible because the construction site is usually unleveled ground with objects, clutter, and cords on the ground. Securing the pathways from any trip hazards is very important.

---

13 Risk in this context means hazard to safety of people and structures.
14 Construction Accident Causes, Injuries & Legal Options – Forbes Advisor.
15 10 Common Injuries And Accidents On Construction Sites – My Case Helper.

2. *Being struck by falling or flying objects.* The extent of the injury depends on the nature of the object (weight, density, and shape), the distance it fell, and the way it struck the worker, including the part of the worker's body. In some instances, it can trigger more falling objects and persons' falls (domino effect).

3. *Electrical Hazards* (Electrocutions): Working with a building's electrical wiring and systems puts construction workers at risk of electric shock. The likelihood of such accidents increases in remodeling and renovation (existing) buildings. Electrocutions alone are very painful and often cause horrible burns. Safety guidelines include items such as shutting off the power before working on an item, testing for live wires, and grounding. Electrocution may also happen because of the lightning, both if the lightning hits the worker directly or hit a structure, equipment, or object connected to the worker with conducting material. This is why insulation is important. Also, stopping the work on site completely mainly outdoors, is a must when lightning starts.

4. *Getting caught in or between objects.* Construction workers can suffer injuries if they get caught in between or underneath materials, equipment, walls, or other objects. This can happen if the worker falls in between objects, or the object falls on the worker. The injury may be broken bones, bruises, lacerations, or suffocation (lack of oxygen).

5. *Equipment and power tools accidents.* Power tools such as power saws, jackhammers, grinders, drills, nail guns, or trowels, create a major risk to all construction workers. There are lots of mishaps that can happen while operating these tools. Accidents happen from faulty equipment, negligent operators, or improper training. Construction workers often lose fingers and limbs from such accidents.

6. *Trench and ground collapse accidents*: Trench accidents can be the most terrifying for construction workers. Getting trapped in a collapsed trench can literally bury a worker alive. Workers in collapsed trenches, or who have fallen into trenches can be crushed or suffocated. OSHA requires employers to provide ladders, steps, ramps, or other safe means of egress for workers working in trench excavations 4 ft (1.22 m) or deeper.

7. *Vehicle accidents* and accidents with moving equipment. In construction sites, drivers/operators of moving equipment such as excavators or forklifts, can hit workers standing or walking in the path of the equipment. They can also hit objects or structures or even fall in ditches, ponds/lakes, or other low areas when not paying attention. The chances of such an accident may increase in crowded and high-traffic / high-noise areas. In construction work on a highway, there is also a risk of being struck by oncoming traffic.

There are many safety measures to prevent such accidents, including equipment sensors and alarms, personal protective equipment (PPE), good site planning, and signage.

8. *Chemical hazards / exposure to toxic chemicals.* Getting exposed to chemicals may have risks to the respiratory system, skin, eyes, and other parts of the body. Toxic chemicals and substances can create some of the worst long-term effects on construction workers. Employers must provide construction workers with proper personal safety equipment such as hard hats, glasses, gloves, masks, and safety boots. In some instances, a protective suit may be required. OSHA has regulations regarding "Lockout/ Tagout," which is a safety procedure that prevents the accidental release of hazardous energy from machines or equipment during maintenance or repair.[16]

---

16 FactSheet3 (osha.gov).

Also, in the past, some building materials contained compounds that were later found to be carcinogenic such as asbestos and lead-based paint. In the past few decades, construction workers worked on abatement of old buildings, either by permanently removing known hazardous material in surface areas or by sealing off hazardous materials (encapsulation), so that it is safe for everyone.

9. *Repetitive motion / vibration / repetitive stress injury*: Some types of construction work, especially labor-intensive, have higher chances of injury that may not surface till days, weeks, or even months later. Common stress injuries include back, leg, and arm strains, as well as rolled joints.

10. *Welding accidents* can cause severe burns, as well as injury to the eyes. The welder, who must be wearing PPE, but protection must include other workers who may come by or pass close to the welding position without such protection.

11. *Crane and hoist accidents* can be very serious, especially large cranes and those with extra-long booms. The accident may happen as a result of the crane collapse or tipping, or if the hoisted object gets loose and falls. Crane can tip because of unstable ground, heavy load,[17] operator error, or mechanical failure. Wind can pose a danger to cranes, so operators have to make careful decisions when operating under wind. The decision depends on the wind strength and direction, as well as the size and weight of the object to be hoisted. Crane accidents kill and injure many people, including those outside the construction site. Large cranes, especially with their long boom, are more vulnerable to lightning strikes, as mentioned earlier in electric hazards.

12. *Fires, explosions, and gas leaks*. Construction sites commonly use toxic chemicals and gases. Leaks, fires, and explosions happen frequently on construction sites, causing injuries such as burns and lacerations. Carbon monoxide can be fatal if not detected early. Chlorine gas is heavier than air, so the leak of large amounts of it can suffocate people nearby. Explosives are used for road and tunnel construction, as well as excavation and demolition projects. They have to be handled by experts and stored in a safe place to avoid accidental explosions. Every site must have fire extinguishers that are suitable for any type of fire, including chemicals.

13. *Structural collapse*: This can happen to both permanent and temporary structures. In situations where the contractor does not provide sufficient strength to scaffolding, walkways, and other temporary structures, or if the load is heavier than originally planned, a collapse may happen causing fatalities, injuries, property damage, and delay. Placement of the concrete mix with a crane and bucket on the formwork of a suspended slab may cause such collapse if not spread out instantly.

14. *Overexertion* due to work under extreme temperatures, whether high or low, or extended work hours. Working under high temperatures may cause heat strokes, while working under low temperatures may cause hypothermia. Extreme cases may cause death.

15. *Noise pollution injuries*. Getting exposed to high noise for extended periods, without using ear protection, may and usually does cause hearing loss among other issues. Some operations are particularly noisy such as explosions, driving piles in the ground, and using jackhammers. Some construction sites have quarries to crush stones, which also can produce high noise. Loud noise can create physical and psychological stress, reduce productivity, interfere with communication and concentration, and contribute to workplace accidents and injuries by making it difficult to hear warning signals.[18]

---

17 The load capacity has to do with the "arm length," which is the horizontal distance from the load center of gravity to the center of the crane tower.

18 Occupational Noise Exposure - Health Effects | Occupational Safety and Health Administration (osha.gov).

In some cases, depending on the site location, there can be other risks such as attacks by wild animals and insects and reptile bites. In unstable regions, there can be risks from wars, terrorism, armed robberies, tensions with certain local parties, or other sources. Some of these risks may need to be handled by the project security management, which is discussed later in this chapter.

There are many other types of accidents that may be classified under one or more of the types mentioned earlier. For example, a demolition accident may be a combination of types of numbers 2, 4, 6, 12, and/or 13. Also, one accident may trigger a series of accidents. For example, a flying object may hit a worker and cause him to fall or even cause a structural collapse, if the object is heavy enough. Also, a fire accident can cause a structural collapse.

We can classify hazards in construction projects into these types:

1. Safety hazards such as slips, trips, and falls, or faulty equipment.
2. Physical hazards such as noise, temperature extremes, or radiation.
3. Work organization hazards such as conditions that cause stress and fatigue.
4. Ergonomic hazards such as repetitive movement, lifting, or awkward postures.
5. Chemical and dust hazards such as all types of dust, vapors, fumes, and materials such as cleaning products and pesticides.
6. Biological hazards such as mold, insects/pests, and communicable diseases.

Hierarchy of Controls[19]:

The Hierarchy of controls is a systematized approach to protect people in the workplace of an organization. It has five levels of actions to reduce or remove hazards, as shown in Figure 12.1, with elimination being the most effective and PPE being the least effective. It includes:

1. Personal protective equipment, PPE is equipment worn to minimize exposure to hazards such as gloves, hard hats, safety goggles, ear plugs, respirators, safety vests, safety boots, welding shields, and face masks. The choice of PPE depends on the type of work the employee is performing. The PPE program should address:
   • Workplace hazards assessment
   • PPE selection and use
   • Inspection and replacement of damaged or worn-out PPE
   • Employee training
   • Program monitoring for continued effectiveness

Employers should not rely on PPE alone to control hazards when other effective control options are available. PPE can be effective, but only when workers use it correctly and consistently. PPE might seem to be less expensive than other controls but can be costly over time. This is especially true when used for multiple workers on a daily basis.

2. Administrative controls: Establish work practices that reduce the duration, frequency, or intensity of exposure to hazards. This may include:
   • Work process training
   • Job rotation
   • Ensuring adequate rest breaks

---

19 Hierarchy of Controls | NIOSH | CDC.

- Limiting access to hazardous areas or machinery
- Adjusting line speeds

3. Engineering controls: Requires a physical change to the workplace to reduce or prevent hazards from coming into contact with workers. Engineering controls can include modifying equipment or the workspace, using protective barriers, ventilation, and more. The most effective engineering controls are part of the original equipment design. They:
   - Remove or block the hazard at the source before it comes into contact with the worker.
   - Prevent users from modifying or interfering with the control.
   - Need minimal user input for the controls to work.
   - Operate correctly without interfering with the work process or making the work process more difficult.

   Engineering controls can cost more upfront than administrative controls or PPE. However, long-term operating costs tend to be lower, especially when protecting multiple workers. In addition, engineering controls can save money in other areas of the work process or facility operation.

4. Substitution is using a safer alternative to the source of the hazard. An example is using a trenching machine instead of having workers manually excavate the trench or using a mechanical scaffolding system instead of plywood boards over wood frame.

   When considering a substitute, it is important to compare the potential new risks of the substitute to the original risks. This review should consider how the substitute will integrate with other elements in the workplace. Effective substitutes reduce the potential for harmful effects and do not create new risks.

5. Elimination removes the hazard at the source. This could include changing the work process to stop using a toxic chemical, heavy object, or sharp tool. It is the preferred solution to protect workers because no exposure can occur.

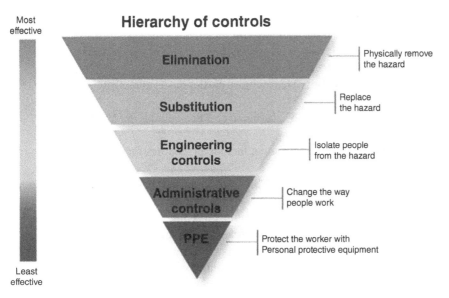

Figure 12.1 Hierarchy of Controls according to the National Institute for Occupational Safety and Health (NIOSH).

Preparing a Safety (or HSE) Management Plan

It is important to define the safety plan and safety policy, and to differentiate between them.

Safety plan is the comprehensive roadmap to achieve the goal while the policies are the guidelines and rules that the authority uses in the implementation of the plan. The plan may come in phases with interim goals, where the policies apply in their entirety but may change or be updated periodically.

A project safety plan in construction is a formal written document that outlines safety procedures, rules, and regulations, to help identify potential hazards and mitigation procedures. The objective of this document is to protect workers from injuries and accidents during the execution of the project. Protecting property is secondary objective of the plan, after protecting people. In most cases, the two objectives are linked.

The plan must be written by experts based on the requirements of the contract and applicable laws, as well as the project conditions and requirements. It must be drafted before the construction project starts; however, it should be updated whenever necessary. Such updates may be necessitated by changes in the law, the project conditions, methods used, or any other factor.

The safety management plan must proceed in a sequence of steps: draft, review, finalize, formalize, train, implement, enforce, monitor, document, report, update, closure, evaluation, and feedback.

The contractor may start with the plan used for a previous project, but it must be adjusted and updated to be compatible with the project and contract requirements, as well as applicable laws and other conditions. The project safety plan must also be consistent with the company's (or organization's) safety policy, which can be more stringent than the law.

Some of the common elements included in a project safety plan consist of the following:

1. The plan goals. It is possible to create interim goals as well, but this requires clear distinction between interim goal(s) and final goal, along with defining stages and their duration.
2. Company safety policy and standards.
3. A description of the project scope, constraints, and specific hazards.
4. A description of site conditions, location, and site-specific characteristics. Climatic and weather conditions also have to be mentioned, particularly when they have an impact on safety.
5. A list of responsible persons by position and role.
6. Safety and emergency contact information. This includes parties onsite and offsite, within and outside the contractor's team.
7. Hazard identification, analysis, controls, and safe practices. This includes the Material Safety Data Sheets (MSDS).
8. Crisis and emergency preparedness plan. This must include procedures for any emergency, natural or not.
9. Required certifications, training, and education. This is an important item that needs continuous monitoring and updating.
10. Compliance management through monitoring and updating procedures. This includes monitoring the execution of the plan, and reporting any noncompliance (violations, infractions, and "close calls") along with the response taken. Also, reporting any event or condition that triggers an update to the procedures.
11. Documenting and reporting procedures. Part of the documentation and reporting is required by law, but it is useful and important to the organization itself for review and planning.
12. Closure, evaluation, feedback, and continuous improvement.

Safety training saves lives, suffering, reputation, time, and money. Train everyone who will be on the job site, especially new hires!

Contractor: Use toolbox talks as safety training opportunities. Involve everyone and use examples and real cases. Emphasize safety in every talk or occasion.

Benefits of a Safety Program:

1. Reduced workers' compensation claims.
2. Reduced expenses related to injuries and illnesses.
3. Reduced absenteeism.
4. Lower employee complaints.
5. Improved employee morale and satisfaction.
6. Increased productivity.
7. Reduction of hidden costs.
8. Reduced insurance cost.

Cost of Safety Violations and Accidents

Workplace fatalities, injuries, and illnesses cost the country billions of dollars every year. In its 2021 Workplace Safety Index, Liberty Mutual estimated that employers in the U.S. paid more than US$1 billion per week for direct workers' compensation costs for disabling, nonfatal workplace injuries in 2018. The U.S. National Safety Council estimated that work-related deaths and injuries cost the nation, employers, and individuals US$171 billion in 2019. Employers that implement effective safety and health programs may expect to significantly reduce injuries and illnesses and reduce the costs associated with these injuries and illnesses, including workers' compensation payments, medical expenses, and lost productivity. In addition, employers often find that processes and other changes made to improve workplace safety and health may result in significant improvements to their organization's productivity and profitability.[20]

Accidents, violations, and even near-misses add a cost to the contractor. When construction accidents are mentioned, people may think of the direct costs such as only medical bills and property damage. In fact, the cost is much more than that. It includes direct and indirect cost items such as:

Direct costs:

1. Medical and Workers' Compensation expenses.
2. Property damage.
3. Equipment damage.
4. Regulatory fines and lawsuits.

---

20 Business Case for Safety and Health - Overview I Occupational Safety and Health Administration (osha.gov).

Indirect costs:

1. Cost resulting from delay caused by the accident. This includes the cost of accelerating the project and/or penalties for missed deadlines.
2. Administrative time spent by supervisors and safety personnel.
3. Replacement and training costs for new or substitute employees.
4. Lost time productivity due to complete work stoppage and post-accident low morale.
5. Damage to reputation. This affects the likelihood of winning future projects.
6. Increase in insurance premiums.
7. Costs of implementing corrective measures.

Indirect costs are believed to be much higher than direct costs. In Chapter 6, Project Budgeting and Cost Management, we gave an example with a comparison between two contractors, with EMR[21] $= 1.30$ for the first one and 0.80 for the other. We calculated the difference in Workers' Compensation insurance between the two contractors to be around \$650,000 for a \$20 million job, which is about 3.25% of the total cost of the project. This alone can make a difference in the bid amount, giving the contractor with better safety record an edge over the other contractor and a higher chance of winning, not to mention the impact on the reputation.

If you think safety is expensive, consider the cost of lack of safety!

Many organizations have compared prevention costs with accident costs, and found out that doing nothing is far more expensive. The US National Safety Council surveyed financial managers, and 60% answered that every dollar invested in prevention saves \$2 in accident consequences.[22]

Environmental Preservation

While scientists are debating major issues such as global warming, it is the duty of each individual and organization to do their part to preserve the environment in every way we can. There are many things the general contractor can do to make a positive contribution, including creative ways to save materials and energy and minimize waste. The EPA has guidelines for the construction industry.[23]

The concept of conservation is as old as humanity itself, but it was formalized in the past few decades as the need for it increased to a level of desperation with the rapidly increasing population of the earth along with their consumption of resources and increased pollution. Environmentalists came up with the concept and symbol of the 3 Rs: Reduce, Reuse, and Recycle. This simply means that we must reduce the consumption of materials, reuse what can be reused, and recycle what cannot be used. This concept is not as simple as it sounds. It must be understood and practiced as a culture, not as a hobby or public relations! The fact is that we buy and consume so many things that we either do not need at all or we need a lot less than we acquire. The power of marketing and advertising, especially in a capitalistic society, has

---

21 Experience modification rate, EMR, which has a national average of 1.00.
22 Construction Safety Investment: An Ethical and Lucrative Decision (ny-engineers.com).
23 Construction and Development Effluent Guidelines | US EPA.

taken us on that route to the point that shopping itself has become a hobby and addiction for many people. This phenomenon has increased with rapidly evolving technology and new gadgets at relatively affordable prices. Add to that, the rising labor rates in Western societies, which made buying new items more attractive than repairing or restoring old ones. The overuse of disposables, such as bags, boxes, bottles, plates, and utensils, has become a huge crisis. Landfills are filling fast, and plastics and other non- and slow-biodegradable materials are filling the earth, land and seas.

Many people resorted to the third option: recycling, which – in the author's opinion – is the least effective option after reducing and reusing. Basically, there are two major concerns with recycling: first, it is expensive. It takes a lot of stages to convert the disposed material into a usable (semi-raw) material, consuming effort and fuel. The stages usually include sorting, cleaning, shredding or melting, processing, and manufacturing. In most cases, these stages do not occur in the same place or country but shipped from one point to another. Keep in mind that not all plastic is recyclable. The second concern is that recycling may give a false feeling that we are doing our "homework" and duty to Mother Earth by just putting our paper, plastic, metal, and glass trash into the recycle bins! This may eliminate the guilt feeling from lack of reducing and reusing.

The concept of the three Rs as we all know and shown in Figure 12.2a, seems to give equal weights to the three functions: Reduce, Reuse, and Recycle. This concept can be improved so it becomes more effective, if we prioritize the three Rs as shown in Figure 12.2b. The weight of these Rs must never be the same. We must first reduce as much as we can, then reuse also as much as we can, and lastly recycle when we need to acquire it but cannot reuse it. We are not perfect, but the more we do, the better it gets for the environment, and consequently for the next generations.

(a)   (b)

Figure 12.2   The concept of the Rs: Reduce, Reuse, and Recycle: (a) Not prioritized, (b) prioritized.

The concept of sustainability must also be observed and practiced. Sustainability means leaving the environment the same as or better than it was before the operations. For the designer (architect/engineer), there is plenty that can be done to improve the sustainability of the project. This includes items such as:

- Using local materials that are efficient, economical, and friendly to the environment
- Finding ways to preserve and even generate power
- Finding ways to preserve and reuse water, including capturing and using water from rain and snow
- Finding ways to reduce the need, effort, and cost of operation and maintenance
- Utilize greenery for shading, increase oxygen output, and even provide fruits and vegetables
- Design with the construction process in mind, in order to help the contractor save materials, energy, and effort

In roads and bridges, there are also ways to help the environment such as using recycled aggregates and/or other local and environmentally friendly materials. What is important in sustainable design for buildings and other structures, is to deal with the design from a comprehensive perspective, taking into account the integration and interaction among the components, and also the long-term impact.

For the general contractor, sustainability means leaving the environment in and around the job site after the completion of the project, same as or better than it was before the operations. It is the responsibility of every individual and organization, but this responsibility is proportionate to their role and authority. It is more than just complying with the law, which may have loopholes, but it is a conviction, culture, and dedication. The EPA has several guidelines and recommendations for the construction industry, including "Sustainable Management of Construction and Demolition Materials."[24]

Environmental preservation for construction projects can be optimized when the project owner, designer (A/E), and general contractor are on the same page, collaborating in the effort. There are many ways to make the project a friend of the environment, both in the construction stage and during its service life. This includes the design itself, materials used, and construction methods.

Site Planning and Security

The contractor must secure the site during and after work hours. Security is protection from any incident or action that may threaten people or property or may adversely impact work progress. Security procedures differ from one situation to another despite having many common items. Having a project in the downtown of a major city or an urban area differs from a project away from cities and people. Site security plan starts with site planning before the start of the construction work. The project site must be planned to manage:

1. Construction activities.
2. Construction support activities such as sawing wood, cutting tile, welding, and bending reinforcement bars.
3. Storage including indoor, outdoor, and special storage (chemicals, explosives, fragile/valuable items, items that must stay away from moisture or within certain temperature).

---

24 Sustainable Management of Construction and Demolition Materials | US EPA.

4. Entrance/exit for delivery and other traffic.
5. Site offices with parking spaces. A "main office" must be clearly identified for visitors.
6. Utilities: drinking water, portable toilets, cafeteria, and lounge.
7. Equipment repair and maintenance shop, if required.
8. Workers' housing and facilities, if required.
9. Emergency evacuation plan.
10. Clear signage for all of the above.

The site plan may and does change during construction based on the changes on the ground and requirements of the projects. Some subcontractors are required only in the early stages of the project, while others may not start till the late stages of the project. Requirements for equipment and stored materials also vary with the progress of the work.

The contractor must visit and inspect the site prior to mobilization for several purposes, site planning is one of them. After that, the contractor creates a site plan that shows:

- The clear boundaries of the entire site and its location in regard to the area, showing access to the adjacent streets. If there is access through someone else's property with permission granted, it must be clearly marked on the map.
- The footprint of the project to be built.
- Area to be dedicated to office trailers, storage, parking, and other facilities.
- Roads and traffic directions inside the site, including emergency roads and exits.
- Connections to public utilities such as electricity, water, and sewage, if to be used. Also, temporary stormwater drainage system.
- Any important structures, points, objects, or issues that may either impact work or get impacted by the work.

In the end, the contractor must always make sure:

1. The site is safe for all workers and other people in and around the site.
2. Every worker has sufficient and efficient workspace.
3. Logistics are efficient and do not interfere with work activities.
4. Storage is located appropriately for the most efficiency.
5. Services and utilities are available to all subcontractors sufficiently and conveniently, such as (space for) site offices, storage, workshops, parking, portable toilets, and utilities.
6. Clear and safe instructions/signage for workers and visitors.
7. Overall, the objectives of workspace management are safety, work efficiency, and optimum cost. Compatibility with local laws and regulations is always a must.

> When making signage for the site or other items, use pictures and symbols as much as possible. Remember that not all workers and other people on site speak the same language. Pictures and symbols are universal language.

There are several tools the contractor may use for the security of the project site such as erecting a fence around the site with gated entrance (access control), employing security guards, providing sufficient lighting, installing burglar alarm system, and installing security cameras. Today's technology in solar power combined with high-capacity small-size rechargeable batteries, made it easier to install cameras anywhere, away from the electrical grid. In addition, and on special occasions, the contractor may utilize drones with cameras. Drones can reach certain locations that are difficult to reach in person, and they can zoom in to inspect confined areas even when it is dark there. The contractor can be watching what the drone camera is showing and saving relevant or important information as pictures or video clips, along with date and location coordinates for future use or records.

Security, like safety, is the responsibility of everyone onsite, but roles must be assigned clearly. So, if a worker observed a security breach or a threat, he/she must act promptly by informing his/her supervisor or the safety officer directly. Security is needed to protect people and property from crimes and other dangers such as theft and vandalism.

Fencing the site and securing equipment does not only provide protection to these assets but also may protect the contractor from lawsuits stemming from possible harm caused by equipment to trespassing people, especially children. This is the case of "attractive nuisance," which is defined as a dangerous condition or object on a property that can attract children and harm them, even if they are trespassers. Examples of attractive nuisances are construction equipment, ladders, scaffolding, and chemical containers. The property owner, or the party in charge of the property such as the general contractor, may be liable for the injuries caused by the attractive nuisance. The concept applies to any property such as homes, subdivisions, commercial, and other types, that have unsecured items that can fit into the description of attractive nuisance such as swimming pools, ladders, and fire escapes.

Exercises

12.1   What do the acronyms HSE and HSSE stand for?

12.2   What is the main objective of an HSE program for a construction company?

12.3   Do a search to create a timeline for OSHA since it was established and through important and landmark decisions.

12.4   What are the key elements of a successful HSE management system?

12.5   Safety management includes proactive and reactive measures. Mention some examples.

12.6   An HSE plan is a document that includes what?

12.7   Safety is the responsibility of both employees and employers. Explain this statement and give some examples.

12.8   Mention some of the common causes of construction accidents.

12.9   Safety issues include not just accidents. Explain this point and give examples.

12.10 How can you classify hazards in construction projects?

12.11 What does the Hierarchy of Controls mean? Briefly explain.

12.12 Discuss how to prepare a Safety Management Plan. Mention the common elements included in the plan.

12.13 What is the difference between a safety plan and a safety policy?

12.14 Mention the benefits of a safety program.

12.15 Mention the direct and indirect costs of safety violations and accidents.

12.16 Mention the benefits of good safety record on Workers' Compensation insurance.

12.17 Conduct a search on environmental preservation and answer the question: Why it is so important?

12.18 What are the drawbacks of focusing on recycling as a primary objective for environmental preservation?

12.19 Meet an architect and a general contractor, separately, and discuss ways to improve the sustainability of a project each one is involved in.

12.20 What is the importance of site planning for a general contractor?

12.21 What are the points a site plan must include?

12.22 You are the CEO of a construction company with a mission to improve site safety, security, and work efficiency. Write a memo to all your project managers instructing them to make sure their sites are well prepared and maintained, itemizing the issues.

13

Project Claims and Other Dispute Management

Introduction

Changes in construction projects are almost inevitable. They happen for many reasons, which may force a change to the original agreement between the owner and the contractor, mainly in the cost and schedule of the contract. This comes in a process that involves claims and change orders (COs). Most claims are resolved satisfactorily between the owner and the contractor, but sometimes the two parties disagree on the assessment of the situation, where we then need claim or dispute resolution. In this chapter, we will address issues related to claims management, both prevention and resolution.

Claims

A claim is a demand by one party for additional time or money. The American Institute of Architects (AIA) defines the claim as a written notice of a demand for payment of money, a change in the contract time, or other relief with respect to the terms of the contract. It provides the notified party a chance to investigate the matter and associated risks. Notice of claim provisions typically require that the notice must be filed within a reasonable time and contain detailed information about the issue.[1]

A delay, in its simplest definition in the construction industry, is the action or condition that results in finishing a project later than stipulated in the contract. A delay can also pertain to starting or finishing a specific activity later than planned. If we put the two terms, claim and delay, together, a delay claim simply means a claim related to a delay. It is important though, to note that a claim is not necessarily a dispute or confrontation. Many claims are legitimate and are routinely resolved to the satisfaction of both parties.

Change Orders

A change order (CO) is a formal written document, signed by the owner, directing the contractor to make changes to the original contract. A change order can be used for adding, deleting, or substituting work

---

1 Notice of Claims: Are You Prepared? (aiacontracts.com)

Project Management in the Construction Industry: From Concept to Completion, First Edition. Saleh Mubarak.
© 2024 John Wiley & Sons, Inc. Published 2024 by John Wiley & Sons, Inc.
Companion Website: www.wiley.com/go/nextgencpm

items. A change order usually, but not always, has an impact on the project's cost and/or schedule. In some countries outside the United States, a change order is called a variation order (VO).

The change order states the agreement of the contracting parties upon all of the following:

1. Change in the Work.
2. The amount of the adjustment, if any, in the Contract Sum, and
3. The extent of the adjustment, if any, in the Contract Time.

Part 1 above describes the change, while parts 2 and 3 represent the impact of the change on the project's cost and time (budget and schedule), as agreed previously in the contract. When changes are made to a project's scope, cost, or time, these changes are formally incorporated into the construction contract using a change order. Baselines for schedule and/or budget may have to be adjusted accordingly.

The relationship between claims and change orders is intimate. Here are the different scenarios in how they may happen:

I. Scenario 1: Contractor faces unforeseen conditions or other situations that potentially qualify the contractor for compensation and/or extra time.
    a. Investigate the situation and assess the impact on cost and time.
    b. Initiate a "Change Conditions" change order to the owner, describing the condition and requesting additional compensation and/or time per the terms of the changed conditions clause of the contract.
    c. If the owner approves, the contractor starts the execution. If the change has an impact on the design, the applicable designer (architect, structural engineer, and/or others) has to approve it too.
    d. The signed change order becomes part of the contract documents. The budget and schedule baselines may have to be adjusted accordingly.
    e. If the owner does not approve, the contractor either abandons the issue or files a claim. This claim needs to be resolved (see Figure 13.1).
II. Scenario 2: Contractor has a condition that potentially qualifies the contractor for extra time but no compensation, such as extreme weather or labor strike.
    a. Assess the impact of the situation on the schedule.
    b. Initiate a change order to the owner, describing the condition and requesting extension of time.
    c. If the owner approves, the signed change order is official and becomes part of the contract documents. The schedule baseline will be adjusted accordingly, and liquidated damages will not be incurred.
    d. If the owner does not approve, the contractor either abandons the issue or files a claim for a time extension. This claim needs to be resolved (see Figure 13.2).
III. Scenario 3: The owner wants to make a change, addition, deletion, or substitution.
    a. The owner, directly or through the designer, sends a description of the proposed change with a request for quotation to the contractor.
    b. The contractor studies the change, if within the broad scope of work, and then determines the impact on the project's cost and time.

Figure 13.1 Process for a change order (CO) for unforeseen conditions.

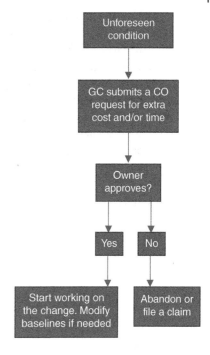

Figure 13.2 Process for a change order (CO) for extreme weather conditions or Force Majeure.

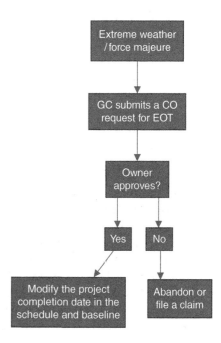

   c. The contractor issues a claim to the owner, informing them of the impact of the change on the project's cost and time as per the changes clause of the contract.

   d. If the owner approves, the contractor starts working on the change. The signed change order is official and becomes part of the contract documents. The budget and/or schedule baselines will be adjusted accordingly. If the change impacts the design, the change order has to be also signed by the applicable designer.

   e. If the owner rejects, the change order is abandoned. Nothing will change. The owner may negotiate the matter with the contractor and perhaps modify the change suggested, so the cost and/or delay are acceptable. In the end, the owner either agrees to the terms of the contractor or abandons the change (see Figure 13.3).

Figure 13.3 Process for a change order (CO) for owner's requested changes.

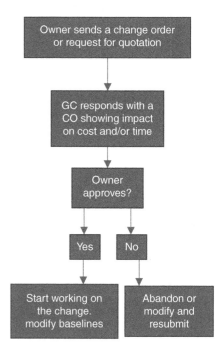

IV. Scenario 4: The owner issues a directive to the contractor ordering a specific change. If the change impacts the design, the relevant designer needs to approve it.

   a. The contractor studies the change and its impact on the project's cost and time.

   b. The contractor starts working on executing the change.

   c. The contractor issues a change order request to the owner, informing them of the impact of the change on the project's cost and time.

   d. If the owner approves, the signed change order is official and becomes part of the contract documents. The budget and schedule baselines will be adjusted accordingly.

   e. If the owner rejects, the change order becomes a claim and needs to be resolved (see Figure 13.4).

Figure 13.4 Process for a change order (CO) for owner's directive.

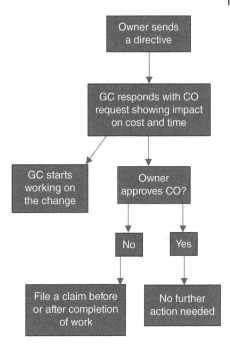

Constructive Acceleration

In all scenarios above, if the owner rejects the contractor's change order request but insists on executing the change with no extension of time, the contractor will be in a dilemma with two options: to stop the work till the dispute is resolved or continue the work and dispute later. The second option is called constructive acceleration, which is a situation that happens when a change occurs, as a result of the owner's request or other cause with no fault on the contractor's side. This change has an negative impact on the project's finish date, but the owner refuses to approve an extension of time. The contractor accelerates the schedule in order to finish the project on time, but will later file a claim with the owner (most likely, after the completion of the project), demanding reimbursement for the cost of the project acceleration.

If the contractor refuses to perform the change order till the owner signs the contractor's claim, this can exacerbate the situation by adding more delays. The better option for the contractor, as defined by constructive acceleration, is to send a letter to the owner informing them that he/she will comply with the work requested but at the same time expressing clear disagreement with the owner's decision not to grant extra time (and possibly money). The contractor postpones the resolution of the claim till after the completion of the project. Documenting everything is a must for the contractor to prove his/her claim.

For example, an owner issues a change order or directive delaying or extending the duration of a critical activity or even non-critical activity but disrupting the contractor's work plan. Other examples include differing subsurface conditions, errors in the design documents, or a delay in a delivery of an item by the owner. There are many situations that may not seem to the owner as causing a delay or disruption, but the contractor sees it otherwise. Constructive acceleration works best for both parties by limiting the damage (delay and tension between the parties) and preventing it from getting worse. It is like "I'll do it now and argue later."

> The term "Constructive Acceleration" underscores the importance of time. If the contractor stops working every time there is a dispute, there will be negative consequences likely to all parties. Constructive acceleration implies: Keep working, document everything, and file a claim later.

What Situations Trigger Change Orders/Claims?

1. Unforeseen site conditions. This includes all unexpected and differing subsurface and underground conditions. There is a wide variety of situations that cover cases, above and under the ground, that were neither mentioned in the contract documents nor could have been seen or detected by the contractor in a normal site inspection. When the contractor starts work, especially excavation, unexpected things may be encountered. This can be different types of soil, underground water, buried objects, old utility cables, or others. Some owners try to shift the risk to the contractor by adding a clause to the contract that the submitted maps and reports are only for the contractor's information and that the owner does not guarantee the accuracy of these documents. Even with such a clause, courts may still allow the contractor in some cases to file successful claims. Generally, if the unforeseen condition was either visible or could have been easily discovered by a construction professional during the due diligence process, then the contractor will get no relief. However, if it was not easily detectable before the work started, then it can be a basis for a claim.

   The contractor is expected to make investigative visits to the site, before bidding, before signing the contract, and before mobilizing. The contractor can also do their own soil and other investigations.
2. Design errors or omissions: These mistakes happen for several reasons such as designer's carelessness, "copy and paste" sections and paragraphs from previous design, poor coordination between documents, and others. If the mistake is so obvious to a professional, like saying the joists are spaced @ 16' (instead of 16") on center, then the contractor can correct it but needs to alert the designer. No change order is needed there. However, if the mistake is not obvious, like mentioning the thickness of the slab-on-grade as 4 inches in one drawing and 6 inches in another drawing or the specs, or giving conflicting numbers for concrete strength, then it needs a request for clarification and perhaps a change order.
3. Owner's delays, such as owner's failure to provide project site on time, late notice to proceed (NTP), the owner's failure to make timely decisions, delay in delivery of owner's furnished equipment, unjustified delay in progress or other payment, and other problems that are deemed to be the responsibility of the owner.
4. Changes by the owner: Owners have the right to change their minds on any matter as long as the request is reasonable and there is no change to the project scope. Contractors typically have the obligation to execute such changes. The timing of the change may make a difference in both cost and time, as discussed in Chapter 5, Scope Management. The name of the design change the owner makes differs also based on the stage when it happens, as shown in Figure 13.5. Too many change orders

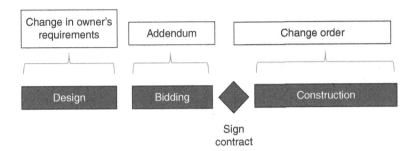

Figure 13.5 The name of design change during the different stages of the project.

based on owner's demand (change of mind) are likely an indication of poor scope definition and/or management.

5. Changes due to market fluctuations and trends: Sometimes a specified product or material is unavailable or a better one is introduced to the market. Such a change may originate from the owner, the designer, or the contractor, especially when a new and better product becomes available on the market.

6. Delays that are not the fault of any contracting party, such as unusually severe weather, will likely qualify the contractor for a claim for time extension but not money. The keyword here is "unusual," which implies "unexpected." So, if the contractor encountered severe weather, but this was normal for that location and that time of the year, then there will be no relief for him/her. If the severe weather was expected but was severer than expected or lasted more than usual, the contractor may qualify for an extension of the difference in days, for example:

   - It usually rains heavily 10 days in the month of August in this location, but there was 13 days of heavy rain this August, then the contractor may be given 3 days extension.

   Other delays include natural disasters, wars, pandemics, and labor strikes.[2] Claims under these conditions do not qualify the contractor for any financial reimbursement. Such losses should be covered under the contractor's Builders Risk insurance.

   In all cases, the contractor should keep accurate records to ensure full reimbursement for any changes.

What is a Force Majeure?

The term "force majeure" (also written as "force majure") came around 1883 from a French origin that means "superior force." It is an alternative to the term "Act of God." In the context of construction projects, this term generally combines three important elements:

1. *Superior*, overwhelming, or overpowering, i.e., cannot be prevented
2. *Unexpected* or cannot reasonably be anticipated or controlled, and
3. Has a *destructive or disruptive* effect on the construction process. The term destructive implies financial losses and the term disruptive implies delays.

---

2 Provided that it is not cause by the contractor's unfair practices.

Types of Delays

With regard to construction schedules, delay events are classified into the following three types:

1. Excusable delay: An excusable delay entitles the contractor to additional time for completing the contract work. Excusable delays usually stem from reasons beyond the contractor's control. These can be:
   a. Delays that are beyond the control of, and are not the fault of, the owner, such as those caused by unusual weather conditions, natural disasters, wars or national crises, or labor strikes. These types of delays most likely entitle the contractor to a time extension but not to monetary compensation (non-compensable delay).
   b. Delays caused by the owner or the designer. These types of delays typically allow the contractor to recover the costs and time associated with the delays and are known as "compensable delays."
2. Nonexcusable delay: By definition, a nonexcusable delay does not entitle the contractor to either a time extension or monetary compensation. This type includes:
   a. Delays caused by the contractor, such as slow mobilization, poor workmanship, labor strikes due to unfair labor practices, problems in the contractor's cash flow, or accidents on the project site caused by the contractor's negligence or lack of preparation.
   b. Delays not caused by the contractor but should have been anticipated by the contractor under normal conditions such as adverse weather conditions that are expected for that location during that time of the year, lack of skilled labor, or late delivery of the contractor's furnished materials and equipment.
3. Concurrent delay: A combination of two or more independent causes of delay during the same general time period may be considered to be concurrent delays. Often concurrent delays include excusable and non-excusable delays. The contractor may try to blame all the delay on the excusable delay.

   It is important to note that just because the additional work may require time to perform, the contractor is not automatically entitled to an extension of time. The contractor may have to demonstrate that the additional work impacts the critical path of the project. The use of critical path method (CPM) schedules and scheduling experts is often necessary to prove or disprove such a claim. Conversely, the contractor may be entitled to extra monetary compensation in lieu of a time extension even if the contractor finished by the original finish deadline. Typically, this extra payment covers the contractor's effort – above and beyond the original work plan – for accelerating the schedule. This is the case mentioned earlier under the title, "Constructive Acceleration."

Documentation

For the purpose of prevention and resolution of delay claims, both the owner and contractor must keep all project documents in a well-organized and easily searchable and accessible manner. One key recommendation is to document promptly because people forget. Today's computer systems and software programs give great solutions for such documentation systems. The information is usually stored in the cloud, so anyone with authority can search and retrieve needed information in no time. Some recommendations are mentioned in more detail in Chapter 14, Project Close-out-Efficiency and Success.

Delay Claims Resolution

Most contracts contain clauses for the resolution of claim disputes. The process usually starts with the simplest and quickest procedure and then, if not resolved, goes into the next level, and so on. Since the court system is time consuming, costly, and complicated, other alternative dispute resolution methods have been used. The following are the methods that are usually used in resolving disputed claims:

1. Negotiation: This is the most direct method for resolving any type of construction claim. In many cases, the "truth" may not be entirely known or acknowledged, and "fairness" is seen differently by the different disputing parties. Negotiation requires certain skills that tend to reduce the gap between the negotiating parties. Every negotiating party needs to know and focus on their own priorities as well as the other party's priorities.

   Although negotiation occurs basically between the disputing parties, independent experts or consultants may be asked to give opinions on certain issues. Negotiation starts with the parties at the project management level, but higher-level officers may get involved if the lower-level team fails to reach an agreement. While complete satisfaction may never be reached, the objective of negotiation is to reach a solution that is acceptable to both parties.

> Negotiation is the simplest way to resolve disputes. It requires certain skills and can make disputing parties avoid other methods that cost more time and money.

2. Mediation: When negotiation does not work, either because of lack of trust or skills, an independent neutral or mediator may get involved to bring the disputing parties together. The mediator can be an individual or a team. The mediator usually starts by explaining the role of a neutral in bringing the parties to an agreement. The mediator must demonstrate neutrality and patience and must collect all of the facts before making any recommendation. Typically, the mediator spends time with each party individually but – if the atmosphere is encouraging – may bring the parties to a meeting. The mediator, like the negotiator, must be skilled in narrowing the gap between the disputing parties and focusing on the positive side of any proposed solution. For example, although the dispute may be on financial issues between a general contractor and subcontractor, the mediator may sense that the subcontractor is concerned about security and future work. The mediator may then convince the subcontractor to accept a lesser financial settlement in exchange for a good and extended relationship with the general contractor.

3. Arbitration: Arbitration is a formal process that is performed by an independent professional arbitrator. It is defined by the American Arbitration Association as a "referral of a dispute to one or more impartial persons for final and binding determination. Private and confidential, it is designed for quick, practical, and economical dispute resolution."[3] Arbitration in the construction industry is usually performed by experts in the industry, such as architects, engineers, or construction management professionals. Although arbitration is usually binding to both parties, arbitrators do not have the enforcement power that judges do. An arbitration award in a binding arbitration can be become an

---

3 A Guide to Commercial.pdf (adr.org)

enforceable judgment by a court of law, if necessary. It should be noted that in many contracts, arbitration is specified as a binding process if negotiation and mediation fail.

Arbitration does not require the employing of lawyers, although it is not unusual for parties arbitrating important matters to employ lawyers. Not being subject to the complicated procedures and restrictions of the court system gives the arbitrator more freedom and makes the arbitration process proceed faster. The powers given to an arbitrator differ from one state to another. Mediation and arbitration can be combined into a two-stage process: the disputing parties select a mediator and agree that the mediator will become the arbitrator if the parties fail to reach a mediated settlement within a specific timeframe.

> Arbitration is "another court system," but shorter with less formalities and not enforceable. It is usually performed by arbitrators who are construction professionals.

4. Litigation (the court system): When none of the previous processes work and the parties cannot reach a settlement, the complaining party may sue the other party in a court of law. The legal process is complex and often involves pretrial discovery procedures, such as interrogatories (written questions to the other party), requests for the production of documents, and pretrial testimony in depositions. When the matter goes to trial, the facts may be decided by either a jury or a judge who may be overwhelmed by the technical details involved in a construction delay claim. Construction litigation is usually complicated, time consuming, and expensive. Many plaintiffs spend more money on attorney's fees than they later recover by the court's judgment.

5. Dispute review board: This method is proactive because it starts at the beginning of the project and before any dispute arises. A dispute review board (DRB) may be used to resolve disputes as they arise on a construction project. The board consists of three members: one representative of the owner, the second represents the general contractor, and the third is independent neutral, chosen by the agreement of the other two members. This third person will chair the board.

   The DRB technique was created to help in the process of investigating disputes by involving the neutral (the chairperson) at the very early stages of the project. DRB members meet regularly, often on the construction site, to become familiar with the progress of the work and to help resolve disputes in their early stages. Unlike mediation, arbitration, and litigation, DRB does not need the usual "let me tell you what happened" from either disputing party since the chairperson is familiar with the work progress. The chairperson will listen to their versions of the dispute and then make the decision.

   Negotiation, mediation, and DRBs are nonbinding. This means that the solution changes its shape – like a piece of dough – during the process until it reaches a form that is acceptable to both parties. No party is under any obligation until the party accepts the proposed solution. Chapter 16, Soft Skills for Construction Project Management, contains discussion on skills needed for negotiation and conflict resolution.

   Contracts often mention the method of dispute resolution that must be used if any dispute arises. Arbitration and DRB are usually mentioned. They can be combined with the DRB utilized first, and arbitration used next if one of the disputing parties does not agree with the DRB decision.

Table 13.1 Comparison among negotiation, mediation, arbitration, and litigation.

Method	Negotiation	Mediation	Arbitration	Litigation
Third party involved?	No	Yes	Yes	Yes
Voluntary?	Yes	Yes	Depends on the contract	No
Costly?	No	None to very little	Moderately	Yes
Takes time?	Little	Little	Moderately	A lot
Involves attorney?	No	No	Unlikely, but possible	Yes
Binding?	No	No	Yes, but not enforceable	Yes
Formal?	No	No	Yes, but simple	Yes, and complex
Resolution formation	Takes several cycles	Takes several cycles	Arbitrator's decision	Judge's decision

> DRB is the only proactive dispute resolution method!

Since the arbitrator does not have the power to enforce its decision, any of the disputing parties can still go to court and sue the other party. However, if this happens and the judge learns about the arbitrator's decision, it is extremely likely that the judge will uphold the arbitrator's decision. The judge will not reject the arbitrator's decision unless there is an egregious error, a conspiracy, or criminal intent on the arbitrator's part, which is highly unlikely (see Table 13.1).

6. Adjudication: For construction claims in the United Kingdom, there is an alternative dispute resolution method known as adjudication. The goal of adjudication is to provide a speedy and cost-effective method of resolving disputes on an "interim basis," allowing the decision to be "enforced pending final determination."

Methods of Schedule Analysis

There are many methods that have been employed to demonstrate the impact of delays upon a project. Most involve sophisticated evaluation of schedules but most analysis uses the dissection of the schedule and isolating the impacting factors. Some of these methods are:

- As-built schedule
- Updated impact schedule
- As-planned schedule
- Comparison schedule
- Accelerated schedule

For this reason, it is extremely important for the contractor to start with an accurate schedule and then update it regularly without any gaps. Adding notes to any incidents or events that impacted or altered activities' dates is also important. This will make it easier and more compelling to prove a claim.

Conclusion

Claims need to be prepared by professionals, so the cost estimator must be involved in the monetary part of the claim, while a scheduler needs to analyze the time delay. Both parties, the owner and the general contractor, use professionals to prepare, analyze, and prove or disprove a claim. This is why accuracy, organization, and professional practices are necessary for any contractor. Winning every dispute may not be the ultimate goal, but it is rather maintaining the integrity, sustainability, and success of the organization that matters.

Exercises

13.1 Define a claim, a delay, and a delay claim.

13.2 Define a change order. What information a change order form must include?

13.3 Discuss the relationship between claims and change orders.

13.4 Classify claims in terms of the contractor's demands.

13.5 What does the term "unforeseen conditions" mean to a contractor?

13.6 What are the unwritten professional obligations of the contractor regarding the site? The answer to this question will help define what conditions qualify as "unforeseen."

13.7 What is constructive acceleration? Why is it better than halting operations till the dispute is resolved?

13.8 The owner has authority to make changes to the project, but there are limits. What are these limits?

13.9 The earlier the change is decided, the less of a negative impact on the project it will cause. Explain this statement and give an example.

13.10 What situations trigger change orders/claims?

13.11 What is a Force Majeure? What conditions must apply for an event to be classified as force majeure?

13.12 What are the types of delays in terms of the contractor's entitlement?

13.13 An owner issued a change order to an activity, extending its duration from 5 to 8 days. If this activity had 4 days of total float, would the contractor be entitled to time extension, monetary compensation, both, or neither? Justify your answer.

13.14 An owner issued a change order that adds more work but insisted on keeping the project's original completion date. Will the contractor qualify for extra monetary compensation in lieu of a time extension? Justify your answer.

13.15 A contractor filed a claim for time extension due to extreme weather conditions. The owner refused the claim citing climate report showing that this area is prone to extreme weather conditions at this time of the year. How would you resolve this claim?

13.16 A contractor filed a claim for time extension and monetary compensation for removing an onsite old utility pole that was not on the site plan submitted by the owner prior to signing the contract. How would you resolve this claim?

13.17 A contractor filed a claim for time extension and monetary compensation for differing site conditions. The soil report submitted by the owner prior to signing the contract said the soil is sand, but during excavation, at depth of around 6 feet, the contractor confronted green clay with a substantial amount. How would you resolve this claim?

13.18 It has been said that the winner of a dispute is usually the party with better documentation. Elaborate on the importance of documentation in the context of proving / disproving claims.

13.19 What are the methods for resolving delay claims?

13.20 What is the DRB method? What advantages does it have over the other methods?

13.21 Compare negotiations and mediation methods.

13.22 Compare arbitration and litigation methods.

13.23 What are the skills needed for negotiation?

13.24 What are the skills needed for mediation?

13.25 An owner and a contractor signed a contract for building a project in Chicago, Illinois, at a lump-sum cost of US$3,550,000. The project is supposed to start in April and finish by the end of the same year. The owner had unexpected problems in financing the project that forced a four-month delay in the start of the project. Luckily, this happened before the contractor's mobilization, but the contractor filed a claim demanding raising the price to US$4,100,000 due to inflation and the fact that the project will be built during less productive time. The owner disagrees with the second cause. How would you resolve this claim?

14

Project Completion, Close-Out, and Beyond

Introduction

The process of project closing out is a bit complicated, as it includes technical, legal, financial, and other aspects. It usually happens in stages that result in the handing over of the project from the general contractor to the owner. However, the process may not completely end at this point, as there will be other issues to settle and resolve. In addition, the contractor needs to assess the success of the project and use "lessons learned" for improvement of future work.

The Termination of the Construction Project

Construction operations in a typical project start slow, pick up and peak, and then gradually decrease until it ends. The entire site reflects this expansion and shrinkage in operations, including the number of subcontractors and workers. There are usually several milestones in this process, depending on the type of the project. We will mention them mostly, but not entirely, in chronological order:

1. Project substantial completion: This is a major milestone, and it is defined as the point when the project reaches the stage of being ready to be used for the original purpose it was intended for. For example, if this is a road project, the original purpose is to have cars drive over it. Once, it reaches this stage, it is said to be substantially complete, even if the landscape work or the light poles installation is incomplete. This stage needs to be certified by the chief designer (architect/engineer) who will issue *the certificate of substantial completion.*
2. Project walkthrough or pre-final walkthrough: It is when the project is completed, so the general contractor and the owner (or their representatives) walk through the project to check all work items and make sure everything has been completed according to the contract documents. It is very common to have the project designer participate in the walkthrough, because the designer is the party that knows best about the project design and, thus, can detect any defects or variations. Defective and incomplete items are compiled in what is called the punch or snag list. It is the list of items requiring immediate attention before the project is completely finished and turned over to the owner. The

Project Management in the Construction Industry: From Concept to Completion, First Edition. Saleh Mubarak.
© 2024 John Wiley & Sons, Inc. Published 2024 by John Wiley & Sons, Inc.
Companion Website: www.wiley.com/go/nextgencpm

owner and the general contractor agree on a timeline, usually a short period like one or two weeks, to have all these items completed and/or corrected.

3. Project final walkthrough: This walkthrough takes place to make sure all the items in the punch list have been taken care of; completion or correction. If, for any reason, there are any items that still need some work, the owner and general contractor will agree on a time to do so. This can happen before or even after the project turn over to the owner.

4. Construction close-out: Construction close-out mostly refers to the physical work on the project and the site, while project close-out extends to legal and financial process that will officially end the project. Construction close-out is a group of activities that includes the punch list, certificate of final completion, and certificate of occupancy. Like *the certificate of substantial completion*, the certificate of final completion is issued by the chief designer: architect or engineer. However, the certificate of occupancy is issued by the local government authority. This certificate is likely to be required by the contract for releasing the final payment.

5. Contract close-out: This stage includes all the legal and financial steps needed to complete the project and its transition from the general contractor to the owner. The general contractor is required to submit certain documents to the owner such as warranty documents, lien waivers, and final payment applications. Upon satisfying all physical and legal requirements, the owner will release the final payment, which usually includes the summation of the last progress payment request and the held retainage.

   The term project close-out includes both construction and contract close out.

6. Site cleaning and demobilization: Upon completion of the construction work, the contractor must move out of the site after completion of these tasks:
   a. Disposing of surplus materials. The contractor must prepare in advance for this step, and plan for how surplus materials will be dealt with.
   b. Removing trailers, equipment, and temporary facilities. Although subcontractors may be required to remove their own trailers, equipment, and materials, the general contractor still has the responsibility to make sure this is done properly and promptly.
   c. Cleaning the site and building from any trash, debris, or other unwanted materials. Special attention must be given to items such as nails, broken glass, and sharp objects.
   d. Repairing any facilities/infrastructure that were damaged because of the construction activities, both on- and off-site.
   e. Removing trash and disposing of it in a legal manner.

7. Start-up and testing. This is particularly important for industrial and commercial projects. In some projects, the general contractor may have the obligation to train the owner's employees on the project's main systems.

Archiving: We cannot overemphasize the importance of archiving. It must be done properly and professionally. The contractor may need certain documents for many reasons such as dispute resolution, tax audit, or for own records that constitute the reference for future cost estimates, schedules, risk plans, and others. This is more of an ongoing process than a stage, but it is particularly important by the end of the project.

Figure 14.1 summarizes the components of project close-out in construction. Industrial projects may have additional steps such as start-up and training.

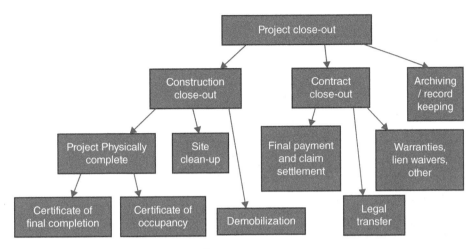

Figure 14.1 Project close-out steps and components.

The Transition Back to the Main (Home) Office

The activities mentioned earlier must be coordinated with the main office that will take responsibility of transitioning and reassigning all resources. The management in the main office needs to deal with each resource category: staff, labor, materials, equipment, and others, in a way to optimize the utilization and minimize waste and inefficiency. The contractor must always watch overhead expenses and make sure the company operates as lean as possible.

The closeout process of a project does not occur at once. It takes weeks or even months, which gives the chance for the main office staff to plan for the transition and management of the resources. It is not unusual for a project manager or other project management team members to spend time in the main office between project assignments, but it can be frustrating for these professionals to feel unutilized. Such periods of time must not be long and can be used for a variety of tasks such as attending continuing education seminars, attending a course in a software program, helping build a database for cost estimating, conducting training for others, or taking a vacation.

Project Management at the Corporate Level

In the past chapters, we addressed several areas of the project management such as planning, budgeting, scheduling, project control, quality management, risk management, HSE[1] management, field management and subcontractor coordination, project administration, and others. Each of these areas can be optimized at a local level, by focusing on it independently. But, the fact that matters most is that any action taken on any of these areas may and does impact other areas. We saw, for example, in the scheduling chapter, how time and cost interact. Quality, in the sense of defining and conforming to the

---

1 Health, Safety, and the Environment.

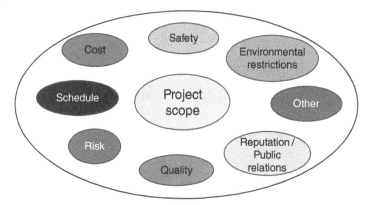

Figure 14.2 Project objective and constraints.

specifications, has a direct impact on both cost and time. Risk management is basically managing potential events that can impact project's objectives, mostly cost and time. Safety is an extremely important aspect of project management, from humanistic, legal, and financial aspects. So, it makes perfect sense to take an integrative panoramic look at all of these aspects combined, as seen in Figure 14.2, for the best interest of the project and the entire the organization.

Let us also remember that the construction company is in the business to make profit, so ultimately everything in the business needs to lead to profitability, of course, within the boundaries of the law and professional practices. While there is management at the project level, there must also be management at the corporate level that balances the needs, priorities, risks, resources, and responsibilities among projects. This corporate-level management looks after the interests of the organization, especially in the long run, which is something project managers may not pay enough attention to. So, while a project may not be as profitable as typically expected, the upper management may allow this situation to go in exchange for maximizing the benefits or minimizing the loss for the entire organization. Some companies sacrifice corporate profit in slow economic times, in exchange for staying in business and paying their overhead expenses. The upper management sets priorities to achieve the best interests of the organization but must also be keen on fulfilling the contractual obligations in each project.

> The project manager must always have a "wide-angle lens" to focus on the objective but also see all the constraints, in proportion to their importance.

For this reason, the company's management must look at every option to explore ways to make work more efficient, minimizing expenses, and maximizing profit. Operating in a lean way increases profit not only directly but also indirectly by giving the company a competitive advantage in bidding and negotiating projects. Companies may also have other competitive advantages such as having a technical niche, a reputation for finishing projects on or ahead of schedule, possessing a good safety record, earning a good reputation for quality products, operating in an environmentally-friendly way, ability to take on risky projects, and others. Combining more than one advantage is possible and will boost the strength and profitability of the company. On the flip side, greed, dishonesty, sloppy and unsafe work will lead to loss of work and profit, especially in the long run, and perhaps to the failure of the company.

Crew Productivity

When we address crew productivity, we have to differentiate between two cases:

1. Crew with major equipment: When a crew operates on major equipment such as an excavator, trencher, concrete placement (crane and bucket or pump), or concrete power trowel, the crew production is mainly determined by the equipment production according to its manufacturer's specifications but also influenced by several factors. For example, RS Means online provides the rate of production for excavation based on the type of excavation and equipment used, as shown in Figure 14.3.

   Figure 14.3 shows different types of excavations with different machines. Production rate is shown in both hourly rate with the description and daily rate under the "Daily Output" column, based on 8 work hours/day. So, the same worker(s) can have higher or lower production under the same circumstances but using different equipment. The numbers must be adjusted based on several factors, such as those listed in the same section on RS Means online and shown in Table 14.1.

   Although these adjustments are mentioned for cost purposes, they apply to productivity at the same rate[2], and hence duration. The contractor may also apply additional factors such as weather and equipment conditions.

2. Crews without major equipment: When a crew operates with no major equipment needed such as in installing blocks or bricks, placing floor tile or carpet, framing, or painting. Productivity is mainly dependent on two factors: The crew (number of workers, their role and specialty, and skill level), and work type and complexity. The contractor has to take other factors into account such as weather conditions and job site congestion. In general, balancing the workforce is a challenge. As discussed previously in the scheduling chapter, resource leveling is performed on major resources to minimize the fluctuation in daily use.

Productivity and Other Improvements

It is generally believed that labor productivity in the construction industry lags behind manufacturing and other industries. The US Bureau of Labor Statistics (BLS) publishes productivity numbers on an annual basis. In the United States and Western countries, labor cost is considered the highest single-item expenditure in any construction project. It averages usually around 40% of the total cost[3] but can go a little more or less depending on the type of project and other factors. Generally, labor time includes

---

2 The adjustment numbers are applied to productivity in the opposite way, as they apply to cost. So, item 312 316 420 360 (marked with an arrow) has a base daily production of 480 BCY (bank cubic yards): with soft soil or sand, we deduct 15% from the cost and add 15% to the production rate. For heavy soil or stiff clay, we add 60% to the cost and reduce the production rate by 60%.

3 This percentage is a rough estimated by the author based on several sources such as:
RS Means online Note R013113-60
What Percentage of Construction Costs Is Labor? (botkeeper.com)
Labor versus material cost in construction: 6 things to keep in mind – Bridgit (gobridgit.com)
The percentage is higher if we include management and administration staff that are part of the overhead expenses.

Line Number			Description	Unit	Crew	Daily Output
312316420010			EXCAVATING, BULK BANK MEASURE			
312316420011			**Common earth piled**			
312316420020			For loading onto trucks, add			
312316420050			**For mobilization and demobilization, see Section 01 54 36.50**			
312316420100			**For hauling, see Section 31 23 23.20**			
312316420200			Excavator, hydraulic, crawler mtd., 1 C.Y. cap. = 100 C.Y./hr.	B.C.Y.	B12A	800.00
312316420250			1-1/2 C.Y. cap. = 125 C.Y./hr.	B.C.Y.	B12B	1000.00
312316420260			2 C.Y. cap. = 165 C.Y./hr.	B.C.Y.	B12C	1320.00
312316420300			3 C.Y. cap. = 260 C.Y./hr.	B.C.Y.	B12D	2080.00
312316420305			3-1/2 C.Y. cap. = 300 C.Y./hr.	B.C.Y.	B12D	2400.00
312316420310			Wheel mounted, 1/2 C.Y. cap. = 40 C.Y./hr.	B.C.Y.	B12E	320.00
312316420360	➡		3/4 C.Y. cap. = 60 C.Y./hr.	B.C.Y.	B12F	480.00
312316420500			Clamshell, 1/2 C.Y. cap. = 20 C.Y./hr.	B.C.Y.	B12G	160.00
312316420550			1 C.Y. cap. = 35 C.Y./hr.	B.C.Y.	B12H	280.00
312316420950			Dragline, 1/2 C.Y. cap. = 30 C.Y./hr.	B.C.Y.	B12I	240.00
312316421000			3/4 C.Y. cap. = 35 C.Y./hr.	B.C.Y.	B12I	280.00
312316421050			1-1/2 C.Y. cap. = 65 C.Y./hr.	B.C.Y.	B12P	520.00
312316421200			Front end loader, track mtd., 1-1/2 C.Y. cap. = 70 C.Y./hr.	B.C.Y.	B10N	560.00
312316421250			2-1/2 C.Y. cap. = 95 C.Y./hr.	B.C.Y.	B10O	760.00
312316421300			3 C.Y. cap. = 130 C.Y./hr.	B.C.Y.	B10P	1040.00
312316421350			5 C.Y. cap. = 160 C.Y./hr.	B.C.Y.	B10Q	1280.00
312316421500			Wheel mounted, 3/4 C.Y. cap. = 45 C.Y./hr.	B.C.Y.	B10R	360.00
312316421550			1-1/2 C.Y. cap. = 80 C.Y./hr.	B.C.Y.	B10S	640.00
312316421600			2-1/4 C.Y. cap. = 100 C.Y./hr.	B.C.Y.	B10T	800.00
312316421650			5 C.Y. cap. = 185 C.Y./hr.	B.C.Y.	B10U	1480.00
312316421800			Hydraulic excavator, truck mtd. 1/2 C.Y. = 30 C.Y./hr.	B.C.Y.	B12J	240.00
312316421850			48" bucket, 1 C.Y. = 45 C.Y./hr.	B.C.Y.	B12K	360.00
312316423700			Shovel, 1/2 C.Y. cap. = 55 C.Y./hr.	B.C.Y.	B12L	440.00
312316423750			3/4 C.Y. cap. = 85 C.Y./hr.	B.C.Y.	B12M	680.00
312316423800			1 C.Y. cap. = 120 C.Y./hr.	B.C.Y.	B12N	960.00
312316423850			1-1/2 C.Y. cap. = 160 C.Y./hr.	B.C.Y.	B12O	1280.00

Figure 14.3 Excavation types and production rates. Source: From https://www.rsmeansonline.com.

Table 14.1 Adjustments for excavation rates.

For soft soil or sand, deduct	15.00%	15.00%
For heavy soil or stiff clay, add	60.00%	60.00%
For wet excavation with clamshell or dragline, add	100.00%	100.00%
All other equipment, add	50.00%	50.00%
Clamshell in sheeting or cofferdam, minimum	11.95	14.89
Maximum	32.00	39.72

productive work, supportive work, and wasted time. Wasted time can be attributed to several factors such as not having the required materials delivered on time and poor coordination with other trades.

In other countries, labor cost can be significantly less, but many other issues can also differ such as labor skill and productivity, labor laws and regulations, benefits, insurance (mainly workers compensation), and labor import process and cost (including living and transportation expenses.)

Improving labor (or crew) productivity is a very important subject, with hundreds of articles and research studies written on it. Labor (or crew) productivity affects both the cost and duration of the project. However, it will be better to address crew productivity from the more comprehensive "work efficiency" concept because the latter addresses crew productivity in addition to other factors such as materials and overhead management. The objective is to make the entire work, onsite and in the main office, technical and non-technical, as efficient as possible.

In this book, we will not get into the details of this wide and deep subject but will instead mention general notes. The objective is to coordinate the resources; labor, materials, equipment, and overhead, within all the constraints, in a way that maximizes productive work and minimizes waste of all types. Here are some tips for the general contractor:

I. Before you start

1. It all starts with planning. Resist the urge to rush, especially when time is crucial. A little good planning can save you time, money, and headache in execution.
2. If possible, participate in the design processes. Add any constructability and value engineering suggestions.
3. Utilize building information modeling (BIM) for clash detection and improvement ideas before you even start executing.
4. Negotiate the contract to suit the situation and make it a win–win situation for all parties.
5. Take on calculated and manageable risks only. Balance your risk: there is no profit without risk, but higher risk than you can afford may lead your organization to trouble.
6. Be proactive in claims and other dispute resolution. Address such methods in the contract.
7. Optimize the schedule using techniques such as Optimum Scheduling[4]. This allows you to perform activities under the best climatic and other conditions and within your constraints.

> Planning is the mother of all work success!

> If you overpromise, you will likely under-deliver!

8. Make sure responsibilities are assigned clearly and fairly. You cannot hold someone accountable unless you clearly define their mission, responsibility, and authority.
9. Visit the project site perhaps multiple times: before you bid, after the award, and before mobilizing.

---

4 Review Construction Project Scheduling and Control by Saleh A. Mubarak, 4th edition, Wiley.

10. When bidding on or negotiating a project in a country or location you have never worked in before, make sure you familiarize yourself with all requirements such as local laws and regulations and availability and cost of the resources. Have a local trusted partner or consultant.

11. Ambition is good, but it has to be realistic. Use the "110% rule," which attempts to gain a 10% improvement in performance as compared to last time. Those who set a target that is unrealistically high, say 150%, are unlikely to achieve that, which will turn into frustration and the feeling of failure.

12. Multitasking is a myth if it means focusing on more than one issue at the same time. We need to plan, organize, and *prioritize*, so we focus on one issue at a time.

II. On the project site

1. Review your own practices, policies, and regulations to make sure you are treating all in a fair manner.
2. Raise the culture of suggestions, constructive criticism, creative thinking, and transparency.
3. Create and encourage team spirit. It is "we or us," not "I or me."
4. Improve the procurement process and supply chain management by allowing the data to flow across departments and make sure everyone obtains all necessary information on time. Pay attention to long-lead and large-quantity items. Have a continuously updated vendors' list, organized and searchable by name, specialty, location, and other criteria.
5. Improve onsite operations. This actually is a group of items, including organizing the site to allow efficient work operations, including support work, smooth and safe flow of traffic and people, efficient materials storage, and clear signage.
6. Adopt an efficient and dynamic materials management system that balances between "inventory buffer" and "just in time" theories. Make sure the right material is available in the right quantity at the right time, and that the total cost of materials is minimized. Make sure all materials are inspected appropriately and reduce materials waste and reuse as much as possible.
7. Keep equipment in good condition. Have on- and off-site maintenance plans, particularly for projects in remote areas, including likely to need spare parts on hand.

> The project management team members are on the same boat. The entire team succeeds or fails!

8. Pay special attention to cash flow past and predicted future. This means money being transferred into and out of the company's business. Failing to track and predict cash flow is a prime reason for the failure of the business.

9. Note the difference between failure of a project and failure of the entire company or organization. Although the contractor must be keen on the success of each project, projects may fail sometimes for reasons that may or may not be the fault of the contractor. However, a strong organization should be able to continue business and succeed even if this happens. If failure of projects becomes repetitive, it is an indication that the organization lacks the ability to manage its projects and will ultimately fail unless it makes a substantial positive change.

10. Improve communications among your team members and with other project stakeholders. Make sure all requests for information and clarification are responded to in a timely manner.

11. Embrace technology: methods, concepts, software, materials, equipment. However, remember that the tool that worked best in one situation may not be the best for another situation.

12. Consult with experts when you do not have enough knowledge, especially in important matters. Make sure the expert you consult with is unbiased and has no motive for or against the item or action being discussed.

13. Utilize cameras, perhaps with time-lapse photography, so you can observe field operations for analysis. This includes direct work, support work, logistics, workers breaks, safety hazards, and more.

14. Reskill the workforce and staff by periodical workshops and continuing education, especially when a new concept, technology, equipment, or other item is introduced. Utilize own talents for training others.

15. Make Safety a priority. A good safety record gives an advantage to the contractor in direct savings, particularly in workers compensation insurance, and in reputation. Accidents have direct and indirect costs, with the indirect cost several times the direct cost.

It has been said that the party that is likely to win any dispute, is the one with better documentation!

---

16. Documentation is an important matter, so make sure it is handled properly by a professional. Information must be organized in a clear way that makes it easy to search and retrieve, yet confidentiality and security are not compromised.

17. Watch your overhead spending. Make sure you provide the necessary items and services but operate as lean as possible.

18. Set priorities for your organization and make sure you have a strategic plan for good and slow economic times. Make sure you have risk management plans at the organization level.

19. Be friendly to the environment by planting and protecting vegetation and preventing any action that damages the environment. Save on energy and utilize clean energy methods when available.

20. Be fair and honest with your workers, employees, subcontractors, vendors, and clients. Treat them as you like them to treat you. Greed is a destructive trait that ruins relationships and businesses. Failing to appreciate a good employee's performance and reflect this appreciation in acknowledgement and rewards may lead to turnovers that cost you more than just money.

21. Good quality pays off in so many ways: customer satisfaction, less rework, and better reputation.

Reputation is like a glass jar: It can be broken only once!

The quality of a product is as good as the quality of its weakest component.

---

22. Avoid long-term fatigue for your workforce. Overtime should be utilized only when necessary and only to the extent needed. In extended overtime, the contractor pays more to get less (per unit of time)! Workers are encouraged and motivated when they see the light at the end of the tunnel, this means working harder for a limited time with the end within sight.

III. After the completion of the project

1. It has been said that the devil is in the details, so make sure you pay attention to these details and do not leave any stone unturned.
2. It is not over till it is over; physically, fiscally, and legally.
3. Make sure you have a proper and comprehensive documentation system that allows you to search and retrieve any information, easily and quickly.
4. Create and maintain a database for your information pertaining to cost and production rates. Make sure you add notes explaining the conditions attached to the numbers so they can be easily used and adjusted for future projects. Also have a database for all subcontractors and vendors, that can be sorted by specialty, location, and other criteria. Both the documentation system and the database must be managed by professionals.
5. Survey your clients to measure their satisfaction. Design the survey detailed enough to separate issues but brief enough to improve the response rate.
6. Also, get feedback from your teams. Be open-minded on any criticism or suggestion. The objective is continuous improvement and no one is perfect!
7. Use mistakes as "lessons learned" to avoid repeating them in the future and use them to improve work. Do not allow them to be used to blame people or for grief or anger. Remember that the past cannot be undone but can be a good foundation for the future.
8. Remember that heavy rain, wind, and other severe conditions, are not a cause for defects in the project. In fact, they help put the project to the real test and detect quality defects.

IV. In the main office

1. Create a project management office (PMO) unit that standardizes and monitors the management of all projects. This unit can range from a part-time staff member, i.e., a person who has other responsibilities, to large unit with specialized professionals, as the company needs.
2. Dedicate a unit: a person or persons, in the company as the technology leader, not in the context of information technology (IT) but rather in tracking advancements in the world of construction. Although tracking and adopting technology is for everyone in the company, this unit will take the lead in this area.
3. Monitor overhead expenses and make sure it is under control.
4. Utilize continuing education opportunity through different means. Also, give opportunity to employees who can contribute to educating colleagues through means such as weekly "brown-bag seminars." An occasional retreat that mixes entertainment with education, is a good idea.
5. Treat your employees fairly. Organizations where employees are always looking for a better place will not likely survive for a long time. The organization must do its best to retain its good employees by giving them an atmosphere, where they can be happy and productive.

Project Success

Project success is measured from different perspectives: the owner, the designer, the contractor, and other parties. They may have a completely different assessment of the project success. For example, an investor decides to build a hotel in a spot that he believes will attract tourists. The investor hires a great architect who created a wonderful design and then hires a reputable contractor who built the project and finished it on budget and on schedule. Unfortunately, the area was not as attractive to tourists as the investor believed, so the hotel occupancy was less than anticipated and the rate of return was disappointing.

The question is: Do you consider the project a success? The answer is: It depends on whose point of view. From a design and construction point of view, it was a success but from the owner's perspective, it was not. In other words, it was a success from design and construction perspective, but not as an investment.

What is Success?

Before we can measure success, we need to define it. In the general meaning, success is the achievement of something desired, planned, or attempted. This may be wealth, respect, fame, education, pleasure, position, or others. So, success is not a goal per se, but it is the achievement of the goal. In the world of project management, project success is the achievement of the set objective within the defined constraints. For example, the objective can be an office building with specific design and specifications in a specific location. The constraints include quality, budget, schedule, and others.

So, what if the contractor achieved the objectives but was late or over budget? And what if this schedule or cost variance was minor or major? This is why we do not consider project success as a "yes or no" but rather a degree between 0 and 100%. It is similar to reviews customers give to certain merchandise or services. What has been observed in such reviews is that the more general the review question is, the more subjective the review will be. To make it more objective and accurate, we break the review into specific categories and request a review on each category separately. For example, instead of a hotel making one review to its guests "How was your stay?," it can ask a few specific questions like check-in, room comfort and cleanliness, staff service, amenities, food, etc.

Success in general is not a simple "yes or no" but rather a degree between 0% and 100%. However, it may become a "yes or no" if:

▶ You are a surgeon and your "project" is the save the patient's life.
▶ You are a candidate for a position in an election or a football team playing the championship game.

Even when failing, one must learn the lesson, so you improve your success chances next time.

We follow the same methodology in measuring project success, but we also give different weights to the constraints as the importance of these constraints varies from one project to another. The author created a simple spreadsheet for measuring success, using these steps:

1. The objective must be well-defined.
2. The constraints are written in a random order, see the "Criteria" column in Table 14.2.
3. Each constraint is given a weight proportional to its importance in this project. The weights have to add up to 100, but if they do not, they can be adjusted later to have a total of 100, see the "Weight %" column in Table 14.2.

Table 14.2 The project success model.

Criteria	Weight %	Score (out of 10)	Weighted score	Comments
Cost	35	8	28	
Schedule	25	10	25	
Safety	8	9	7.2	
Environment	4	7	2.8	
Change orders	8	6	4.8	
Claim disputes	7	8	5.6	
Relationship with other parties	5	8	4	
Public image	3	9	2.7	
Satisfied employees/workers	5	8	4	
Total	100		84.1	

Table 14.3 The category grade classification.

Legend	Range	Color code
Very Good	9–10	
Acceptable	7–8	
Marginal	5–6	
Not Accepted	<5	

4. A grading system must be developed to assign a grade, 1–10, to the performance in each constraint, see the "Score" column in Table 14.2. The cell is given a combination of grey shades and pattern compatible with the grade, as shown in Table 14.3.
5. The grade in each constraint is multiplied by its weight, see the "Weighted Score" column in Table 14.2. For cost, it is $35\% \times 8/10 = 28\%$. The sum of these multiplications will be the degree of success to the entire project.

The user can set guidelines for the score in each category, see for example Tables 14.4 and 14.5.

The overall grade of the project can be compared to a ranking, also set by the contractor, similar to the one in Table 14.6.

In addition, the contractor can also set "triggers" where if one category goes below a certain threshold, it is deemed "not successful," regardless of the overall score, such as if the safety or cost score is below 5 out of 10. Also, if the project is building a stadium for an important event, and the contractor failed to deliver it on time, it can be considered "not successful," regardless of the overall score or other criteria.

Advantages of the model:

1. Simplifying the decision by breaking it down to individual criteria.
2. Making the evaluation more objective by creating a scale for evaluating each criterion.
3. The model is customizable and adaptable by not only changing success criteria but also assigning the appropriate weights to suit the project priorities.

Table 14.4   Cost grade system.

Cost	
Outcome	**Grade**
Finished on or under budget	10
Cost overrun <1% of budget	9
Cost overrun 1–2% of budget	8
Cost overrun 2–3% of budget	7
Cost overrun 3–4% of budget	6
Cost overrun 4–5% of budget	5
Cost overrun 5–6% of budget	4
Cost overrun 6–7% of budget	3
Cost overrun 7–8% of budget	2
Cost overrun 8–9% of budget	1
Cost overrun >9% of budget	0

Table 14.5   Schedule grade system.

Schedule	
Outcome	**Grade**
Finished on or ahead of schedule	10
Finished late 1–3 days	9
Finished late 4–6 days	8
Finished late 7–9 days	7
Finished late 10–12 days	6
Finished late 13–15 days	5
Finished late 15–17 days	4
Finished late 18–20 days	3
Finished late 21–23 days	2
Finished late 24–26 days	1
Finished late >26 days	0

Table 14.6   Project evaluation scale.

$\geq$85%	Project successful
70–85%	Project partially successful
$\leq$70%	Project not successful

> ► Success is not a course you take in school.
> ► Success is neither an art nor a science.
> ► Success, in most cases, is not dependent on a single factor.
> ► Success, in most cases, is not a 0% or 100%
> ► Success is not a "gut feeling." It can be measured!
> ► Success is not a "copy paste" experience!
> ► Success does not come by luck or coincidence.
> ► Although success may vary from one project to another, a good contractor should see project success rates going and staying up consistently.

Mitigating/Adjusting Factors?

Sometimes, things do not go as planned for a variety of reasons. It may be issues beyond expectations and above our control. This may result in a lower-than-expected success rate for the project. The "success percentage" and numbers cannot give the total picture without analysis and explanation. We need to be fair to ourselves and others, so the success numbers have to be read and interpreted within the context of the circumstances. On the other hand, sometimes even with good success numbers, we can say "we could have done better" if the conditions were better than expected.

Success of the Project versus Success of the Organization

Although a good businessperson would like to win in every transaction or project, a wise businessperson would look at the overall picture, making sure it is profitable and it is going in the right direction. There is risk in every project, so it may happen for a company that takes on many projects, that one or two projects may not rise to the level of expectations. Sometimes, the contractor's primary objective in a project, is to establish presence or reputation, or to survive through tough economic times, with modest profit expectations. In this case, the success of the project must be measured with this objective in mind.

For the success of the organization, the goal must be set such as making an overall net profit percentage, like 5%. Such a process must be coupled with the strategic planning and risk management for the organization. This must be a vigilant and dynamic process, responding to market changes and fluctuations on one hand, and maintaining the company's strength, growth, and resilience on the other hand.

Service After the Completion of the Project

Figure 14.4 shows the stages of the project throughout its lifecycle. The contractor's main work is in stage 2, the construction of the project. However, the contractor may also be involved in the preconstruction services, either formally by owner's appointment or internally, such as in the case of design-built contracts.

After the completion of the project, there will be a period when the contractor will be responsible for any defects, breakdowns, or other types of customers complaints. This can be a challenging period for the contractor because most of the work has been performed by subcontractors who may not be available.

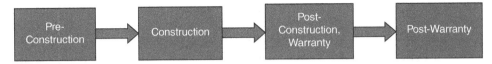

Figure 14.4 Stages of the project.

Sometimes, there are issues where subcontractors' specialties intermingle together. For example, electrical and mechanical (HVAC[5]) are intimately linked. Roof water leak could be caused by one of several reasons that fall under different trades. Also, some items such as the air conditioning systems and appliances, have manufacturer warranty, which covers manufacturer's defects, but not installation defects or user's abuse.

One more complication is the fact that not all the project components have the same period of warranty. In Florida, for example, there is a law that provides most buildings with a 10-year mandatory warranty on the structural system.

In general, issues that appear after the completion of the project, fall under one of the following:

1. Clearly under warranty,
2. Clearly outside the warranty, either after the expiration of the warranty period or the problem happened because of user's fault or abuse.
3. In a grey area.

The contractor who is keen on having a good reputation, would always honor the after-construction service, including and most importantly warranty issues. When the matter is in a grey area or even clearly outside of the warranty, the contractor may offer help, which can be doing the repair as a courtesy, giving a discount on the repair fees, or recommending someone reputable to do the repair. The best advertisement for the contractor is the word of mouth[6] from satisfied customers.

Exercises

14.1   What does a project's substantial completion mean? How different it is from final completion?

14.2   What does "project walkthrough" include? What does it mean to the owner or to the general contractor? Is there one final walkthrough or could there be more than one?

14.3   What does construction close-out mean and what does it include?

14.4   Is construction close-out and contract close-out synonymous or they are different?

---

5 Heating, ventilation, and air-conditioning.
6 Nowadays, the term "word of mouth" is a metaphor. This happens, in most cases today, as online review in both numerical and word evaluation.

14.5 What are the general contractor's obligations after the construction close-out regarding the site?

14.6 The project is ending in about two months. As a project manager, how would you prepare for it? Discuss your actions regarding the workers, staff, equipment, surplus materials, trash, and the site itself.

14.7 Now, you are the CEO of the construction company for the project in Exercise 6. How would you handle the transition between individual projects and the main office? Each project manager is communicating with you as if he/she is your top priority, but your top priority is the welfare of the company.

14.8 Talk to a construction project manager about lessons learned:
   a. Any mistake happened in the past that they made adjustments to make sure it will not happen again?
   b. Any area where they discovered ways to improve performance?
   c. Any new technology applied that helps improve performance in any way?

14.9 Talk to a construction project manager about tips that may:
   a. Increase the productivity,
   b. Reduce the cost,
   c. Improve safety, and
   d. Improve labor conditions and morale.

14.10 In the previous two questions, investigate all the improvements that were implemented. Ask the question: do any of these improvements require prior preparation? Focus on planning and early preparation.

14.11 Talk to a construction project manager about the documentation system:
   a. What type of system do they use?
   b. How do the main office and projects communicate?
   c. Any offices in a different country? If yes, any challenges?
   d. What do they do for cyber security? Any issues with hackers?
   e. Did they ever go back to old archives to resolve a dispute? How valuable is to keep accurate and searchable information?

14.12 How do you define project success?

14.13 How does the success of the project differ among project partners: The owner, the designer, and the contractor?

14.14 A project for a World Cup stadium is to be built with a very strict deadline schedule. The "success criteria" and their weights are as follows:

Criteria	Weight
Cost	20
Schedule	40
Safety	6
Environment	7
Change orders	5
Claim disputes	5
Relationship with other parties	5
Public image	8
Satisfied employees/workers	4
Total	100

The project finished with this information:
- There was a 2-day delay.
- There was a 1.4% budget overrun.
- One minor accident.
- No environmental pollution or violation.
- A few change orders and very few disputes.
- There was some tension between the owner and general contractor (GC) but in the end, everything was fine.
- The public image of the project was very positive.
- Many workers and employees complained about work conditions, but the GC management was able to resolve most of these complaints.

Using the provided Excel sheet, estimate the project success grade. Use the provided tables for grading cost and schedule, and your own educated opinion for other criteria.

14.15 In exercise 14, assuming an unforeseen condition delayed the contractor by 10 days. Since the project finish deadline is very strict, the owner demanded project acceleration so the project could be finished on time. This led to a claim by the contractor which made the budget overrun 3.8%. Calculate the project success grade. Do you believe there were any mitigating circumstances that we must take into account when calculating project success?

14.16 An office building project is to be built with a very strict budget. The "success criteria" and their weights are as follows:

Criteria	Weight
Cost	45
Schedule	20
Safety	10
Environment	5
Change orders	5
Claim disputes	5
Relationship with other parties	5
Public image	2
Satisfied employees/workers	3
Total	100

The project finished with this information:
 - There was no delay.
 - There was a 0.9% budget overrun.
 - One major and two minor accidents.
 - Minor environmental violation that led to a government fine.
 - A few change orders and very few disputes.
 - The relationship between the owner and GC was mostly fine.
 - The public image of the project was acceptable.
 - Workers and employees were satisfied with no major complaints.

Using the provided Excel sheet, estimate the project success grade. Use the provided tables for grading cost and schedule, and your own educated opinion for other criteria.

14.17 Redo exercise 16 with 3.2% cost overrun.

14.18 A company decided to expand its operations in a new region. It finished its first project there on time. The client was happy but there was barely any profit. How do you look at the project success from a corporate perspective? (Do not use the Success Model spreadsheet.)

15

Project Administration and Corporate Management

Introduction

Managing a project has more than one side. We addressed the "field side" where construction operations are going on to physically build the project. The other side is in the office, both on the job site and in the main office. There is where much of the work happens; before, during, and after the construction work, and we may call it project administration. However, this administration occurs continuously at both the project and corporate levels, with the objective of providing projects with the necessary resources, technical and financial support, documentation, and other services. Projects are temporary by their nature, but this administration is permanent despite the changes and fluctuations to respond to projects' needs, corporate plans and policies, and market changes.

What does Project Administration Cover?

Although the titles and functions of administrative departments vary from one organization to another, we can combine all these functions under one title: services to the construction operations and staff. Here is a summary of the departments or services:

1. Human resources: This includes all services needed for the employees such as the process of hiring, layoff, transfer, and retirement planning. It includes payroll, benefits, medical insurance, scheduling vacations, logistics, and other services to the employees.
2. Finance and accounting: This includes managing the organization's finances (accounts payable and receivable and cash flow), taxes, and other finance and accounting matters. Its main function is to make sure the organization has financial resources sufficient to sustain the operations.
3. Procurement: The role of this department varies from one organization to another. It can be responsible for acquiring materials, equipment, temporary installations, and supplies. Its responsibility can also include the acquisition of labor and staff.
4. Communications and information technology (IT): This department provides computer services and IT solutions to all projects and employees. It can be considered the nervous system of the company where all information and communications are managed.

Project Management in the Construction Industry: From Concept to Completion, First Edition. Saleh Mubarak.
© 2024 John Wiley & Sons, Inc. Published 2024 by John Wiley & Sons, Inc.
Companion Website: www.wiley.com/go/nextgencpm

5. Documentation and archiving department where all information and documents are filed and managed. This department may be combined with the IT department.
6. Legal department that looks after the legal matters and perhaps contracts issues.
7. Marketing, media, and public relations: This department takes care of collecting and preparing the materials necessary for marketing and public relations.
8. Training: This department takes care of all training matters, technical, and other. Training is important, especially when a new computer program, technology, or other system is implemented.
9. Preconstruction studies and services and post-construction services: This is a technical department that provides two types of services: pre- and post-construction. Preconstruction services include feasibility studies and planning. Post-construction services include repairs, warranties, and follow-ups.
10. Senior level management: This department is like the brain of the organization. It provides directions to where the organization is heading, strategic planning, investment, and risk management at the corporate level. It also deals with emergencies and makes the final decision on important matters.
11. Research and development department: Large organizations may have a research department that conducts research on methods, materials, and technologies. It may also exchange information with technical teams for data analysis and provide testing and feedback on technologies, methods, materials, and other matters for possible use. Many large companies partner with universities and professional societies on research.
12. Other services such as security, cleaning, maintenance, and facilities management.

Project administration is the "infrastructure" of the construction company!

Construction companies vary in size and management style. So, the departments or services mentioned above also vary in shape, content, and size. Some departments may not even exist, as the services may be outsourced. Also, in most small companies, the same department or even person will be doing more than one function. Generally, technical issues must be handled by professionals. Unjustified cost cutting may backfire and cost several times what it would cost to hire a professional. Take, for example, IT issues and cyber security: the lack of a competent professional may jeopardize the company's records and data, including confidential information.

Balancing services with their quality and rate of return is important for the company. This balancing must be an ongoing process, where the administrative, operational, and overhead expenses are reviewed and revised based on the demand and effectiveness of these expenses. As an example, the cost of unsuccessfully bidding on a project can be high, as the contractor dedicates a team to study the project, do the cost estimate and critical path (CPM) schedule, and prepare all bidding documents. This expense counts with the main office expenses. This is a normal scenario when the rate of successful bids out of all bids is reasonable and acceptable. This is why the contractor must study the project before making the decision to bid. The decision to bid or not to bid is crucial and risky. Not bidding may represent missing an opportunity, while bidding has a risk of losing money with unsuccessful bids. This is why the contractor must have a bidding strategy with lessons learned. This strategy must be periodically reevaluated. Factors considered in the decision to bid were discussed in Chapter 3.

The Main Office and Projects Management

The main office acts like the hub for all projects. Regardless of the individual organization structure (projectized, functional, or matrix), the main office is the permanent component of the organization with channels extended to all projects. These channels convey resources, money, and information both ways. The main office closes the project's channel after its closing and opens a new one for the new project. The responsibility of the main office staff is delicate, as there is sometimes a "tug-of-war" situation among project managers, where each is pulling needed resources, but the main office has limited resources, so it needs to allocate them as efficiently as possible.

> The leadership in the main office must balance its relationship with regional offices:
>
> – Too much control may annoy the management of these regional offices and may be counterproductive.
> – Lack of sufficient control may lead to chaos and disorder in the company.

Enterprise Resource Planning (ERP) Programs

ERP stands for enterprise resource planning. This is a term many organizations use to refer to a comprehensive computer system that encompasses all the core business processes needed to run a company: finance, human resources, supply chain and inventory, services, procurement, and others. At its most basic level, ERP helps to efficiently manage all these processes in an integrated system that is sometimes described as "the central nervous system of an enterprise," as shown in Figure 15.1.

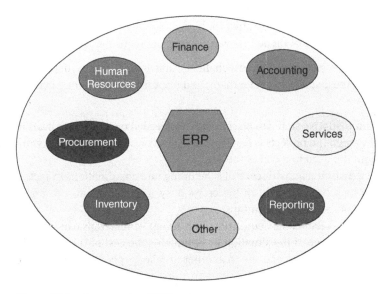

Figure 15.1 An example of ERP system.

An ERP software system provides the automation, integration, standardization, and intelligence that is essential to efficiently run all day-to-day business operations. Most or all of an organization's data should reside in the ERP system to provide a single source of information across the business.

Multi-Project Management

It is very common for a contractor's organization to be executing several projects at the same time. These overlapping projects may differ in type, size, location, or other attributes. They also may compete for the contractor's limited resources. The contractor needs to prioritize them in a way that fulfills the needs within the constraints of time, cost, resources, and perhaps others.

Competition or Collaboration?

Having more than one project under the responsibility of the same contractor is both a challenge and an opportunity.

The decision to take on a new project has to take into consideration several factors such as the company's bonding capacity, risk (at each project level and then collectively at the organization level), and availability of resources. In some cases, the contractor teams up with another contractor for one or more reasons such as:

1. Raising the bonding capacity to allow them to take on the project,
2. Each contractor has a technical specialty or niche that complements the others' domains and specialties. Together they cover project's needs,
3. The main contractor is not local. Teaming up with a local contractor, who does not have the capacity to independently take on the project, creates an opportunity for both.
4. Sharing the risk if the risk is too high for one of them.

Sharing resources among the company's projects, can also be a challenge and an opportunity, depending on the location of the sharing projects (the distance between them) and resources required for each project. Here are some factors to consider:

1. Some of the management and administrative staff can be shared if not needed on a full-time basis in one project, and transportation between the projects is feasible and economical. In some cases, work can be performed remotely without the need to travel to the site.
2. Crews can be shared, if distance is reasonable, with careful scheduling and coordination. In fact, if conditions are favorable, this can be a blessing to keep the crews busy. This applies also to major pieces of equipment such as excavators and concrete pumps.
3. If the distance between the projects is suitable, the acquisition and storage of materials can be coordinated. The contractor may take advantage of the situation by optimizing the cost of the materials. Also, the contractor may borrow materials from one site to another or take surplus from one site to another.

4. Although the cash flow must be prepared for each project independently, the contractor will also look at the cash flow for all projects combined. This allows the contractor to move the company's liquidity where needed most.

   The main office overhead expenses are usually distributed over the projects in proportion to amount of consumption of the resource or based on duration and financial volume. Generally, the percentage of overhead expenses out of total expenses tends to decrease with larger companies managing many projects as compared to smaller companies with fewer projects, although there are exceptions. The interpretation of this observation is that large companies utilize their resources, human and other, more efficiently by sharing them among projects.

5. Prioritizing is extremely important. In most situations, you must not lay down issues horizontally, i.e., at the same level of importance and/or urgency. You need to rank them vertically; first matters first! The contractor will always be joggling attention and resources among projects and even within each project, based on "priority." The priority's criteria differ from one project or situation to another. It usually stems from:
   a. The needs as specified in the contract such as project completion date, a set budget, or environmental or safety constraints.
   b. The contractor's own interest and constraints, which include profitability, needs of other projects, or plans for future projects.
   c. The consequences of the decision, like giving priority to project A over project B for a needed resource. This may result in consequences such as penalties and/or bonuses, or other consequences, negative or positive, based on the contractor's evaluation. For example, the contractor expedites the work on a project because he wants to finish the work there before extreme weather conditions occur or to start work on another project.

6. Several aspects of the project management plan may be shared among projects, especially when they are in the same geographic area, such as risk register. The contractor can take advantage of an overlap in the effort needed for the projects, in planning and execution.

7. Larger construction organizations are more likely to have a "project management office, PMO" or similar structure, to standardize the process of managing the projects, and provide them with needed staff. This can be coupled with "functional project organization," where all or most of the PM's team members belong and report to their functional managers in the main office[1].

Managing the Starting and Closing of Projects

The fluctuations in a typical construction company are more, in both extent and frequency, than most companies in other industries. As a project is about to start, the contractor must plan and prepare all the required resources in type, amount, and timing. On the other hand, when a project is about to close, the

---

1 This topic was discussed in Chapter 1.

contractor must plan and prepare for what to do with its resources. Ideally, you would like to take these resources from the project that just finished to another one that is starting. This happens with two major challenges:

1. The timing may not be optimum for the contractor. Take, for example, projects 1 and 3 in Figure 15.2. There is a 2-month gap between the final completion date of project #1 and the start date of project #3. Will the contractor keep all the resources, especially staff and equipment, even though they are not being utilized? Of course, there are other resources such as materials (surplus from previous project or construction materials such as formwork and scaffolding) and temporary structures (trailers and others) that need to be managed.
2. The type and amount of resources needed for the succeeding project may not match those released by the preceding project.

So, the contractor must manage all these resources in the most efficient possible method:

1. Resources that have high usage or high likelihood of use in the near future, must be available and ready.
2. Resources that can be acquired on a temporary basis, the contractor must make sure they are available when needed. For example, hire temporary workers or rent equipment, upon need.
3. Services that can be subcontracted or outsourced. This includes needs in both the project site and the main office.
4. The items above also change with the situation of the contractor. For example, a temporary worker may become a full-time employee if there is an extended need, and the employee demonstrates competency.
5. Many construction companies have their own facilities to store materials and equipment, possibly with facilities for maintenance and repair. Having such facilities depends on economics and whether the overhead expenses of such facilities are justified.
6. This resource management at the organization level impacts the organization's cash flow, which must be monitored carefully. The contractor must have a balanced and educated common-sense attitude in the spending and borrowing policy. Those who take too conservative or too liberal approaches will later pay the consequences.

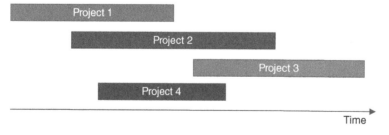

Figure 15.2 Running several projects in an organization.

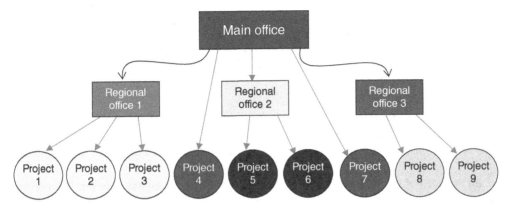

Figure 15.3 Main office, regional offices, and projects.

Main Office and Regional Offices

Many large and international organizations have a main office and also regional offices. The main office can still directly manage some projects, like projects no. 4 and 7 in Figure 15.3.

The main office and regional offices: centralized functions?

The relationship between the main office, on one side, and the regional offices and projects, on the other side, must be defined in a clear way. Lines must be drawn in defining roles, responsibilities, and authority. Even with the possible presence of regional offices, certain functions will still be done in the main office. The upper management has to define its policy regarding centralization and the degree of autonomy the regional offices and projects have. Regional offices may help standardize projects in its region. For example, if an American company has several projects in the Middle East and North Africa (MENA), its regional office there will adapt to the standards, regulations, culture, and markets of that region.

Liability is a major concern to the company, so it has to balance centralization versus autonomy not only from an administrative and financial perspective but also from a legal perspective as well. Documentation and archiving (record keeping) is another matter that may have to be centralized and standardized.

When a company plans to do operations in another country, it may decide to establish a new company in the new location, that is legally a separate entity even though it shares a lot with the mother company. Sometimes, this is done as a legal requirement, when the local authority requires that the company working on a certain project must be local.

Exercises

15.1 What does project administration mean and what does it cover?

15.2 How does a major construction company handle the human resources requirements and their transition to starting projects and closing projects?

15.3 The main office has many expenses to the construction company, such as rent, salaries and benefits, utilities, and others. Some people suggest these are unnecessary expenses (waste), while some of the employees themselves say they are overburdened, and they would like to see more employees added. How would you analyze the situation and make a recommendation to the upper management?

15.4 You are the owner and manager of a midsize construction company. You have a highly qualified and loyal scheduler, but his work fills only 50% of his time as a full-time employee. You do not like to see him leave but you cannot justify his salary for 50% of work only. What would you do?

15.5 You are the head of the IT department in a major construction company. The president of the company sent you an email expressing interest in acquiring an ERP program. Create a plan to find the needs of the company and match them with the best available program in the market.

15.6 Your company uses excavators in almost all projects, yet you do not own any excavators but rent them on as-needed basis. You are thinking of changing this practice and believe you will save money by purchasing two excavators now (and perhaps more in the future). Make an analysis and mention the pros and cons of renting the excavators versus owning them.

15.7 Your company signed a contract to build a project, but you do not have a project manager available. Your choices are:
a. Hire a new qualified project manager.
b. Assign the responsibility to Joe, a qualified project manager who is responsible for on-going project not far away from the proposed project.
c. Give an opportunity to Jose, a young assistant project manager, who shows signs of maturity.

What would you do? Justify your answer.

15.8 You are the president of a construction company. You like to keep control of your staff, so you lean more toward the functional organization. Mark, one of your project managers, is a solid believer in the projectized structure. He is always arguing with you for control of his staff. He likes to be the "CEO" of his own project. Sometimes, he even makes decisions for hiring and laying off that are supposed to be under your authority. You decided to meet Mark and have a frank and decisive discussion with him. Write your perception of the scenario during this meeting.

15.9 You are a consultant, and you were hired by a construction company to help make a decision regarding opening a regional office in a location where you have two on-going projects and possibly more coming in the near future. Currently, the main office runs operations there, but it is over 1000 miles away from these projects. Write an investigative analysis that will explore the new option as compared to the status quo. This analysis will help the leadership make the decision.

15.10 You are the president of a construction company. The main office has a large area of land, most of it used as a parking lot and storage for materials and equipment. Your operations manager suggests building a garage on the lot that will do the maintenance for all the company's vehicles and equipment. It will cost around $500,000 to build the garage and require hiring a staff of three. On the other hand, it will save lots of money and time in keeping the company's vehicles and equipment ready and well-maintained. Write an analysis that will explore the new option as compared to the status quo. This analysis will help the leadership make the decision.

15.11 Your main office has unused space. A person approached you to rent part of this unused space. You are concerned about privacy and security because the tenant will share some common facilities. Analyze the situation, mentioning the factors that will make this proposal acceptable to you. Also, mention the factors that you cannot accept.

15.12 It looks like the economy is coming to slow times that may last for several years. You know you cannot survive unless you take major steps including tightening the belt and saving on items that are not basic. You like to downsize but keep operating (bend down till the storm passes away.) Write a plan showing in bullet points what you propose.

16

Soft Skills for Construction Project Management

Introduction

Managing a construction project requires several types of skills. Just because the project has a technical nature does not mean the project manager has to be an expert in every technical specialty. Also, this does not mean that technical skills are the only or even main required qualification. We have seen brilliant engineers and architects who failed to manage an engineering or architectural firm. This is because managing an organization requires more than technical skills. This applies to all types of projects. In the construction industry, the project manager's knowledge and skills are described as mile wide and inch deep. The project manager needs three groups of knowledge and skills, as shown in Figure 16.1:

1. Technical in construction specialties,
2. Technical but supportive such as IT, accounting, finance, and legal, and
3. Non-technical such as leadership, communications, negotiation, and other skills can be combined under the title "soft skills," "interpersonal skills" or "people skills."

The project management team members also need the same combination but with different components and proportionalities, depending on the position and the role each member has. It is very

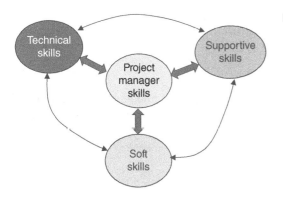

Figure 16.1 Project manager's skills.

important to know and understand what it takes to do the job of the construction project manager, including but not limited to:

a. Supervising the construction work on site; both technical and administrative, on site and in the office,
b. Coordinating among subcontractors and other work teams,
c. Anticipating and following up with own team members for tasks needed for the successful completion of the project,
d. Communicating with other project stakeholders, mainly the owner/client but also the designer, the project management consultant (third party), vendors and suppliers, related government agencies,
e. Balancing several competing interests and requirements: Own organization's interest, the client's demands and satisfaction, own team's welfare, etc.
f. Managing risks and emergencies,
g. Dealing with conflicts; within own team as well as external parties,
h. Making decisions, technical and other; sometimes on the spot,
i. Acting as a leader and motivator,
j. Keeping an open mind for feedback and criticism,
k. Being adaptive to change and new concepts, technologies, materials, and methods, and
l. The responsibility becomes more challenging and complicated when working in an international environment where the project stakeholders and the workforce come from different cultures and backgrounds.

What are Soft Skills?

Soft skills are personal traits and skills that directly relate to how well a person can effectively communicate and interact with others. People vary in nature and character in how much they naturally possess these skills. However, people can learn and improve these skills through a host of means, including, but not limited to:

- Attending professional courses and seminars,
- Reading books and articles on soft skills topics[1], and
- Learning from personal and others' experiences.

> It has been said: It is not what you say... it is how you say it!

Importance of Soft Skills

Several studies showed that our success at work or in life depends more on emotional intelligence than intellect[2]. Emotional intelligence sits at the center of all soft skills. In fact, the author believes that it is the common thread among most, if not all, soft skills.

---

1 There are many "self-help" books that one can read, enjoy, and benefit from.
2 EQ versus IQ Which is Most Important in the Success or Failure of a Student? By: Andrei Cotruș, CameliaStanciu, and Alina Andreea Bulborea. Procedia - Social and Behavioral Sciences' Volume 46, 2012, Pages 5211–5213. https://cyberleninka.org/article/n/226446.

Figure 16.2 Soft skills where the personal and professional sides complement each other.

Soft skills work on the person's entire life cycle, personal, and professional. In fact, each side complements and enhances the other. There is always a connection between one's professional and personal life. Those who can handle their job efficiently are happier and more successful, which reflects on their personal life, and vice versa.

The objective of acquiring soft skills is:

A. Better performance in your job including and most importantly how to interact with other people.
B. Improve the quality of your life, which in turn, helps you perform your job even better, as shown in Figure 16.2.

What Types of Soft Skills Do I Need?

The title "soft skills" is an umbrella that combines a large number of topics. It may not be easy for the organization to decide not only on the topics of interest for training, but also on the depth of coverage, method of training, and other criteria. Basically, the challenges are:

1. Which topics to choose from among so many?
2. What sources (book, trainer, and video) should we use?
3. How to integrate this new knowledge in our professional and personal lives?
4. How do you put different types of training in a panoramic synergizing perspective? Figure 16.3 emulates the integration among the different types of soft skills training in one's life.
5. How to evaluate the acquired skill/knowledge in short and long-term basis?
6. How to update these skills with the continuously evolving knowledge?

Figure 16.3 Soft skills topic in the overall picture. Source: Toncsi/Adobe Stock.

In this chapter, the author will cover a variety of soft skills topics that are believed to improve the performance of the construction project manager and his/her team members. The coverage of these topics will be limited in depth, only to shed light on them.

Time Management

Time management is the ability to use one's time effectively and productively, especially at work. Time management is one of the most important skills because if you can manage your time, you probably can get things done. In most cases, it is not a matter of lack of time; it is rather a matter of organization and prioritization. Time management and good organization skills help your brain: retain more and perform better!

> If you can manage your time, you can very likely get things done.

We can summarize time management by two functions: organization and prioritization. Having a good system for organizing tasks, things and data is essential for time efficiency. Any organization or person must create a system where everything (physical or digital) has a place known to them, so if they need to use that item, they can find it and access it quickly. Being unorganized not only wastes time but also adds frustration which, in turn, may reduce brain power and work efficiency. Having a "to-do" list is part of this organization, with many formats possible. New technologies, including software and cell phone apps, can help in creating such lists and setting reminders.

Prioritization is the other important component. Most of us do not have enough time to do everything we want, so we must prioritize tasks, putting the important ones at the top of the list.

In addition to the unorganized people, there are other people who cannot prioritize tasks easily such as the perfectionist and the one with too much empathy. The perfectionist insists on doing everything, important or not, in a "perfect way," thus, wasting precious time[3]. The other personality is the one with too much empathy. Such a person may not be able to say, "I'm busy now" to a friend or colleague who wants to share a personal story, or even put time limit for the story. Such a person may be considered passive, and therefore, ineffective at making good decisions. Both characters mentioned here are good if these traits are within quantitative and qualitative limits.

Some people are too passionate about work to the point of doing too much work at the expense of their health or family welfare. If you do not take care of your own and family's well-being, sooner or later you will not be able to do your work well. One of the best examples of putting logic over emotions in prioritizing tasks is when a flight attendant gives safety instructions before a flight: In case there is a need for oxygen masks and you have a child, put on your oxygen mask first *before* you help your child.

Classifying tasks according to their importance and urgency is crucial for time management. We can utilize a form of Eisenhower Matrix, as shown in Table 16.1. Once you put all tasks in their four quadrants, you will be surprised how much we mess with our tasks' priorities, which reduces our output efficiency.

---

3 Also, those who are diagnosed with obsessive compulsive disorder, OCD.

Table 16.1 Eisenhower matrix with task classification according to importance and urgency.

	Urgent	Not urgent
Important	**1. Do first/do now** – Crises and accidents – Meeting with your manager on important matter that cannot wait – Fix gas leak – Bid on a project that you really want (with deadline very close)	**2. Schedule** – Important ongoing projects – Hire new staff member – Book dentist appointment for non urgent matter – Exercise, take a vacation – Bid on a project that you really want (with deadline far away) – Visit parents (with no urgent issues exist)
Not important	**3. Delegate** – Meeting in a few minutes to discuss an IT issue (delegate it to the IT manager) – A phone call from a salesperson on a project you are planning to bid – Watch basketball game starting soon – Attend an online seminar with little importance to you	**4. Do last or don't do at all!** – Read Facebook feed – Watch TV – Time wasters – Going through your junk mail (unless you are missing an important email) – Restoring your antique car

When the "to do" list becomes a group work assignment, there are online programs, such as Clickup, that allow to organize work tasks with titles, description, assignment, dates, ball-in-court, reminders, and more. It allows for people to communicate and collaborate on work tasks. They can work on a desktop or cell phone.

Essential Time Management Skills

Time management needs several skills, according to experts, such as[4]:

1. Goal setting
2. Prioritization
3. Self-awareness
4. Self-motivation
5. Focus
6. Decision-making
7. Planning
8. Communication skills
9. Questioning and challenging
10. Delegation/outsourcing

---

4 Source: https://www.coachingpositiveperformance.com.

11. Coping skills
12. Stress management
13. Working effectively with others
14. Record keeping
15. Organization and filing
16. Patience
17. Forgiveness

Each of the skills mentioned above helps in a way to make the brain focus more. There are a lot of things that occupy the brain, thus compromising its ability to remember, think, analyze, and process information. Items on the "do not forget" list, if not written down, can play a role in impeding brain power. They are like heavy load on your pickup truck bed or like unnecessary files cluttering your computer memory or hard disk. Once you "empty" these items from your brain into external medium; paper or electronic, your brain can focus better. Emotions play a major role in this matter too. This is why major decisions must not be taken when the person is in a state of emotional outburst or excitement. If we take forgiveness, the last-mentioned skill earlier, for example. When you hold grudges and negative feelings against others (sadness, anger, fear, bitterness, envy, jealousy, and revenge), they will remain in the background of your brain, practically reducing its power and agility, and likely biasing your decisions.

Basically, emotions belong to either past (events happened already) or future (the unknown/uncertain), with a strong connection between them (certain events that happened to us drive us to do something about them.) There are several ways to deal with these emotions and rationalize them. For past events, we can expand the Sunk Cost principle[5] to cover emotions. Acceptance is a great strategy that allows us to put the past to rest and focus on the future. While we cannot reverse or change that past, we can influence the future. This gives us an additional benefit, the objective and rational analysis of past events is that we can learn lessons, improving our performance, and minimizing the likelihood of repeating our or even others' mistakes[6].

> There are two basic elements of time management: organization and prioritization!

Do We Really Perform Better Under Pressure?

Many people like to leave tasks till the last minute, under the claim "I perform better under pressure", as the case shown in Figure 16.4. While it may be true that most people tend to perform faster, but not necessarily better, when threatened by consequences, this tactic does not always work and may backfire and have negative consequences. It may be okay to go with that principle for trivial issues but not in matters with serious consequences such as project management activities.

---

5 Sunk cost is a cost that has already been incurred and cannot be recovered.
6 More discussion on this topic in the "Emotional Intelligence" section in this chapter.

Figure 16.4   Work under pressure.

But is Stress Good or Bad for Us?

There are many studies on stress and its impact on us, with most indicating that when stress comes at low level, perhaps with occasional sporadic jumps, it is not harmful. In fact, it can be a stimulant and booster of brain power[7]. As a clinical neurophysiologist put it "Stress is not all bad for your brain. In fact, moderate stress can actually improve brain performance by strengthening the connection between neurons in the brain. This helps to improve memory and attention span in order to make you more productive overall. This is why some people tend to perform "better under pressure."[8,9]

However, when stress comes at high sustained level, it can be a killer. Stress management is a key to a balanced, successful, and happy life. There are many causes for stress: personal, family, work, society, national, and global. They may relate to work, health, finance, legal, social, and other issues. We may have a degree of control or influence over them, while this may not be true for others[10]. There is a correlation between time management and stress management, as people get stressed out when

---

7 https://www.health.com/condition/stress/5-weird-ways-stress-can-actually-be-good-for-you#:~:text=It%20Helps%20Boost%20Brainpower,Shelton%20said.

8 https://premierneurologycenter.com/blog/6-ways-stress-affects-your-brain/#:~:text=In%20fact%2C%20moderate%20stress%20can,Dr.

9 https://www.healthline.com/health/benefits-of-stress-you-didnt-know-about.

10 In this regard, the line between "can" and "cannot" may not be so clear. In fact, in many situations we think we cannot do anything about them, when indeed we can, at least partially. Some matters are more societal than personal such as environmental preservation, but even societal matters start with individual initiative and then become collective. When a solid group of people induce a change, it will have better chance in spreading locally or even globally. This is an interesting subject that falls outside the scope of this book.

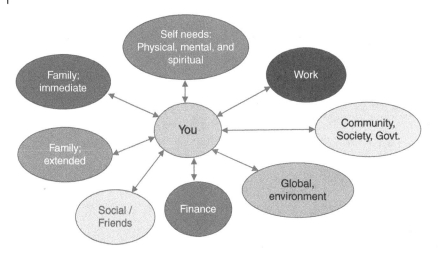

Figure 16.5 Obligations of the individual.

there is a need to do something but there is no sufficient time available. This is why planning and good time management help tremendously in reducing stress. But of course, both time and stress management have other factors impacting each. Stress management is not covered in detail in this book although it is an important area. It is a specialized area that needs in-depth discussion by professionals. Figure 16.5 shows the obligations of the individual in society. These obligations are correlated among each other and compete amongst them.

> Your brain is like your garage (or computer hard drive): No matter how big; it will soon be filled with clutter.
>
>   Once you organize it, not only there will be more space available, but it will be much easier and faster to remember, work, and process!

A Few Tips on Time Management

1. The "To-Do" list:
   a. Using a specific format is good in order to organize and standardize information.
   b. You may create a master "To-Do" list and a sub-list for today. The master list needs to be updated on a daily basis, with "today's list" produced after the update at the beginning of each day, particularly workdays.
   c. The list must contain needed information in concise way, so the user does not need to look for such information somewhere else when needed.
   d. Estimate not only the time to perform the task but also the preparation and/or wait time. For example, you need to replace the floor tile in a house. The activity itself may require five days, but

ordering the materials may need 10 extra days. Even personal tasks such as having your annual physical exam or doing maintenance to your car, usually require an appointment that may need to be scheduled in advance.

    e. Using the cell phone, perhaps synchronized with your computer, for such a list is a good way to set reminders. The reminder will take into consideration the time needed for preparation. For example, if the task is "attending a webinar" in your office, a 15-minute reminder is sufficient. If the task is attending a physical meeting somewhere 40 minutes driving-distance, then a one-hour reminder may be required. Make sure you have a backup for the information, especially important items, in case technology fails.

2. Planning is very important. It not only saves time and cost but also reduces stress and chaos.

3. Do not leave a task unfinished unless you either cannot finish it now, or you plan to finish it at a certain time for a good reason.

4. Find your most productive time of the day to focus on important tasks. Usually, the morning is the most productive time when most people have full physical and mental energy.

5. Utilize "idle/dead time." For example, while traveling or waiting, you can think, plan, and communicate. If driving, it is not advisable to be distracted but you can listen to audio books or relaxing music or may be able to communicate if this is done without compromising safety. When using public transportation, the person can do almost any task. Some exercises may be performed while doing work in the office, especially when hands are not required for work.

   So, multi-tasking is possible if only one activity requires mental effort or concentration.

6. Taking stretching and relaxation breaks during work reenergizes the person and boosts work efficiency. These breaks may also be good for reflection and contemplation.

7. Watching your time is important because many people may get drifted during meetings and conversations or when immersed in a task.

8. Documentation and well-organization are important. Many tasks require information and/or documentation that may not be handy or even available.

9. Set your personal boundaries. Learn when and how to say no to others or even to yourself!

10. The project manager, or any person in a leadership position, may depend heavily on an assistant or secretary in organizing work tasks and setting reminders. This is normal and acceptable, but a backup plan must always exist, in case the assistant becomes absent from the scene, suddenly or by arrangement, for whatever reason.

Change Management

Change management is the planned and systematic approach that includes dealing with the transition or transformation of organizational goals, rules, processes, or technologies. Change always involves people in the organization but may also involve other resources and assets such as technology, processes, equipment, materials, finance, buildings, and/or infrastructure.

Change management incorporates the organizational tools that can be utilized to help individuals make successful personal transitions resulting in the adoption and realization of change.

Figure 16.6 The transition in change management.

Change may happen at the level of:

1. Individual,
2. Group: Family, school, organization, community,
3. Society, or
4. Global.

It can be optional or forced. It may happen in one-step (sudden), multi-steps, or gradual. In most cases, there is a transitional period that leads to the permanent planned stage, as shown in Figure 16.6. In both the multi-step and gradual change, there must be a plan to get to the goal, and a mechanism for measuring progress. Take for example emission limits governments put with a future target date.

Reasons for Change

It is natural and inevitable for organizations to experience change throughout time. It can happen for many reasons such as:

a. Change in leadership/personnel: Every leader has his/her own vision, methods, and character. The change the organization faces when the leader changes may be minor or major, depending on the personality of the new leader as compared to the previous one. Some are described as revolutionaries, adopting the moto (rock the boat!), while others prefer the status quo, adopting the moto (don't rock the boat or if it ain't broke, don't fix it!).
b. Change in mission and/or expectations of the organization. It may be caused by economic expansion or going into a new market, or for other reasons. Many systems and tools the organization uses fit, or are best for, certain situations. When the situation (time, location, or other) changes, these systems and tools may need to be changed as well.
c. Change in technology or equipment which requires the staff to learn how to use and deal with the new software, equipment, or system. This has become more prevalent these days as technology is rapidly changing. Keeping up with, or even ahead of, technology is a requirement for leadership in the industry, and sometimes survival.
d. Societal/environmental: This includes societal shifts such as movement to combat air pollution and use of clean renewable energy, recycle materials, increase safety and security requirements, or automation. Such changes may be mandated by the government or just market trends.
e. Natural/forced (significant events) such as the COVID-19 pandemic, which restricted the movement of people and goods for over two years, and the long-term impact lasted even longer. Also, the climatic changes and increase of seismic and volcanic activities in certain areas, promoted the

invention and/or implementation of methods that make structures more resistant to wind, temperature, earthquakes, and other destructive phenomena.

In this rapidly changing world, "doing business as usual" may not be enough to stay in business. In fact, it may be a reason for failure as happened with business giants such as Kodak and Nokia! In the past, there used to be giant companies, so solid and strong to the point that people used to say, "this company is not going anywhere," meaning that there is no fear that this company will fail or go out of business. This concept is completely obsolete these days, as we witness the demise of many giants, most likely because they did not keep up with changes and technical developments. This, of course, does not mean that not keeping up with technology is the only reason for the failure of construction (and other) companies, but it is one main reason.

Humans, by nature, are not perfect. This is good news because it means there is no ceiling for improvement!

The Decision to Change

The leadership of the organization needs to answer a few questions before making the decision to make the change:

- What? The answer must describe the change.
- Why? The answer must focus on justification for the change.
- Facts and assumptions on the change such as roles, methods, timing, cost, stats, etc.
- Short- and long-term goals.
- The plan for the change with expectations and requirements (How to get there).
  - Obstacles and possible solutions.
  - Alternative plan(s), also known as plan(s) B/C

The leadership must study the decision before implementing: Is it a must or an option (upgrade, new opportunity)? Perhaps doing a pilot study is a good idea, if circumstances allow for such a study. The leadership must also measure all expectations and outcomes, cost of change versus cost of not making the change. Possible alternatives have to be considered and studied objectively.

Reaction to Change

People differ, by nature, in reaction to change. In general, there is a degree of resistance to leaving one's comfort zone. When change is mandated by the government or leadership of the organization, people react differently to a wide range of attitudes. Some are enthusiastic while others are resistant, and most are somewhere between enthusiasm and resistance. We can classify people in their reaction to a mandated change to:

A. The catalyst/disciple: The one who is enthusiastic about the change, who leads toward achieving the change, and helps in pulling and pushing others.
B. The follower/self-propelled: The one who goes along with the change on their own without resistance.

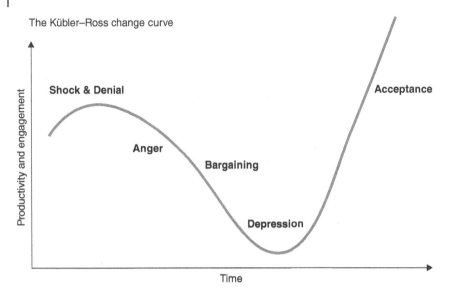

Figure 16.7   The Kübler–Ross Change Curve.

C. The slacker: The one who needs to be pulled or pushed, but eventually will go along with the change.
D. The stagnant/stubborn: The one who remains consistently against change despite the help. Such a person has no place in the organization!

There are theories and experts' opinions on the reaction to change including the Kübler–Ross Change Curve[11,12]. It describes the general reaction to the change imposed on the employees of an organization. According to this curve, the reaction goes through the stages of shock and denial, anger and frustration, bargaining, grief or depression, and finally acceptance, as shown in Figure 16.7. The reality, as the author believes, is not a clear-cut case. It depends on many factors such as the nature of the change, the character of the individual, and the efficiency of the management plan. The matter also depends on whether employees believe this change is absolutely mandated with no choice for them, or it is a case where protest and "negotiation" may influence the management's decision.

Motivational Enhancement Therapy (MET), which focuses on human motivation for change, has a model with six stages of change[13]:

1. Pre-contemplation
2. Contemplation
3. Determination
4. Action

11 https://whatfix.com/blog/kubler-ross-change-curve/.
12 https://www.cleverism.com/understanding-kubler-ross-change-curve/.
13 According to the model created by DiClemente and Prochaska (1998), see What Is Motivational Enhancement Therapy (MET)? (positivepsychology.com)

5. Maintenance
6. Relapse

Of course, the sixth stage, relapse, is a stage that you hope will not happen, but it does happen in a percentage of the population that went through the change, especially in the addiction case. So, we must prepare for it particularly in the fifth stage, maintenance.

In the aftermath of COVID-19 pandemic that started in early 2020, people were impacted globally. There were information, recommendations, and decisions at the: International, national, state, local governments, and individual organizations. But, of course, any decision at a certain level must not negate or contradict a decision from a higher authority. The "area of choice" for individuals and organizations was changing continuously with varying gray area of "not sure" or "do not know." It was a stressful and often confusing situation for many people and organizations, regarding matters such as:

- Social distancing and quarantine,
- Sanitization,
- Wearing face masks,
- Working in the office or virtually from home,
- Vaccination.

In addition to the impact of the above items on productivity, construction companies faced more challenges such as spiking prices for essential commodities such as lumber and labor. Sometimes, it was a question of availability rather than cost. The general atmosphere forced these companies to make changes, temporary or permanent. People's reaction ranged from going beyond instructions from authorities and their superiors, to the opposite including protesting, challenging, and even rebelling against orders. This type of global and high-impact change affects the lives of everyone, and the way people do business.

> Every human has capability to change his/her habits and characteristics.
> - It is more than what many of us think!

Strategies for Change

The leadership of the organization must start the change process with the right foot. As mentioned earlier, the change may be either a mandate (by the government or a governing body) or optional. For the mandatory change, we skip the vetting process and jump to preparing the plan and execution. As for the optional change, like acquiring new software to replace an existing one or moving the main office to a new building, all viable options have to be studied before a decision is made, as shown in Figure 16.8.

Here are some tips for evaluating the change:

1. The leaders must get the "buy-in" from a group[14], inside or outside the organization, who will be the management's support in the implementation process, at least until things settle down and the people

---

14 The author calls this group the "solid nucleus."

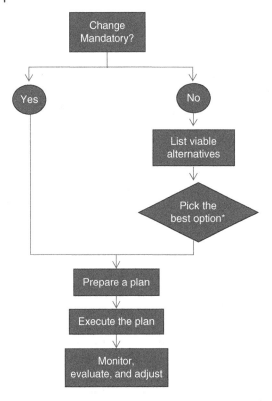

Figure 16.8 Change process choice.
*Including the option of not making any change.

in the organization accept the change and become familiar with it. The members of this "Solid Support Group," as shown in Figure 16.9, must:

a. Understand the new change well,
b. Believe in the benefits of the change whole heartedly,
c. Are enthusiastic and willing to convince and help fellow members,
d. Serve as the catalysts for the change.
e. Be the management's line for defense/offense.

2. Doing a pilot study, if possible, is a good idea. For example, if the company is contemplating the acquisition of a new ERP system[15] or accounting software, it may assign the evaluation to a few experts and have them install and test the proposed software. In fact, most software vendors allow the testing of their software for a period of time with no obligation. At the end of this testing period, the management will get the feedback from the testers group, which will help the management to make the final decision.

---

15 Enterprise resource planning (ERP) is the integrated management of main business processes, helping organizations run their businesses more efficiently and effectively. It comes usually in the form of a software program that is integrated and utilized through the entire organization.

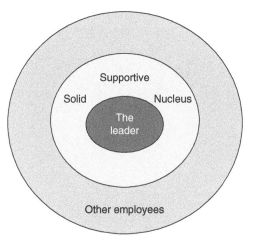

Figure 16.9 The change layers.

3. It is extremely important for the management to make the transition as smooth and painless as possible. This includes conducting workshops or seminars for all impacted employees and providing instructional materials. Also, designating certain employees as experts to help others when needed.

4. It is also important for the management to make it clear that it is fully committed to the change, and that there is no backdown. The change will go as planned with no "if's, but's, or maybe's." Those employees who refuse to accept and embrace the change have no place in the organization. The management must balance between firmness/decisiveness and compassion.

5. After the implementation of the change, the management must monitor the results and conduct follow-ups to ensure success. It can do so by face-to-face interviews, observations, and/or surveys.

   Some types of change take a long time to show results. The management should be patient while observing at the same time. A seed takes years to become a fruit-bearing tree, but you can – and should – monitor the seed growth stages to make sure matters are heading in the right direction, as shown in Figure 16.10a. Keep in mind that the change may not be "linear." It is possible that it starts slowly and then accelerates, as shown in Figure 16.10b. It is important to realize this point, so we do not get frustrated if we do not see quick results. We may take care of the "seed" for couple of weeks without seeing any external results. As we mentioned in the MET stages, pre-contemplation and contemplation take time but may not be visible before we get to the "determination and action" stages. However, management must make sure there is progress toward the goal. Otherwise, it will be a waste of time and resources if it waits for the full period and then gets no results.

6. The management needs to define and measure success because, in most cases, it is not a simple yes or no, but there is a degree in-between. Such a measure may be repeated several times periodically.

7. Change, generally, is difficult and goes against human nature, in many cases. For this reason, it is important to create a supportive and encouraging environment, whether this change is personal or professional. The key to success is discipline and expanding one's comfort zone and adaptability.

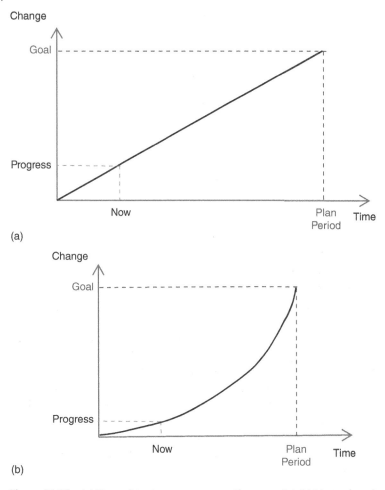

Figure 16.10 (a) Measuring change progress – linear model. (b) Measuring change progress – non-linear model.

> Vision without action is merely a dream. Action without vision just passes the time. Vision with action can change the world.
>
> Joel A. Barker

In a situation where the author was involved, a large government organization decided to replace an outdated project management system in 2002. The author was one of a three-member team selected to design a replacement system. The new system was professionally designed and successfully tested. The upper management announced the new system to all employees of the organization. A training seminar was offered with many sessions. All departments' directors required their staff to attend. There was a six-month overlap between the old and new system, with a clear and loud announcement that we will "pull the plug" on the old system at the end of the six month time limit.

In the beginning, there were many employees who either complained about the new system or just bluntly wished to keep the old system. We tried to comfort them and help them overcome any difficulty with the new system. By the end of the six-month overlap, the number of complainers significantly decreased, and the old system was turned off as planned and on schedule. Shortly after, the new system became the backbone of the organization's management of hundreds of projects, with high job satisfaction from almost all employees.

Leadership and Teambuilding

Leadership is the art of motivating a group of people to act toward achieving a common goal. In a business setting, this can mean directing workers and colleagues with a strategy to meet the company's needs and goals.

In business, leadership is linked to performance. While it is not solely about profit, those who are viewed as effective leaders are those who increase their company's bottom lines. Individuals in a leadership role must meet profit expectations set by boards, or else they may be terminated[16].

A lot was written and talked about the difference between managers and leaders. There are two important questions to ask in any organization:

- Can we combine both (leadership and management skills) in one person?
- Do we need both types in the organization?

Management is a must, so we cannot replace it with leadership alone, but a good manager also carries the traits of a leader who:

- Cares for his/her subordinates, listens to them, and makes them feel valuable.
- Follows the rules but makes sure these rules are fair and applicable; both in the general and special cases.
  - It is not only the law that matters, but also the spirit of the law!
- While doing "business as usual," always looks for advancement and improvement of the status quo.
- A good manager is a visionary too; proactive and not reactive!
- Always questions "why?" and goes beyond the surface of matters.
- Dynamic, creative, and is not a "prisoner to the textbook."
- Has an element of rebellion or defiance. This implies not accepting things that are unfair or do not make sense, even when "following the rules." This rebellion/defiance must be in a constructive and controlled manner.

In the author's opinion, in any organization, the presence of a leader or few leaders is healthy. They "rock the boat" when needed. However, too many leaders in the same organization may cause chaos and have a negative impact, as each leader may be pushing in a different direction. This is why it is best to have managers that have leadership characteristics.

---

16 Leadership Definition by Susan Ward https://www.thebalancesmb.com/leadership-definition-2948275.

Leaders: Born or Made?

While there are people who seem to be naturally endowed with more leadership abilities than others, people can and do learn to become leaders by improving particular skills. There are situations when a person is "forced" to become a leader. For example, an abrupt departure of a CEO forced the company to assign the position to another person in the organization. Such assignment usually happens on an interim basis, but when this interim leader demonstrates high qualifications and good fit for the position, he/she may stay in that position on a regular basis. The situation may happen in many other scenarios:

- The leader of a country departed by death, resignation, or other, and someone else was chosen to lead the country, often immediately.
- A project manager for a general contractor left his/her position in the middle of a large and critical project.
- A sports team loses its leader due to injury just before the playoffs or the championship game.

There are many examples in history of people who successfully stepped into a leadership position unexpectedly and without much preparation but managed to demonstrate strong leadership beyond the expectations of many people. Perhaps, they possessed traits and qualities that helped them step into roles of leadership, but it did not show because they were not given the opportunity before. Or, perhaps, the pressure of the new role helped get their traits and qualities out on the surface. This does not negate the fact that leadership qualities can be improved through various methods of learning. There are a lot of tips and tools a person can learn that make their performance better, those in leadership positions, and others.

> A good leader has courage and discipline. The balance between the two traits defines a great leader! Resilience is another trait that makes the leader stay alive and well when unexpected unpleasant events happen.

Leaders versus Managers

A lot has been said about the comparison between leaders and managers, such as the comparison in Table 16.2. However, such comparison should not lead us to believe that the characters of leaders and managers are mutually exclusive. Best leaders are those who are good managers, and best managers are those who carry leadership traits.

In motivating your own team, the project manager must follow a carefully balanced approach. (S)he should better follow the "110% Rule" which says: Try to challenge yourself (or your team) to be 10% better than last time. If you do not challenge yourself, you will not see any improvement. However, if you challenge yourself unreasonably, like demanding a 50% improvement next time, you are not likely to achieve it, and perhaps get disappointed and frustrated.

Another important point is that managers always have formal positions that assign specific responsibilities to them and give them specific authority. Leaders can be anyone in any position in the team or organization, but of course, the impact of the leader gets bigger as his/her position gets higher. For

Table 16.2 Leaders versus managers.

Managers	Leaders
Focus on things	Focus on people
Do things right	Do the right things
Doing what is right	Knowing what is right
Follow the rules	Follow the spirit of the rules and try to fix the rules
Rigid	Flexible
Plan	Inspire
Direct and control	Motivate and inspire
Delegates	Empower
Concerned most about superior's opinion	Concerned most about own team's welfare and satisfaction
Problem solvers	Innovators
Mostly reactive	Proactive
Concerned about the positions and ranks	Break the barriers among positions and ranks
Like formalities	Dislike formalities
Focus on short-term goals	Strategic look
Listen to others	Participate with others
Feared by subordinates	Liked, respected, and trusted by subordinates
Business as usual, follow trends	Anticipate, lead trends
Conservative in risk-taking	Willing to take calculated risk
Maintain status quo unless forced to change	Embrace change

More details in "The 21 Irrefutable Laws of Leadership" by John Maxwell.

example, in the general contractor's case, the leader can be the CEO/president of the company, the project manager, the superintendent, the foreman, or any other person or persons. It is also possible for someone at a lower rank position to demonstrate leadership character, so their superiors realize that and give them bigger role to utilize their character in positively impacting others toward the organization's goals.

One important leadership trait is the willingness to take risks when needed. Sometimes, the leader gets into a situation where they must take the stand they believe is the right one, even though it may be against the organization's rules or culture. Such a stand may cost the leader their position with the organization, but the leader is willing to take this risk anyway. These situations distinguish true leaders from those who "play it safe." Those who play it safe prioritize their positions over principles, whereas true leaders recognize that the position is worthless if it comes at the expense of principles and values. It was easy after the civil rights movement succeeded to change the rules to protect all sectors of society, for someone to say, "I support civil rights for everyone." But true leaders are those who stood for it even when it was not politically correct or legal.

Team Building

The most important thing about a successful team is that the members only use the pronoun "we" and not "I" in team matters, especially taking the credit. You may be a brilliant engineer, accountant, IT specialist, or other, but if you are not a good team member, your talent is not being utilized efficiently. Team members complement each other. In fact, each one is like a piece of a puzzle that, when put together, makes the right and complete picture.

There are some qualities of the effective team:

1. Share information openly among team members.
2. Members participate in all team's tasks, except when a specific assignment is given to a member or certain members, for a specialized matter that ties with the work of the other team members.
3. Encourage each other. No opinion is silly or ridiculous. The team, not individual members, takes the responsibility and credit for all ideas. Validation is a good communication tool. It does not necessitate agreement, but merely that a person can feel encouraged through validation.
4. Use all of the team's resources. Nothing is mine or yours, among the team's resources.
5. No lines are drawn along specialty work or ranks. Everyone is valued and respected.

Different opinions, discussion, and criticism are encouraged but once the team adopts an idea or decision, the entire team moves on. No more "we should've done so," "See? I told you!", or "I'll go with you, but still not convinced." The leader must be fair and objective in explaining the adoption of a decision after the discussion, even though may not be able to convince every member.

Building the team must be a careful process because any change in the composition of the team or its leadership may return the team to the forming stage. It is important for the leader to delegate responsibilities to team members as much as possible but stay in the picture, as shown in Figure 16.11. During

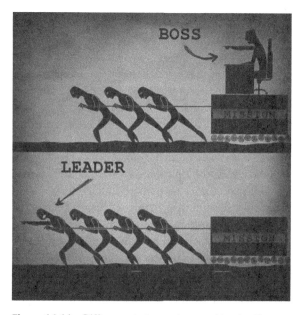

Figure 16.11 Difference between boss and leader. Source: rudall30/Adobe Stock.

work, there should always be breaks for reflection and review. Adjustments are taken, whenever and wherever needed.

The team leader must build trust and cooperation among the team members, understanding and utilizing the talents and expertise of each one of them. Using motivation and balancing compassion and firmness, the leader sometimes finds out that a certain member does not fit well in the team. This can be either from a lack of qualifications, a negative attitude, or an ambition from the team member that takes him/her away from the team objectives. At this point, the leader will need to take the decision to let the team member go and find a replacement. The leader must do this as professionally, quietly, and smoothly as possible so it will not negatively impact teamwork.

Professional Democracy

In all projects, the project manager can not be an expert in all areas of the project. This is normal and very common since the project includes many specialized areas, as long as he/she is familiar with these areas and has the qualifications to manage and lead the project. But in specialized technical matters, an employee who is two or three tiers in rank below the project manager, may know a lot more than the project manager. So, will this employee have the courage to correct or even contradict a statement made by the project manager? And if yes, will the project manager accept such opinion with an open mind? In many cultures, this is a difficult situation, and the employee may find him/herself intimidated and lacking the courage to speak. It may even get worse if this employee praises and endorses the superior's opinion, even though he/she knows for a fact that it is not the best opinion for the situation.

A good leader will not only allow such contributions from subordinates, but also encourages them to speak out freely and praises them for good ideas. On the other side, the employee must convey such criticism in the most appropriate and professional manner. For example, if the manager decided to buy new personal computers for the team. The IT specialist in the team does not agree with the choice of the manager, so he/she will articulate his/her opinion on these points:

a. Thanking the manager for thinking of such an update, affirming the need for new and powerful computers.
b. In a table format, the features, characteristics, and cost of the system proposed by the manager are compared to other system(s) that the IT specialist believes is (are) better.
c. Leave the conclusion to the manager, but the numbers and facts in the table speak loudly.

In a different example, the city is trying to solve a traffic jam problem at a major intersection. The mayor suggested creating an underpass for one of the two intersecting roads, so it goes under the other road at the intersection. Thus, the traffic continues uninterrupted on both roads. A city engineer realized that an underpass is not a good idea in that location because of the high water table[17]. Sooner or later, there will be water leak problems. In addition, there is a safety concern for flooding in case of heavy rain. The city engineer explained to the mayor and committee:

a. Indeed, there is a traffic issue at that intersection. Thanks to the mayor for paying attention to this issue and trying to find a solution.

---

17 The water table is the boundary between the unsaturated zone and the saturated zone underground. Below the water table, groundwater fills any spaces between sediments and within rock.

b. There are two solutions to this problem: An underpass or a bridge. Both are "good" solutions that will eliminate the traffic jam at the intersection.
c. The underpass can be the better solution in areas where the water table is not that high, and heavy rain is not expected. However, this is not the case here (the engineer can focus on the potential safety and maintenance issues with the underpass solution in this location).
d. The bridge is a solution that has the same advantages as the underpass, namely eliminating the traffic jam and keeping the traffic flowing in both directions and is not impacted by water leaks or flooding.

However, professional democracy does not negate the presence of a hierarchy. On the contrary, a hierarchy is needed for the good decision-making process. An employee may express a technical opinion on a matter to the leader. Even though the employee is correct from a technical perspective, the leader may decide to go with a different opinion due to a strategic goal that the employee may not see or know. The important thing for the leader is to listen to the different opinions, analyze them objectively, and then make the decision that is in the best interests of the organization.

Conflict Management

A conflict is simply a disagreement between two parties on an issue (or issues) relating to a matter of mutual interest. Conflicts in general are a normal and inescapable part of life. They occur periodically in any relationship. However, a conflict is also an opportunity to understand opposing preferences and values. The roots of the conflict may exist even if the conflict itself does not surface. In some cases, the conflict seems to arise from unworthy cause, but the roots of the conflict may go much deeper than the apparent cause.

In many instances, good things come out of conflict resolution. There are many benefits of conflict resolution, especially when the solution makes a win–win situation for the parties, perhaps not exactly the way they saw it prior to the resolution. In fact, this is precisely the mission of the mediator: To shed light on other fruits of the resolution beyond and outside the original cause of the conflict. Conflict resolution helps reduce stress and tension at the conflicting parties, retain good employees (when a conflict happens between an employer and employee), and improve relationships in general.

In construction projects, conflict may arise between two entities such as the general contractor and the owner. It can arise between two subcontractors, between the general contractor and a subcontractor, or

Figure 16.12 Conflict management and resolution.

between a contractor and a vendor. Conflicts may also arise within the same organization such as between the project manager and the main office staff or between two departments or people in the same organization.

Conflict management is the process of limiting the negative aspects of conflict while increasing the positive aspects of conflict. The aim of conflict management is to enhance learning and group outcomes, including effectiveness of performance in an organizational setting. The steps and tips discussed below are beneficial in both cases: when the reader is one of the conflicting parties or when mediating between two conflicting parties.

The steps to conflict resolution can be summarized as[18]:

1. Identify the source of the conflict. This requires listening and "digging deep" by the mediator to explore any areas, in the past or present, that have a direct or indirect impact on the conflict. As mentioned earlier, there may be roots for the conflict deeper than the apparent cause.
2. Look beyond the incident. In many cases, the conflict may be about financial matters, but there are other issues that can be of interest to the conflicting parties or one of them such as future relationship with the other party or reputation. In some cases, the incident was minor, but it escalated because it transformed into personal attacks. The mediator must be skilled in isolating the issues.
3. Request solutions from both conflicting parties. Such solutions may have to be adjusted or smoothened through the discussion with the mediator to make them more acceptable to the other side.
4. Identify areas of common acceptance in the suggested solutions. The mediator has to be skillful and delicate in tailoring a solution out of the suggested solutions that is likely to work out. The mediator will need to identify the "green, yellow, and red areas" for each of the disputing parties. Once this process is completed, the mediator has created a solution both disputants can support.
5. Strike an agreement. The final stage of the agreement may require some tweaking and final touches from the mediator before the handshake occurs. Making the agreement formal (written and signed) depends on the situation. In some cases, this may be necessary while in other cases, it is not. The agreement should be celebrated as an achievement for the conflicting parties as well as the mediator.

> We do not seek conflicts; we try to prevent and resolve them. But when they occur, they can be an opportunity!

When the dispute is long and complicated, use these steps:

1. Breakdown the dispute into segments or phases.
2. Stop and pause if matter escalates or tones turn higher.
3. Take a break after each round. Perhaps have a light meal and do some entertainment.
4. Reevaluate the issues during the break. Make any needed adjustments.
5. Celebrate every step toward the resolution.

---

18 The original source is https://www.amanet.org/articles/the-five-steps-to-conflict-resolution/ with some editing from the author.

Conflict management requires several skills[19], such as:

1. Communication Skills
2. Emotional Intelligence
3. Listening/Empathy
4. Creative Problem Solving

Active listening is being able to summarize what the person is trying to communicate and advocate. At the same time, the mediator may express empathy by validation, mostly by facial expressions and nodding the head. This does not mean "I agree with you," but rather means "I am listening." This shows respect as well.

In relationships among people, we sometimes use the principle of "credits and charges." This principle applies in a continuous relationship between two people or organizations, who like to maintain this relationship. When one of them gives in to the other side in a dispute or a decision with different preferences, he/she earns a "credit." This credit may be used later when the other side gives in to you in another matter, as if you are making a "charge," using your earned credit. Although this is purely hypothetical and there is no tangible measure of these "credits and charges," the principle works in the subconscious of the two parties in reciprocating concessions rather than trying to win every dispute, debate, or disagreement. This does not necessarily mean one-for-one, but it simply says that in most cases of differences, it is hard or even impossible to know or agree on the absolute "correct" opinion. So, they will move on with one party giving in to the other. This works only if both parties are open-minded, flexible, fair, generous (or at least reasonable), and keen on maintaining the relationship with the other party.

Effective Negotiation

The simple definition of negotiation is a discussion between two or more parties, aimed at reaching an agreement. A successful negotiation requires the parties to come together and hammer out an agreement that is acceptable to both/all.

It is interesting when looking at skills required for effective negotiation, we find a large overlap between these skills and skills needed for other missions such as conflict and change management, communications, leadership, problem-solving, and emotional intelligence. In fact, negotiation can be a component of conflict management because it takes good negotiation skills to resolve a conflict.

In business, you don't get what you deserve. You get what you negotiate!
https://www.amazon.com/Business-Life-Dont-Deserve-Negotiate/dp/0965227499

Negotiation is needed for almost everything. In fact, we may not realize that little children "negotiate" with their parents either for a treat or to escape a chore or punishment. We negotiate when we buy a car, house, or other items. Negotiation is not always about money. A general contractor may negotiate a later

---

19 https://www.thebalancemoney.com/conflict-management-skills-2059687.

finish date with the owner or some terms of the contract. A defense attorney may negotiate the fate of his/her client with the judge. Negotiations take more serious consideration when they are between leaders of countries, in peace or war.

Top Effective Negotiation Skills[20]

1. Communication: This means going beyond and around the literal meaning of the words said and heard. This includes the ability of the mediator / negotiator to:
   a. Understand the tone, body language, and allusions of the parties.
   b. To express certain feelings to them, hinting and alluding in "unsaid words." Bossy words and expressions must be avoided.
   c. Ask the right questions and extract the necessary information.
   d. Know when to speak, when to listen, and when to just stay silent.
2. Active Listening: It is important for the mediator to listen attentively, without interrupting or putting the words in the mouth of the speaker. It is okay to show empathy by facial expressions, but the mediator must avoid showing any bias (for or against) with such expressions such as making "I don't believe you" face[21]. Taking notes occasionally is a good idea for keeping thoughts, but without making the other side feel uncomfortable.
3. Emotional intelligence: This is arguably the foundation of all soft skills. It will be explained later in this chapter, but basically, it is the ability to understand and control own emotions and read and influence others' emotions.
4. Preparation/planning: This is a must in every project or complicated task. Planning allows the person to sort out related ideas, obtain the facts, list priorities, and prepare a plan, likely with alternatives.
5. Problem Analysis and Solving: This is the ability to analyze, dissect, and break down complicated problems into small and isolated issues. It also helps distinguish between facts and emotional/subjective wishes. Analysis may also go into assessing these issues in terms of importance to parties (putting value to each) because negotiation may, and usually does, happen on the basis of individual issues (I give some items in exchange of getting others.)
6. Value Creation: This can be a part of "problem-solving." It means the ability of the negotiator to explore and find value in solutions either suggested by self or the other party. For example, if you are selling an old building that needs many repairs, you can also point out and highlight the strategic location of the building and/or its historical and symbolic value. For a general contractor trying to convince a subcontractor to sign a contract at less than the sum requested, the general contractor can create value in establishing a long-term relationship with him or with the owner. This is the art of shedding light on areas that the other party either did not think of or did not value much.
7. Decision-Making Ability: Negotiation may drag for a long time, so the good negotiator must know when to stop negotiation, with or without a deal. Making a decision may not be easy since the decision is likely to bring in gains and losses. Negotiation should be neither total victory nor submission, although it is possible that one party's gains are more than the gains of the other party. The decision

20 For more, read "You Can Negotiate Anything: How to Get What You Want", by Herb Cohen and "Getting to Yes: Negotiating Agreement Without Giving In", by Roger Fisher, William L. Ury, and Bruce Patton.
21 A funny and embarrassing situation faced some American businessmen when traveling to Japan in 1960's. In attempting to sell goods or services, the Japanese host would nod his head. The American businessman would think "He just nod his head, this means yes. We got a deal!" When, indeed, this only meant "Yes, I am listening!"

entails that the successful negotiator puts the gains and losses in one basket and asks the following questions:

a. Is this a win situation for me, overall?

b. Is this the best deal I can get under the circumstances?

c. What are my gains and losses if I reject the deal?

8. Ethics and Reliability: The negotiator, while looking for own interest, must be honest and fair with the other parties. It is extremely important to avoid lying, hiding related issues[22], misrepresenting the facts, or cheating. The negotiator must be reliable in the sense that he/she can back up any statement and fulfill any promise he/she makes. Effective negotiation is also about collaboration and teamwork to reach a deal that maximizes benefits to all parties.

The Power of Persuasion

In many situations, the project manager gets into a situation where he/she is trying to convince a client (current or potential) or other party to strike a deal or accept an agreement. We talked about negotiation earlier, but in many cases, the project manager has the opportunity to make a presentation to lay down his/her case. Such a presentation can be the key to the deal. His/her performance can be the difference between getting the deal and not getting it. The presentation is more important in the case of competition, where a few others also do their own "show", and the client will eventually pick one of the candidates. The client's decision boils down to the question: Why should I hire you and not others?

The presentation can be either in-person or virtual (online). Although the tips may differ between the two types (in-person or online), but it basically includes tips in two groups: The materials and the performance.

> When the contractor acquisition is not based solely on price, there is room for the contractor to convince the owner to select them. This hinges on the ability of the contractor to convince the owner that they can build the project in the best manner, focusing on owner's priorities and matching them with the contractor's experience and skills. In short, giving the owner the highest value for the project and his/her money.

The materials include the presentation itself (e.g., PowerPoint) in addition to other support materials such as brochures, video clips, hand-outs, or even samples. The performance must utilize all available tools: the voice tone, body language, eye contact, facts, humor, relevance, and interaction. A mistake some contractors make is to focus on (and brag about) their previous projects and work that does not relate to the client's specific needs. Bragging about previous work may be justified, and even demanded, only within the context of the client's demands and constraints. For example, you may have a client who has a project with a very strict deadline, another client may have a tight budget, or a third one with strong environmental convictions. The most important thing for the contractor is providing the maximum value to the client. This is similar to the case of a job applicant who boasts

---

22 This is why when doing an in-court testimony, the person is required to tell the truth, all the truth, and nothing but the truth. Telling part of the truth is a distortion that is equivalent to lying.

about being able to speak four languages, but all what the potential employer needs is someone with strong accounting skills!

The presentation usually has limited time, so the contractor (or the candidate) must be ready and must utilize the time efficiently and wisely. There are subtle signs that may turn the client off:

- If you start late, it may indicate that you will be late for your project or that you don't care about others' time.
- If you are unorganized or run around looking for something frequently or even occasionally, it indicates that you do the same thing at work. Math errors are grave mistakes.
- If you treat your assistants or other people disrespectfully, it may leave a bitter impression.
- If you make any statement that either you cannot backup or be contradicted later, it may harm your credibility. It has been said that credibility is like a glass jar, it can only be broken once!
- If you do not look professional and talk as a professional, then you may not be taken seriously.
- Convince the client by focusing on the advantages of your products and services, not by trashing your competitors.

Emotional Intelligence

Emotional intelligence as a concept and component existed for centuries with different names and titles given by different scholars. These scholars and researchers discovered that a person's success in life, both personal and professional, depends on factors that are far more than scientific achievement and IQ. A common thread among these successful people is the ability to communicate and interact with others well. In 1995, Daniel Goleman, a science reporter for the New York Times, published his book "Emotional Intelligence" which instantly became very popular. Like brush fire, this subject spread around the world; both as a stand-alone topic and as a part of many other topics such as project management.

Being a "book smart" is not enough to do the job!.

Many experts suggest that emotional intelligence is more important than IQ (or "book smart") for success in life. In one study, researchers have shown that our success at work or in life depends on Emotional Intelligence 80% and only 20% of intellect[23].

Emotional intelligence is the ability to perceive, interpret, demonstrate, control, evaluate, and use emotions to communicate with and relate to others effectively and constructively. We use both "EI" to denote emotional intelligence and "EQ" as a measure of emotional intelligence, as compared to IQ, intelligence quotient.

The steps to emotional intelligence, as many experts suggested, consist of four steps / components:

1. Self-awareness
2. Self-management
3. Social awareness
4. Relationship management

---

23 https://www.sciencedirect.com/science/article/pii/S1877042812021477.

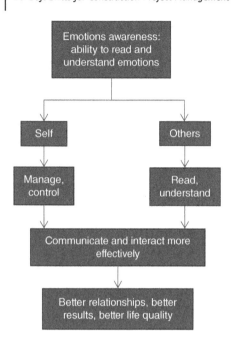

Figure 16.13 Stages of emotional intelligence.

The first step implies understanding self-emotions, which includes knowing and acknowledging each type of emotion that affects me, understanding the causes that stir each type, and why. In the second step, I will try to manage and control the emotion. Doing so does not imply suppressing the emotions but rather managing them in a positive and productive manner. Muhammad (the prophet of Islam) once said "The strong (man) is not the one who can overpower others (as in wrestling); rather, the strong man is the one who controls himself in a moment of anger." The third step is to understand and have awareness of others' emotions. This is a difficult step because of the wide variety of people's characters. Once you successfully complete (or let us say have a grip on) this step, then the final step is to be able to interact with and influence others' emotions, especially those who are in constant contact with self. This is summarized in Figure 16.13.

Understanding others' emotions and knowing "how to push their buttons" is a big advantage when dealing with others. With such talent, you can settle disputes, negotiate, resolve conflicts, and be more approachable and likable, in a better and more effective way. A project manager may understand the personalities of his/her staff, subcontractors, and clients and deal with them differently based on their characters. This makes this project manager more successful in both motivating and minimizing negative issues.

There is a misconception that we need to manage and control only negative emotions such as anger, hate, sorrow, envy, jealousy, frustration, and depression. But positive emotions must also be managed and controlled too, such as joy, love, confidence, pride, respect, admiration, affection, and altruism. The main objective is to always put logic in control of emotions, not the other way around.

If you are one of those who say, "I manage myself well, but someone made me mad when (s)he did so and so.", then you really don't have good control over yourself. Control shows at tense, not normal, times!

Emotional intelligence starts with thoughts, and then feelings, and finally behavior:

Thoughts → Feelings → Behavior

This is easier said than done, but people can train and improve their emotional intelligence, EI. One of the most important keys to improving EI is discipline and willpower. There are training programs that aim at strengthening awareness, critical thinking, decision-making, and willpower. One of the training tactics is to think of the consequences of possible actions. For example, a general contractor is having a confrontation with a subcontractor, that started over a 2-day delay caused by the sub, but escalated farther. The GC is furious and thinks of firing the sub. However, firing the sub at this stage of the project may cause the GC further delays and extra cost. Succumbing to anger emotions will carry a high cost to the GC. The better option is likely to keep the sub and focus on finishing the project on time. After the end of the project, the GC can evaluate his/her relationship with the sub and decide whether he/she will use that sub again.

Another example of EI is a contractor who had a claim dispute with an owner over a $15,000 change order. The contractor can pursue the claim with a good likelihood of winning but realizes that this may ruin his relationship with the client. He realizes that maintaining a good relationship with this client is worth a lot more than $15,000, so he concedes the claim battle, despite alluding to his right in the claim (earning a symbolic credit), but focusing on the desire to continue a relationship with this client, that is prosperous for both.

Emotional intelligence can be used in many different ways in your daily life. Some different ways to practice emotional intelligence include[24]:

1. Being able to accept criticism and responsibility.
2. Being able to move on after making a mistake.
3. Being able to say no when you need to.
4. Being able to share your feelings with others without offending them.
5. Being able to solve problems in ways that work for everyone.
6. Having empathy for other people.
7. Having good listening skills.
8. Knowing why you do the things you do.
9. Not being judgmental of others.

IQ+EQ = Leadership

Those who cannot read and control their own emotions are controlled by their emotions! Which makes them susceptible to be controlled by others.

• Avoid being controlled by others when these people, including your opponents and adversaries, know your weak spots, so they may "push your buttons".

---

24 https://www.verywellmind.com/what-is-emotional-intelligence-2795423.

Emotional intelligence and effective communication are closely correlated. An article[25] on Forbes website, mentions five ways you can communicate with more EQ:

1. Listen and reflect before responding.
2. Acknowledge and affirm.
3. Use your empathy skills.
4. Gather information through reality testing.
5. Do not take things personally.

It is amazing how much negative energy you can defuse from the other side, whether you are one of the disputing parties or a mediator. Remember that listening to others is for understanding what they say, not just to win an argument. You may hear things you do not like or disagree with. Do not interrupt to "correct" the speaker right away. You can take notes for rebuttal but remember that not every point is worth arguing or even mentioning. You even show courtesy to the other side after they finished taking by politely asking "Are you finished?" or "Can I speak now?" Staying calm and objective tremendously helps in defusing rising emotions and softening the other side.

Communication in the international environment may be a bit more complicated. The basic concepts stay the same, but some practices differ due to different cultures. There are expressions that are commonly used in certain cultures, while they are offensive in other cultures. Body language also differs vastly from on culture to another. Not to mention laws, rules, protocols, and professional practices. So, if a company is going to do business in a new market, it is encouraged to do its homework before getting down to negotiation. Having a local person who is familiar with both cultures is a great advantage.

Work for the Best Outcome:

One of the key elements in making a decision in a tense moment is not to think of theoretical situations and what "should be," "should have happened," or "supposed to be," but rather what can be done at this moment. For example, a subcontractor caused a 3-day delay in the project. The GC's project manager was upset somewhat but did not raise the issue because he managed to absorb it, as he had enough "project float." Three months later, the project got behind schedule for other reasons and the GC had to accelerate the schedule in order to meet the contract finish date. In a meeting with all subcontractors who still have work remaining, the project manager was bitter at the subcontractor who caused the 3-day delay three months before. The project manager should be thinking now about options available to accelerate the schedule now, not what already happened and cannot be reversed. This is similar to the "Sunk Cost Principle," which was mentioned earlier in time management. The subcontractor's earlier delay may be mentioned casually if the subcontractor does not cooperate in the acceleration effort.

The most important aspect of emotional intelligence is to focus on the outcome, maximizing the benefits and/or minimizing the losses. This is precisely what "putting logic in control of emotions, not the other way around" means. Take for example a situation when the contractor and the owner are locked in a bitter dispute over a change order submitted by the contractor and rejected by the owner.

---

25 https://www.forbes.com/sites/forbescoachescouncil/2021/11/30/say-it-with-more-emotional-intelligence-five-communication-tips-for-leaders/?sh=42ac12402733.

The contractor can allow his emotions to escalate and take over his actions. He would then tell the owner "I will stop all work until you sign the change order," or make a better and more logical decision of continue the work while postponing the resolution of the dispute till later. Here is a quick analysis of both scenarios:

- Scenario 1: The work will stop for perhaps a long time with potential losses in cost and time, in addition to the stress and strained relationships. Even if the contractor wins the original dispute, there could be another dispute over the consequential delay. The situation is likely to be a lose-lose one.
- Scenario 2: The dispute reached a deadlock, but the general contractor decided to continue work under the "constructive acceleration" term, which means that continuing the work does not imply the concession of the contractor regarding the standing dispute, but rather reserving the rights for pursuing the dispute till it is resolved later. The contractor is keen on documenting everything so he can use it to back up his claim later. Thus, the contractor minimized and contained the potential losses. Most importantly, prevented the consequential delay from happening, and possibly minimized any deterioration in the relationship with the owner (since the matter did not escalate).

The contractor, in following the second scenario, did not use the "absolute right" mentality but followed the "best outcome approach" that likely leads to best results for everyone. We all know many cases where a person made the wrong decision under the pretense "it was a matter of principles". Well, in the majority of these cases, you will find out that it was emotional thinking that led to the wrong decision, resulting in losses. This is not a call to give up on principles, but rather a call to look at the expected consequences and sort out your priorities. It is like when a cashier at a convenient store with a robber has pointing a gun to his head and demanding the money. From a "principal" point of view, the cashier must not give the money to a thief, but looking at the realistic outcomes, he had only two choices: Give the money and save his life or lose the money and his life together!

Another issue in emotional intelligence is that your actions must be independent of your opponent's actions. They may not be pure reactions! So, if someone yells at you or does something illegal or shameful against you, you don't need to retaliate the same way, even when you are convinced that it is your right to do so. Use the same principle: Contemplate your options and select the option that maximizes your benefits and/or minimizes your losses.

General tips:

1. Focus on outcomes: You do not want to win the battle but lose the war...
2. In life, there are factors we cannot control (hereditary/genetic, environment, "luck," etc.) but there are certain factors we *can* control.
   - Positive people focus on what they can control rather than complain about what is out of their control.
   - Even factors that you believe are beyond your control, they may be influenced by concerted and persistent effort!
3. All powers we have, physical and mental, are on a "use it or lose it" basis, so you better use them.

Do not be penny-wise and pound-foolish!

Professionalism and Ethics

This topic is not part of the soft skills, but it is part of the requirements of the project management team members, particularly the project manager.

Professionalism is a combination of competence or skills expected of a professional and the display of behavior or attitude while at work. A professional is the one who performs his/her own work duties in a satisfactory manner, technically and ethically. Professional associations and societies publish their own professional standards and codes of ethics, that members have to abide by.

The Project Manager

The Project Manager is the captain of the ship that sails from point A to point B. He/she is responsible for what and who is on the ship, what happens on it, what resources it needs for the journey, and how to manage the risks that will face this ship until it reaches its final destination, safely, on time, and on budget. More importantly, the "captain" must manage all the ship workers; assigning their roles and responsibilities, making sure they are doing their jobs correctly, and resolving any conflict among them. The skills required from the project manager are the same three groups mentioned in earlier this chapter: Technical in the construction specialties, technical but supportive, and non-technical (mainly soft skills).

The project manager has professional obligations, as shown in Figure 16.14, towards different parties, as shown in Figure 16.15. This requires a delicate balance, without allowing the focus on one party / side to result in neglecting other parties / sides. In addition, the project manager has needs and obligations in his/her personal life that must be fulfilled as well.

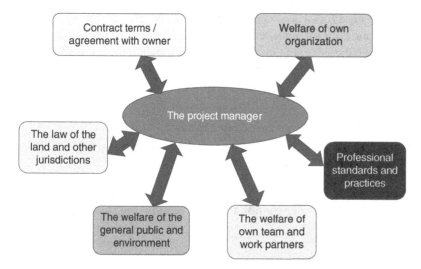

Figure 16.14 The project manager's professional obligations.

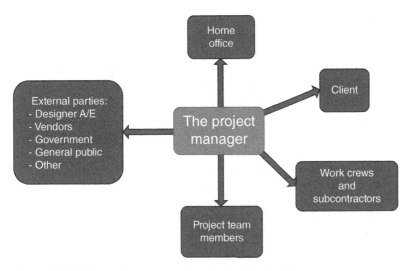

Figure 16.15   The project manager's professional network.

Dealing with Difficult Owners

One of the challenges that faces the project manager is dealing with different personalities, especially when it comes to the owner. Here are some types of owner's personalities with a brief recommendation on how to deal with them:

1. The negotiating / demanding / greedy owner. This type of owner not only looks for the lowest price, but also free favors, sometimes even for work not related to the project. It may be okay for the project manager to do a small task at no extra cost as a goodwill gesture, but it is important not to allow this to become a pattern. The contract must have a clear scope description, so anything above and beyond, must be processed as a change order.

2. The "crack the whip"/ "I want it now" owner. This can be either for the end of the project or for work progress in general. The project manager must assure the owner that it is the contractor's responsibility to finish the project on time according to the signed contract. The project manager must also convince the owner in a nice way, not to interfere with work progress or judge the completion percent complete from the appearance of the incomplete project. It can be frustrating to have the owner present always or frequently on site and giving instructions.

3. The picky / hard-to-please owner. Again, the project manager has to use the written contract as a reference. Attempting to satisfy the owner is something any businessperson would like to do, but to a limit, especially when this satisfaction goes beyond the work required in the contract and takes a toll on the time and cost of the contractor.

4. The cheap / "penny wise pound foolish" owner. This can be a dilemma for the project manager when the owner is trying to tighten the upfront cost instead of getting the highest value. This matter should be discussed before signing the contract, so the owner makes the appropriate decision, balancing their own needs and budget. But the issue can come up again in the case of change orders. The best

course of action the project manager can take is to lay down the options for the owner, along with their pros and cons and the cost. After that, the matter is completely in the owner's court, except in matters where safety or legal boundaries are at stake.

5. The oblivious / too busy / "I trust you man" owner. Although the project manager does not like to see the owner continuously interfering with work, he/she does not like to have the owner completely absent from the site. It is professionally appropriate and customary for the owner to have a presence on the site, even when it is occasional. If the owner still does not show up nor has a representative, the project manager should periodically send the owner a report, perhaps weekly, summarizing the work progress, preferably with pictures. Such reporting puts the ball in the owner's court and stands as a witness stating that the contractor is updating the owner constantly.

6. The intrusive / bossy / "poking nose in everything" owner. This can be a dangerous situation especially when the owner gives instructions directly to subs and workers. The owner must be informed in a polite yet firm manner that the contract and professional practices prohibit them from giving instructions to anyone on site and that the communication must be only via the contractor's project manager.

7. The suspicious / paranoid owner. The project manager must work on this matter in two parallel lines: earning the trust of the owner and documenting everything in an easy-to-retrieve way. The project manager must not show signs of anxiety or anger. By behaving normally and providing convincing answers, the project manager is likely to be able to contain this attitude.

8. The owner who does not follow rules or even the law. This is a dangerous behavior when the owner, under whatever motive, does not care about the rules and the law. The project manager must quickly and swiftly draw the line and refuse to do anything unethical, illegal, unsafe, or in violation of professional standards.

9. The procrastinator / "do not worry, I will do it soon" owner. This is a frustrating matter because the owner has their own timely obligations such as signing change orders, processing progress payments, and conducting certain inspections. The project manager must give the owner reminders using documentable means such as text messages and email. When this behavior from the owner continues, the project manager must send an official message / email alerting the owner to the consequences of this behavior, in both time and money.

10. The forgetful / unorganized / "no I did not say that" owner. The project manager has to be keen on having everything in writing. So, if the owner asks the project manager on the phone to make a change, the project manager must ask the owner for a written request. If such a request does not come, the project manager can instead send an email to the owner mentioning the phone (or face-to-face) conversation and the owner's request, asking for confirmation. Documentation is always important, but it may be even more important in such situations.

11. The irresponsive / "too busy" owner. Like the previous character, the project manager must use written communication, especially for important matters, with the owner. It is a good idea to make the message as brief and concise as possible, and perhaps making important information (cost amount or deadlines) in bold or caps. It may also help for important meetings and other events, to send a reminder to the owner a day or hours before.

12. The "out of this world" / unrealistic / impossible owner. There are owners with fancy, but unrealistic imaginations and those who make unrealistic requests like performing a one-month work in two days or changing a work item after its completion. The project manager should reply in a calm way, giving them the realistic and available options.

13. The "ignorant but does not realize it" owner. This is truly a dilemma when the owner gives an opinion on a technical matter that they have no knowledge in. It is inappropriate for the project manager to tell them that in a direct way: "you do not know this matter", but can instead explain the matter in the simplest layman's terms, providing the realistic and available options.

14. The arrogant / "I only give orders" owner. The project manager is recommended to use a formal approach with such an owner in correspondence and any other business matter. Official correspondence comes in generic, professional, and unbiased format that will minimize the impact on the owner's character. This puts the owner and the contractor as two partners at the same level in this contract, each with their own expectations and responsibilities.

15. The hesitant / "frequently changing mind" / unpredictable owner. The project manager is recommended to follow a two-step approach with this type of personality. First, explain the available options along with their pros and cons. Second, give the owner the opportunity to make a decision based on the given information, and then make sure the decision comes formally in a written and signed document. Any change after that comes with a price to the owner. In this business, there are no "free returns or cancellations."

16. The old-fashioned / "I hate technology" owner. We still have people who refuse to go with today's technology. Even when they have it, they do not know how to use it properly or just do not like using it. Such owners may not read their email or pay attention to electronic notifications. It is important for the project manager to discuss this matter with the owner and reach an agreeable resolution. Sometimes in important matters, the project manager may follow the email with a text message or a phone call: Please check your email!

17. The talkative / "let me tell you a story" / too friendly owner. This type of owner may be a little entertaining but takes too much valuable time and hinders the project manager form doing important duties. The project manager can find a good moment to interrupt the talk and gently apologize for not being able to continue. A little bit of socialization can be a good way to strengthen relationships, but the line must be drawn when such socialization takes important business time.

18. The nagger / moaner / obnoxious owner, who calls you frequently for little or no reasons, perhaps several times for the same issue. Even though this is not a serious issue, it may annoy the project manager and take time from his/her busy schedule. It can be dealt with, by the project manager, in a nice professional manner, comforting the owner, informing them of the project status, and assuring them that the contractor is making all the effort to complete the project according to the contract terms. It is okay for the project manager not to answer the phone sometimes, especially outside work hours. Always, written messages (text, email) are preferred.

Keep in mind that some of these characters can be combined. For example, there could be someone who is talkative, frugal, and does not follow the rules! We can also add organizations that are difficult to deal with, for a variety of reasons such as frustrating bureaucracy, poor decision making, poor management, poor communications, and those with financial difficulties. Here are some general recommendations for the project manager to deal with all types of clients:

1. Instead of saying no, lay down the available options along with the pros and cons and cost of every option. Support your talk with facts and numbers. Let the owner make their decision <u>after</u> being educated on the issue.

2. Draw the line when a request is unacceptable legally, ethically, or professionally.

3. Be professional and respectful but expect respect from the other side.

4. Insist on following proper procedures and full documentation. When receiving verbal request from the owner, in a physical or virtual meeting, or on the phone, do not be shy to ask for a written request from the owner, confirming the order, especially for important matters.

5. Be keen on conducting weekly meetings where many of the owner's inquiries can be addressed and responded to. Meetings minutes must documented and sent to all parties.

6. Do not be embarrassed to say, "I do not know" but follow with "I will find out and let you know as soon as possible."

7. Do not give a promise for a deadline or any other matter unless you have high confidence in fulfilling this promise.

8. Keep the communication channel always open with the owner. Be precise and concise. Use pictures and/or other attachments when useful. Owners can always respond by requesting more information when needed.

9. Always sharpen your skills: Both technical and soft. This gives you a better negotiating and convincing power.

10. While you are looking to perform today's work, keep in mind the long-term relationship with the owner and your reputation, and think of how much it means to your company.

11. Realize and acknowledge the diversity in people's personalities, including project managers themselves and your team members. We are not psychologists or behavioral scientists nor in the business of changing people's personalities and behaviors. We are simply trying to communicate and do business with others in the most effective and efficient manner possible.

The Project Management and Technology

The project manager and team members must always be on the cutting edge of technology, taking advantage of it to do the work as efficiently as possible. However, it is also important to conceptually understand how things work and not blindly depend on technology. For example, scheduling or cost estimating programs may give output that you must be able to interpret, judge, and then accept or reject. Sometimes you may not have instant access to technology and communication means, or technology may fail for whatever reason, can you still function effectively? Successful project managers are those who utilize technology but do not allow technology to drive them. They have to be in the driver's seat, not in the passenger or back seat.

Exercises

16.1 Do research on the topic of "soft skills": What does this term mean and what topics are usually included in it?

16.2 What are the three groups of knowledge and skills a typical project manager needs?

16.3 Construction is a technical field, so why does a project manager need to have other non-technical skills?

16.4 You are the owner of a construction company. One of your cost estimators approached you expressing his/her interest in filling a project manager's position. He/she studied for this position and obtained a PMP certificate. However, your own observations show a lack of "people skills" on his/her side. He/she gets into many arguments and confrontations with others. How do you respond to him?

16.5 The project manager has many duties and responsibilities. It takes a lot of skills to do the job of the construction project manager. Mention some of these duties and responsibilities.

16.6 How do you define time management?

16.7 If time management is summarized in two words, what would these two words be and why?

16.8 Jack, a cost estimator in your company, is always late on appointments, frantically rushing to do things, and complaining about "too much work and not enough time." He is an honest person, but you think he is trying to cover something he does not want you to know. By looking at other employees, including Jack's predecessor, with comparable workload, you believe he has sufficient time. How would you handle this situation before it adversely affects the company?

16.9 Prepare Eisenhower Matrix and fill the four quadrants (avoid the items mentioned in the example in this chapter.)

16.10 Time management needs several skills, mention eight of them.

16.11 Lisa is a good employee, but she likes to leave things till the last minute because she believes she performs better under pressure. This practice upsets and stresses you. In a recent incident, Lisa missed the deadline for an important task because of a family emergency. You believe that she should have finished this task earlier, as she had enough time. How would you handle the situation and convince Lisa to do work tasks as early as possible, rather than leaving them till the last minute?

16.12 Is stress good or bad for us? Discuss this issue in detail.

16.13 Discuss the obligations in the life of a project manager: how they interact and compete with each other. How can the project manager manage them all effectively and efficiently?

16.14 What is change management? How do people change? Is change obligatory or a choice?

16.15 What are the steps for a change at the corporate level?

16.16 Classify people in their reaction to a mandated change, in society or a corporation.

16.17 Discuss the Kübler–Ross Change Curve using a practical example.

16.18 You are the CEO of a construction company. Your company is still using old methods for cost estimating, accounting, and other functions. You decided to acquire an ERP system which will bring substantial change to the 100+ employees. Write a plan, itemizing your steps with a time-line. Discuss each step: what are you planning to do and if there is a plan B?

16.19 Discuss the "Solid Support Group" the president of a company must form before implementing any major change. What is the role of this group?

16.20 The plan you created in Exercise #18 takes a full year. How would you measure progress? (Hint: use Figure 16.10 a and b.)

16.21 You are the owner and CEO of a major construction company. You attended a presentation by a project management software company. You were impressed and decided to acquire the soft-ware despite the objection of your IT manager who does not believe this program is user-friendly enough. You went ahead with the decision and gave everyone one month to use the new system. Your IT manager is doing her best to help with the transition despite her earlier objections. Three weeks after the implementation of the new system, two of your project managers walked into your office, expressing their dismay with the new system. They conveyed this feeling on behalf of their teams. Now you are at a major intersection:
   a. Keep the new system and do not retreat,
   b. Bow to the pressure and go back to the previous system,
   c. Ask your IT manager to find a better alternative, but, in this case, there will be a transi-tional period that you must address.
   Discuss your options along with their pros and cons. Justify your final decision.

16.22 What is leadership? Is leadership an antonym of management? Explain your answer.

16.23 Give an example of a leader (in any field). What are the criteria that made him/her a leader in your opinion?

16.24 Project managers must possess leadership skills, why?

16.25 One of the traits of leaders is the willingness to take risks in the pursuit of their goals. Taking the risk means being prepared for all possible consequences. Search for a story of a leader who took a major risk. Was it worth it?

16.26 Some leaders are described as dictators or stubborn, while they are getting a lot of praise by others. How would you justify these conflicting opinions?

16.27 Are leaders born or made? Explain your answer.

16.28 What are the qualities of an effective team?

16.29 The leader must treat his/her team members with respect and trust. What does this mean? Are there limits?

16.30 You are a team leader. Joe, one of your team members, likes to brag about his contributions to the team. You heard him once saying to a colleague: The team would fail without me! You need to have a meeting with Joe. How would you handle the situation?

16.31 What is the concept of professional democracy? Does the problem usually come from the timid employee who is afraid of expressing their honest opinion for fear of retribution, or from the boss who does not create the atmosphere of such opinions?

16.32 What is a conflict? How do conflicts arise? Can conflicts arise among any of the contracting parties?

16.33 The term conflict generally has negative connotation, but do conflicts always bring only negative consequences?

16.34 Define conflict management.

16.35 What are the steps to conflict resolution?

16.36 You are a mediator in a long and complicated dispute. What are the measures you take in this situation that may not be taken in a minor dispute?

16.37 What are the skills needed to resolve disputes?

16.38 Why do we need negotiation skills? Why would a construction project manager need such skills?

16.39 What are the skills needed for effective negotiation?

16.40 Even though the dispute may be on financial issues, the negotiator or the mediator may focus on other issues, such as what? Explain how and why.

16.41 Why do many people consider communication as the foundation of all skills? Is listening part of communication with others?

16.42 What does the expression "power of persuasion" mean? Explain your answer with an example.

16.43 Define emotional intelligence, EI. How does EI help a project manager, both with their own team members and with external parties?

16.44 What are the four components (steps) of emotional intelligence?

16.45 Emotional intelligence focuses on managing emotions but there is a difference between managing your own emotions and "managing" others' emotions. Elaborate on this point.

16.46 Emotional intelligence starts with thoughts, and then feelings, and finally behavior. Give an example.

16.47 You are a project manager of a major construction company. You had a meeting with an owner to inform him of a delay resulting from underground conditions that you believe entitle you to an extension of time. The owner was angry and insisted that no extension of time will be granted and reminded you of the liquidated damages mentioned in the contract. You do not accept the owner's opinion but do not see the argument with him during the meeting as resolving the issue, if not making it worse. Use your emotional intelligence to resolve the situation. Keep in mind that this is an important client and that your company is looking to forward continue the business relationship with him.

16.48 "Work for the best outcome" is an emotional intelligence rule. Give some examples. You can use the situation in Exercise 47 as one example.

16.49 Mention four different ways to practice emotional intelligence in your daily life.

16.50 There are many types of owner's difficult personalities the project manager may have to deal with. Mention one of these types and create a scenario where you (the project manager) use your emotional intelligence to manage the situation and get things under control.

17

Construction in the International Environment

Introduction

Teams in the construction and many other industries have become diversified and often international. We may have design, construction, and consulting companies contracting for projects outside their original territories. So, it is normal in many cases to see several companies from different countries or regions, collaborate on one project, especially larger ones. Even within the same company and same project team, it is normal to see employees from different nationalities, cultures, and backgrounds. Managing a project in such an environment can be a major challenge, far more than just language issues. But it can also be rewarding by introducing team members to other cultures and broadening everyone's mind.

> Even if you do not leave your location, the "international environment" will come to you. These days, you rarely miss diversity in any team!

What does International Environment Mean?

An international environment may include some or all of the following:

1. Different nationalities,
2. Different languages,
3. Different religions and sects,
4. Different cultures,
5. Different laws, including, but not limited to, building codes, labor laws, import and export, taxes, entry and residence visa, and financial transfer,
6. Different professional standards and practices,
7. Different labor, materials, and equipment markets,
8. Different units of measure,
9. Different in any other aspects.

Project Management in the Construction Industry: From Concept to Completion, First Edition. Saleh Mubarak.
© 2024 John Wiley & Sons, Inc. Published 2024 by John Wiley & Sons, Inc.
Companion Website: www.wiley.com/go/nextgencpm

Managing Construction Projects Internationally

As a contractor, general, prime, or sub, intending to take on a project in a new and different geographic location, you need to do your homework on stages:

A. Before you decide to bid or negotiate the project,
B. After the decision to bid or negotiate the project,
C. During the bidding or negotiation process, and
D. After signing the contract and before mobilization.

The homework, in this context, is a process of investigation in stages. Every stage is supposed to dig deeper and wider to learn more about the site, the project, the market, and all conditions surrounding and impacting the project. The process is progressive because it costs more as the investigation gets more intensive. So, it is an economic balance, just like the investigation any contractor does on a project before bidding, to answer the question "Is it a good idea for me to bid or take on this project?" But the investigation gets more important, and likely more expensive when the project is in a different location or country.

The decision to bid or take on the project in other locations depends on factors such as:

1. The contractor's company's strategic planning and the desire to expand to new locations. The company may have either decided on its own to explore a new market or received an invitation or tip from the project owners. This applies to design companies (architecture/engineering), contractors, and project management consultants. Generally, the company has to be careful in the approach due to uncertainty factors.
2. The nature of the project, and whether the company has desire, expertise, and capacity to take on this project. It is common to team up with other companies for large projects, where every company has its own niche but all together make a complete team.
3. Availability of required resources. This is a major point as things may differ significantly from one market to another. In some markets, labor is unavailable locally and has to be imported from other countries. The contractor may have to be responsible for finding, transporting, housing, training, and providing other services to the laborers. As for materials and equipment, not only do markets differ but also official rules for import and customs differ too.
4. Lack of any legal or other obstacles such as laws preventing or limiting doing business in that location. This includes situations such as embargoes, restrictions on export and import, or financial transactions.
5. In most countries, the local authorities require a local partner (anchor) with any foreign designer or contractor. This requirement serves several good purposes: it provides the foreign company with a local partner that should be a guide to local regulations, practices, and markets. It also assigns liability to the local partner during and after the completion of the project in case the foreign company is no longer there. The contractor must also be familiar with local laws and regulations regarding partnering with foreign entities or allowing foreign entities to operate.

After the contractor makes the decision to bid or enter into negotiations, more homework needs to be conducted:

1. Preparing a project management plan. The plan was explained in Chapter 4 (The Planning Phase), but the plan here takes more importance and expands more since the project is in a new territory. Perhaps within the project management plan, these issues have to be included:

a. Risk management plan.
b. Logistics, both within the project area and between the project and home office.
c. Communication system and protocol.
d. Currency issues such as exchange rates and methods.
e. Dispute resolution methods.

2. Collecting information for cost estimates and resources.
   a. Visiting the project site and collecting information on it: terrain, soil types, any existing structures, vegetation, or objects that need to be removed, plus any other observations.
   b. Checking with the design documents and the local codes, regulations, and restrictions.
   c. Checking on availability and cost of required resources, including import and customs laws and regulations.
   d. Checking on government labor laws, regulations, and practices, such as maximum work hours per day and week, required breaks, overtime laws, and labor rights.
   e. Learning the culture, especially what is related to doing business and the work environment.

3. Studying the weather and climate is important for several reasons, most importantly estimating both activities' duration and cost. Considerations must be made for temperature, precipitation, humidity, and dust, on all aspects of the construction process such as the storage for materials and equipment.

   It is also important to know the local typical work schedule (hours/day and days/week) and holidays and any seasonal occurrences or events such as heavy rain, monsoons, and hurricanes.

   Some places may have other restrictions. For example, the pilgrimage season for holy cities in Saudi Arabia, when construction must come to a complete halt. Other places may have restrictions during an annual event such as the Rio de Janeiro Carnival in Brazil, or a one-time event such as the Olympic games.

4. It is strongly recommended that the contractor performs constructability and value engineering (VE) studies. Materials and methods in the new location may differ from what the contractor is used to in his/her own base, including automation and the use of modular units. The process is to study feasibility first, and then optimization. What worked in the contractor's base may not work in the project location, and what was optimum solution in the contractor's base may not be optimum in the project location. The contractor's main control is over the construction process but may be able to influence the design through VE study or even better if he is involved earlier either in a design-build contract or in construction services.

   The contractor must think "out of the box" in terms of materials and methods. Some materials are more readily available and cheaper than others in certain locations, such as lumber, concrete, brick, natural stone, and steel. Local markets also differ in the varieties of materials, brand names, and prices: what is local in the United States or another specific country, may not be "local" somewhere else. Even a matter like sand and aggregates for the concrete mix and the formwork materials, may differ based on location. The contractor must always study materials and methods that best suit the project and its location: The climate, natural resources, local markets, social economic situation, and culture. The objective is to suit the locale and optimize the utilization of resources.

   Automation is another issue that the contractor must think of: Is it a good idea in that project? If yes, how, and how much? In a case in a project in a remote and underdeveloped area, the project manager needed to do large amount of excavation. He needed large excavators that were not available locally. They had to be imported with high costs for transportation and delivery. In addition, the official local customs rules would add cost and time to the process. Instead, the project manager hired 300 local villagers with simple shovels and baskets. Luckily, the soil was soft, so there was no problem with hand excavation. The project manager's decision saved time and money plus had a positive impact on the locals.

Sometimes, the situation may entice the contractor toward more automation. This can happen for a variety of reasons such as lack of local skilled labor, harsh conditions on site, speed of work, or purely economic. In this case, the contractor may use modular units that require minimum installation effort on site. Every situation has to be evaluated on its merit, considering all impacting factors, in order to allow the contractor to choose the optimum solution for that situation.

> What worked in the contractor's base may not work in another project location, and what was optimum in the contractor's base may not be optimum in the other project.

> The "default" you are used to it, in rules, materials, and methods, may not be the default somewhere else!

English is not English!

In some international work environments, there might be employees who speak English as a mother tongue. They could be from the United States, Canada, England, Scotland, Wales, Ireland, Australia, South Africa, New Zealand, or other parts of the world. The difference, when communicating among them, is not only in the accent and pronunciation but also in the jargon and meaning of some terms and idioms. Even technical positions such as cost estimator and scheduler may carry other titles in other parts of the world such as quantity surveyor (QS) and planner. The term "main contractor" in the UK is equivalent to the American tern general contractor and the bid is a tender in the UK. The author worked in a company with over 35 nationalities and faced many situations of misunderstanding, some were funny, others were embarrassing, while others had serious or potentially serious repercussions.

Sometimes, the name of the company or a product of the company does not have a good meaning in the local language, so it would be a good idea to change it.

Difference in culture affects many aspects, such as:

1. Language, dialect, accent, pronunciation, spelling, and meanings of words and terms.
   Even when people come from countries that speak the same "language," English or Spanish, for example, terminology and expressions differ. This may cause misunderstanding that leads to embarrassing and awkward situations.

> When you mention "football" to American and European people, you are talking about two completely different sports!

2. Body language and hand gestures differ vastly, including "personal space" and ways people use to greet each other. In some countries, like Saudi Arabia, people sometimes sit on the ground, but they

must cross their legs and not have their feet or shoes facing others. Also, serving coffee has sophisticated rules: the server pours one sip into a little cup and hands it to you with his right hand. You must consume the first cup because it is insulting to refuse it. Later, if you do not shake the empty cup, it will be filled again without asking you till you shake the cup. This is just a small sample of cultures around the world that we must be aware of and respect.

Other cultural issues include dress codes which must not violate local law or culture. Also, in many locations, a foreign company must not engage in political statements or activities.

3. The way the date is written: In the United States, the date is written as month/day/year, while everywhere else it is written as day/month/year. So, a date written as 11/4/2023 can be interpreted as the 4th of November or the 11th of April. For this reason, and especially when you have employees from different cultures, it is important to adopt a clear policy regarding the date format. The best way to do it is to write the month in letters, perhaps the first three letters. So, you can clearly distinguish between 11 Apr and 4 Nov.

4. Measurements can be in the SI[1] (metric) or imperial system. Sometimes, there are terms that could be interpreted in more than one way. For example, there is the metric gallon (5 liters) and the imperial or American gallon (3.78 liters). The metric ton is 1000 kilograms, while the imperial or long ton (or tonne) is 2460 pounds, which is $\approx$1016.05 kilograms, and the US ton = 2000 pounds $\approx$ 907.2 kilograms. Even the term "billion" means 10^9 in the United States and 10^{12} in France.

Also, there are units in the United States that some trades use, while it is not known somewhere else such as the square SQ for roofers (= 100 square feet) and the board-foot BF for framers and carpenters (the volume of 1 foot by 1 foot by 1 inch), and often written as MBF, where M stands here for 1000. This leads us to another confusion: while M is used to indicate 1000 units as in Roman numerals, M is used in the field of electrical and electronics to indicate a million, as in the Greek word mega.

Another important thing for the contractor is the compatibility and integration of the materials. For example, the plumbing fixtures' connections have to match the plumbing network outlets. If the system is on one type of unit (SI, for example), everything in that system needs to be on the same measurement system. Also, the contractor may need to consider is the difference in the electric current between countries. The United States is one of very few countries in the world that uses 120 volt/60 Hz as the main household current, although major appliances may run on a 240-volt line and also have different outlets.

5. The "standard" work schedule may differ around the world: hours/day and days/week; also, holidays: religious, national, and other. Not all countries start their week on Monday and have Saturday and Sunday as their weekend. "National Day" differs from one country to another. Also, while Christmas comes on 25 December in Western countries, it comes on 7 January in other countries. Thanksgiving holiday is not the same in the United States and Canada. The same thing can be said for Labor Day (first Monday of September or 1st of May). In fact, even the term "holiday," while it means an official day(s) off for all in the United States, it means a personal vacation in the United Kingdom.

6. Cultures differ in so many ways which may and do affect communications and the way people do business with each other. For example, bureaucracy and punctuality (or lack of) can be part of the local culture. This may be frustrating to the contractor sometimes. In some countries, corruption,

---

1 The International System of Units, known as SI (from French Système International), is the modern form of the metric system.

bribery, and nepotism are common. Also, racism and unfair practices, especially for laborers may be encountered. The contractor needs to make sure that they are not involved in any illegal, unethical, or unprofessional act. The organization can create its own "work culture," but there is practically a limit on your influence on societal culture in terms of both depth and breadth.

7. Logistics can differ by laws, regulations, and customs. In some countries, cars drive on the right side of the road, so the steering wheel is on the left side of the car, while it is the opposite in other countries. This point must be taken into consideration when transporting vehicles to that location. Other points must be taken into consideration when acquiring equipment and vehicles in another country:

    a. To check with local laws for requirements and regulations such as environmental restrictions on fuel consumption and exhaust emissions.

    b. To make sure there are agencies in that location that can provide maintenance and spare parts for these equipment and vehicles.

    c. For complicated equipment such as tower cranes, make sure there are qualified operators.

8. Currency can be a complicated issue. It is not unusual for the company to deal with two or even three currencies in the same project. The contract may be in one currency, US dollars or Euros, while the pay for local workers and materials is in local currency. The contractor must become aware of local financial regulations, including transactions and transfer of money. Borrowing money rules and regulations may differ according to the location. This includes interest rates and borrowing and payment terms.

9. When parties involved in the project – designer, owner, general contractor, and project management consultant – are in different countries, setting up a virtual meeting can be a challenge. Time zones may be vastly different, and some participants have to go out of their way to accommodate others. In many situations, there is a need for an answer to an urgent question or other information between two parties in different countries. It may be off work hours for the receiving party, but the requesting party cannot wait. The company needs to establish a protocol for regular and emergency communications.

10. There are several standard contract templates around the world but in every region, there are popular ones. In the United States, the most popular construction contract forms are AIA[2] and ConsensusDOCS. In Europe, the Middle East, and North Africa (MENA), the popular forms are FIDIC[3] and NEC[4]. For methods of dispute resolution, there are also methods used in some, but not all, countries such as adjudication in the UK and dispute resolution board (DRB) in the U.S. In addition, there are common practices which differ from one country or situation to another. For this reason, when the contractor signs a contract in another country, he/she must read the contract carefully and enquire about items not clearly mentioned, if any.

    Some of the practices that may differ from one location to another:

    a. The percentage of retainage (also called retention), if any, and its requirements. The most common percentage for retainage withheld from progress payments is 10%, but it can be reduced to

---

2 AIA is the American Institute of Architectes.
3 The International Federation of Consulting Engineers is an international standards organization for construction technology and consulting engineering. The FIDIC acronym came from the French title: Fédération Internationale Des Ingénieurs-Conseils,
4 New Engineering Contract, created by the UK Institution of Civil Engineers.

5% or even stopped at the project's midpoint in some large contracts. In some countries, the contractor gets only half of the retainage upon the successful completion of the project. The other half comes after the expiration of the basic warranty period. Such regulation was made to protect owners in case there is a defect in the project that is covered by the builder's warranty and the contractor did not act professionally to fix it. On this note, the general contractor's warranty on the project in terms of coverage and length also differ by location.

    b. Advance payment is a percentage of the contract sum, usually between 10% and 30%, paid by the owner to the contractor upon signing the contract. This practice is common in countries where labor is imported. This payment covers the contractor's expenses to acquire laborers, such as labor agency fees (in the country of source), government fees such as entry visa, medical check-up, and labor permit, transportation to the project site, housing, food, basic services, and training. Expenses may also include other nonlabor items such as importing equipment and mobilization. In other words, advance payment covers all expenses the contractor has to pay before the construction of the project begins. The owner recovers the advance payment by deducting the same percentage from progress payments.

    c. The reimbursement of materials purchased by the contractor, but not installed yet, is another point that the practice differs if not specifically mentioned in the contract.

11. Insurance and bonds also differ in names, coverage, premiums, and other regulations. Contractors everywhere have to provide required bonds and insurance, but they must check with local laws and regulations to make sure not only they comply with the law and contract stipulations but also that they are covered and protected for any incident. Having a local legal expert is recommended.

12. Security for site and personnel is important, as some locations may not be safe. It is the responsibility of the contractor to provide security for the entire site, particularly the personnel.

The "Background Paradigm" Syndrome

Many people grow up in a certain culture and they get used to it to the point where their traditions and practices become the norm for them. Consequently, traditions and practices that do not conform to theirs are deemed unusual and even "wrong." One example given earlier is driving on the right or left side of the road (and then having the steering on the left or right side of the vehicle.) It is funny when a person tells another one: "You folks drive on the wrong side of the road", when indeed there is no such thing as right (meaning correct) or wrong in these situations. While some practices can be judged using scientific and logical arguments, other matters have no absolute right or wrong. It is a matter of societal, local, or even personal culture. The problem comes when two or more different cultures coexist, and each side looks at the other side as unusual or wrong.

Ignorance can be our worst enemy, depending on how we react to it. For example, "I saw one of my colleagues eat strange food or perform weird prayers." I have two choices in such situation:

– I can react negatively by making assumptions about my colleague's food or prayers, without knowledge of that culture. Such assumptions are rarely accurate, so I will likely have the wrong idea about my colleague's food or prayers. Even worse is when I take actions based on these false assumptions.

– I can react positively either by observing without making assumptions or by gently and respectfully approaching my colleague and enquiring about the food or prayers. The second option must be done I must ask in a careful way that will not give any impression of disrespect.

The more you know about others; their culture and language, the more your horizons will expand and the less likely you will have misunderstandings or embarrassing situations with them or even others. The wider the variety of cultures you know, the more choices you will have, so you can compare and choose. One caution though is not to generalize from one simple observation or experience. Diversity exists within each nation or group, so just because one person has certain tradition, belief, or behavior, it does not necessarily mean that all people from that country do so.

> Just because you are used to doing things in a certain way, it does not necessarily mean that those who do it differently are wrong. Be open-minded to learn about other methods, traditions, or opinions.

So, when a company operates in multinational environment, the "clash of cultures" is expected, but with wise leadership, it can be converted into an advantage: learning about others and the exchange of knowledge, which may lead people to get out of their local thinking and embrace better methods and concepts that they were not aware of before. Whether you agree with others' methods and thinking, mutual acceptance and respect are a must for a successful team.

After the Completion of the Project

Typically, in any construction project, the general contractor has to honor the warranties on the project, which may be different in extent and expiration date for different components of the project. For example, each of the following may have a different basic warranty period: The building structure, the air conditioning/HVAC (heating, ventilation, and air-conditioning) system, the appliances, the floor tile and carpet, and the insulation. Even within the same item, the warranty may cover certain but not all issues. It may cover parts but not labor. In addition, the owner may pay extra to extend or expand the warranty period. In all cases, the general contractor must be responsible for fixing any issue covered by the warranty, directly by his/her own team or indirectly by hiring a third party.

It is customary in many countries to have collateral or financial assurances from foreign companies, so they do not leave and abandon the projects under their responsibility. Many companies honor their responsibilities because they are keen on keeping a good reputation and have the opportunity to do more business in that location.

Exercises

17.1 "Contractor: Even if you don't travel the world, the world will still come to you!" Please explain this statement in the age of multiculturalism.

17.2 What are the factors that encourage a contractor to do business in other countries or regions?

17.3 What recommendations do you have for a contractor who wants to do business in a new country or region?

17.4 You are a general contractor with projects spread across three states (Florida, Georgia, and Alabama). You have an opportunity to take on a project in Puerto Rico. Write a plan to prepare for this project.

17.5 In Exercise 17.4, elaborate on the importance of:
    a. Visiting and inspecting the site.
    b. Having a local partner or consultant.
    c. Building the project management team.
    d. Studying the local market (labor, materials, and equipment).
    e. Language and culture issues.

17.6 In Exercise 17.4, the owner wants to sign the contract with you. You are not familiar with this contract template. The owner is not willing to use another template. What would you do?

17.7 Meet with a contractor or a project manager who has worked internationally. Ask them about the major challenges they faced and how they overcame them.

17.8 A conflict arises between two of your team members. When you listened to each of them, you had a feeling there was a misunderstanding due to cultural differences. How would you deal with the situation?

17.9 You are an American contractor, and you are planning to take on a project in one of the Caribbean islands. While in a meeting with one of the locals there, you asked him "You drive like us or on the wrong side of the road?" He did not like your question. How would you ask the same question in a more appropriate way? What does this have to do with the "Background Paradigm Syndrome?"

17.10 A method that worked in one region may not be the best for another region, why? Explain with an example.

17.11 Ask an international contractor about the challenges due to differing laws of import for materials and equipment.

17.12 Ask an international contractor about the challenges due to differing cultures and languages among the workforce. Discuss the importance of signage with pictures and symbols: can such signs be universal?

18

Construction and Evolving Technology

Introduction

The evolution of new concepts and inventions has been happening since the beginning of humanity on this planet, from learning how to set fire to plant vegetation, to the steam and internal combustion engines, to automobiles and aviation, to computers and the internet and wireless communications. Many concepts come to improve or replace previous ones, sometimes even concepts and items that we thought will live forever. Technology and science proved only one thing: Evolution and inventions have no end and no ceiling. Many of us also agree that the pace of advancement has accelerated in the past few decades to the point that the same generation may see several cycles of technology for the same item, electronics and cell phones are just examples.

> Change, in science and technology, is the only "constant"!

Construction, one of the oldest professions in history, is no different from the rest of the industries and professions, in being part of this ever-continuous improvement and evolution. Constantly, we hear of new concepts, theories, materials, equipment, systems, software, or other items that make the construction industry faster, safer, more economical, more efficient, environmentally friendlier, more resistant to adverse conditions/more durable, friendlier to users, or better in any other way.

Along with all these developments, new challenges arose for contractors, such as:

1. Which one of these gadgets is best for me/my case?
2. Is it cost effective, and does it add value?
3. How can I train my staff/workers to learn how to use it?
4. What if I make a large investment in this technology, and then shortly after, it is replaced by a newer technology?
5. What if I do not acquire it, but my competitors do?

It is not easy to answer these questions. The contractor, or any business owner, will always face such questions and must decide based on several factors, most importantly the merits of the situation. Rushing to acquire every new gadget and technology may not be a good strategy, nor is the adoption of one tool

Project Management in the Construction Industry: From Concept to Completion, First Edition. Saleh Mubarak.
© 2024 John Wiley & Sons, Inc. Published 2024 by John Wiley & Sons, Inc.
Companion Website: www.wiley.com/go/nextgencpm

for every situation. Even worse than that, if the contractor adopts the mentality of "We have been doing business this way for 25 years, and we are not going to change!"

> These days, doing business as usual may not be enough to stay in business. You must keep up with technology and utilize it in the most effective way.

In this chapter, we will cover this topic from a general perspective, not necessarily mentioning every new concept, gadget, or tool. In fact, we cannot list all inventions and innovations, simply because there are so many of them, and also because they keep evolving. So, by the time this book is in the hands of the readers, there will be new inventions that did not exist at the time of writing the book. Our objective is to have the awareness and mindset that we, construction professionals, need to keep pace with technology, be aware of new advancements and inventions, and pick what works best for us and gives us an edge in the industry.

New items (inventions, innovations) in the construction industry vary in nature and usage. Here are some categories:

1. Concepts and theories.
2. Design authoring software (CAD and BIM).
3. Construction and building materials and systems.
4. Construction tools and equipment.
5. Construction support systems such as formwork systems.
6. Communication devices and systems.
7. Layout and Reality Capture tools.
8. Non-destructive testing devices and methods.
9. Prefabricated and modular units, 3-D printing, and the process of automation.
10. Environmental concepts and inventions to save energy or make operations environmentally friend-lier, such as solar and other renewable energy sources. Also, the use of recycled and other environment friendly material.
11. Safety systems.
12. Other concepts, materials, devices, or others that help the construction industry in any way.

This chapter also hopes to inspire engineers and construction professionals to follow and lead in both appropriately applying new technologies, as well as inspiring and innovating new inventions[1]. When you have imagination, inspiration, knowledge, and creative thinking, the sky is the limit in what you can do.

How Technology Impacted Projects' Design and Construction

There is an interesting exchange that is happening in the relationship between the participants of the construction industry and the industry itself. The participants include:

a. Academia: Schools of architecture, engineering, and building/construction management. This includes also vocational institutions.

---

1 This can be either inventing totally new technology or discovering new applications to / modifying an existing one.

b. Professional associations and societies.
c. Owners' organizations including public agencies.
d. Building materials and equipment manufacturers, distributors, and vendors.
e. The computer software, hardware, and electronic gadgets industries.
f. Other industries that may have impact on the construction industry, such as communications systems, security systems, solar and other renewable energy, manufacturing (modular units), and logistics.

The advancement in one area/industry induces changes in other related areas/industries. For example, in the 1970s and 1980s when personal computers were in their infancy, engineering schools used to teach computer programming[2], so students could write their own programs. Later, a number of software programs were professionally developed, so schools switched to teaching these application programs. Many functions were performed manually such as architectural and engineering drafting, and then the focus shifted to computer applications for the same functions. Computer-aided design, CAD[3], was developed by the software industry. Schools followed closely with the CAD advancement with continuous collaboration between education, industry, and software developers. Building information modeling (BIM) was another concept developed by the industry and received and implemented by educational institutions.

On the other side, many educational institutions are continuously conducting research and developing new concepts, materials, and other ideas, that trigger the industry to develop, implement, and/or commercialize. The interaction among the participants/stakeholders continues. It is important to note here that educational institutions include different types:

a. Institutions with academic education only.
b. Institutions with academic education and research.
c. Vocational education institutions and trade schools.
d. Professional (continuing) education.
e. Research institutions such as CII[4], CPWR[5], CSI[6], ASTM[7], NIOSH[8], NCCER[9], and others.

Research institutions usually collaborate with other entities that support their research financially, technically, or otherwise.

The chart in Figure 18.1 simplifies the relationships among participants: educational institutions and support industries are the major participants with the construction industry. Other participants include:

– Government agencies, federal, state, and local, with jurisdiction that relates in any way to the construction industry. This includes, but is not limited to, issuing building codes and legislating laws

---

2 Such as Fortran, Basic, C+/C++, and other computer languages.

3 The acronym CAD stands for computer-aided design or computer-aided drafting.

4 The Construction Industry Institute CII - Home (construction-institute.org).

5 The Center for Construction Research and Training CPWR | A world leader in construction safety and health research and training.

6 The Construction Specifications Institute Home - Construction Specifications Institute (csiresources.org).

7 Started as The American Society for Testing and Materials but later goes only by ASTM International ASTM International - Standards Worldwide.

8 The National Institute for Occupational Safety and Health.

9 National Center for Construction Education and Research National Center for Construction Education & Research - NCCER.

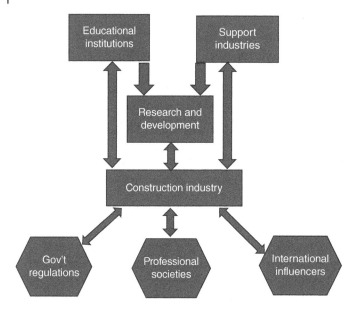

Figure 18.1   The technology development cycle for the construction industry.

dealing with environmental preservation, hazardous materials transport and storage, liability and warranty, labor, and others.
- Professional societies that regulate the profession and set standards.
- International influencers, including governments, professional organizations (national and international), educational and research institutes, private industries, and even markets.

How much and what technology must be implemented in the construction industry?

This is arguably the most important question to consider. In every situation, contractors find themselves with several choices, some of which are more "advanced" than others. Should the contractor always choose the most advanced option? Not necessarily. It depends on several factors including the project design, local market, type of client, options available, cost, and time consideration for each option. For example: The choice of materials. This includes both the availability and cost of local and imported materials as well as the suitability of these materials to the local climate and culture. The choice of methods and equipment also differs based on the situation, including the options of automation and outsourcing.

The contractor must go through two steps: feasibility and optimization. In the first step, all options are laid on the table and vetted for feasibility. The infeasibility may be physical, legal, contractual (e.g., unacceptable large duration), prohibitive cost, or other. Those options that are deemed feasible are then ranked in order of cost, time, and/or other important criteria. Sometimes, there are other factors that may influence the decision such as aesthetics, environmental impact, public relations, the company's strategic plan, and long-term relationship with the client. What is certain is that there is no one tool that is optimum for all projects of all types in all locations. Tools that provide the most added value to the project should be chosen to achieve the desired outcomes.

> Choosing the tool before defining the objective is like putting the carriage in front of the horse!
>   There are many tools available, but you need to define the objective first, and then pick the tool that best fits the situation.

Risk and Cost of Technology

Many technologies, when first come down in the market, carry risks with them such as "bugs" or low benefit/cost ratio. Risks also include long-term effectiveness and side effects. On the other hand, technology vendors may be willing to give financial incentives to those who take the risk and try their product early. The cost, in this context, includes the initial cost (equipment, software, training, etc.) as well as operating cost (maintenance, membership fees, upgrades, more training, etc.). However, there is another potential cost: Reverting cost, which is the cost to switch either back to the product / method used before implementing the new technology or to another competitor. This action may actually cost more than the initial cost, as it may be a source of confusion and frustration for employees.

> It has been said: No pain, no gain!
>   In any business, including construction, no risk, and no profit. However, risk must be taken only when carefully calculated.

However, we defined risk earlier as a threat and opportunity. Contractors who adopt and master a new technology may have an edge over others who lag behind. In fact, many manufacturers and developers like to partner with companies in the industry for pilot studies and "Beta testing," which can be a win–win situation for both parties.

> Two proverbs are not valid anymore:
> - Don't rock the boat!
> - If it ain't broke, don't fix it!

Technology and Construction: Push or Pull?

The interaction between the construction industry and technology is a mix between push and pull. In some cases, a new technology is introduced with an impact on the design and/or construction processes. In other cases, a challenge in design or construction work leads to an invention, as the proverb says: Necessity is mother of invention. Skyscrapers, for example, could not have been introduced or gone to such heights if elevators had not been invented first. Skyscrapers received a huge boost in the 1960s from the brilliant tubular designs by Fazlur Rahman Khan[10]. It resulted in stronger but lighter and more flexible buildings. Khan's design proved that the strength of the building is in its flexibility, rather than the rigidity as used to be commonly believed. However, this flexibility resulted in wider oscillation at the top of the buildings, which created a problem requiring a new solution. The solution was installing dampers to reduce this oscillation. With heights reaching and exceeding 100 stories, other challenges arose in both the design and construction, such as the elevators, the plumbing system, the scaffolding, and the extensions of tower cranes. Concrete creep was another issue that needed to be researched because it happens in the long term, and if not considered, it may have severe consequences on building elements such as doors and windows. The technology of thermal, moisture, and sound insulation has been improving steadily.

Earthquakes pose a great challenge to structures. Traditional structural design that resists earthquakes of high severity, is costly and may not provide significant protection from damage. In Japan, where strong earthquakes are frequent, engineers came up with a creative design that puts sliding seismic isolators between the foundation and the superstructure of the building, as seen in Figure 18.2. Thus, in the case of an earthquake, the foundation will shake but the building will stay relatively stable[11].

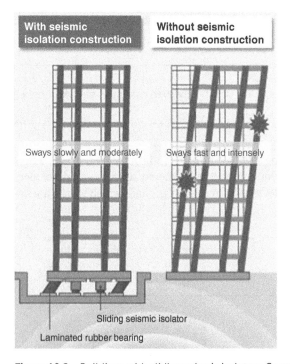

Figure 18.2 Buildings with sliding seismic isolators. Source: https://web-japan.org/trends/11_sci-tech/sci110728.html.

---

10 https://en.wikipedia.org/wiki/Fazlur_Rahman_Khan.

11 https://web-japan.org/trends/11_sci-tech/sci110728.html.

Such technical innovation can save the lives of millions of people and trillions of dollars. It came as a necessity after so many devastating earthquakes. Such innovative solutions also required "thinking out of the box" and not surrendering to the status quo and the mentality of "we can do nothing about it!"

Long bridges over water or land and tunnels posed new challenges that required creative solutions that may have induced new technologies. For example, the Marmaray tunnel in Istanbul, Türkiye[12], goes 8.5 miles at a depth of about 200 feet under the bed of the Bosporus Strait and is resistant to earthquakes known to rock that region. Interestingly, Türkiye also claimed the title for the longest suspended bridge in the world: The "1915 Çanakkale" bridge, opened in 2022, has the longest central span (2023 meter = 6637 ft) of any suspension bridge. The Fehmarnbelt tunnel, connecting Denmark and Germany, a €10 billion project, is expected to be the longest (18-kilometer-long) underwater tunnel and the longest combined rail and road tunnel in the world when it is completed in 2029, significantly reducing travel times between Central Europe and Scandinavia.

The challenges are not only physical (taller, longer, bigger, stronger, etc.) but also in other aspects such as the speed of construction, carrying out the scope with very low budget, or others. In China, several buildings were erected in record time: A 30-story hotel building was built in 15 days, and a 57-story skyscraper building was completed in 19 days only, both with high resistance to seismic forces. However, the execution time may not include prerequisite tasks such as planning and manufacturing. Such superfast projects heavily depend on modular units, which were discussed earlier in other chapters, but certainly, some type of new technology was involved in the design, manufacturing of the modular units, and the erection. Fast execution is needed in many situations such as in natural disasters when suddenly, a large number of people need housing. Also, when a main artery; a highway, bridge, or other structure, collapses and needs a quick repair or replacement.

Architects and designers have been flying high with their imagination and inspiration for attractive, creative, and even unusual designs. Structural and other engineering designers are challenged by such designs and must respond appropriately to make them a reality. The author once described the relationship between architects and structural engineers as "architects like to fly high with their designs, but structural engineers hold their feet and bring them back to earth!" But, without such challenges and provocations, the design process would stagnate without improvements and innovation.

This discussion is not limited to buildings only but includes all types of projects that combine aesthetics and functionality, but often with cost limitations. This includes roads, bridges, tunnels, dams, airports, and more. Landmarks are no longer rare or a possession of the king of the land. They exist almost everywhere and are owned by ordinary people. Every day we see more creative landmark projects that amaze us. Such projects could not have been done without creative solutions, not only utilizing the latest technology, but also inducing and inspiring technology for new inventions.

New Technologies in Design and Construction

When we talk about technologies in construction, we must include related technologies in the design phase. There are new technologies that helped in the design phase for both purposes: the design process itself and the planning for construction, including the constructability study, review and approval process, value engineering (VE) studies, and the transition from design to construction.

---

12 The new name for Turkey.

In the past, all drawings were done manually. Not only this was very time- and effort-consuming, but the process of modifying or correcting errors was also time- and effort-consuming and frustrating. Copies of these original drawings were produced as blueprints[13]. Later, computer-aided drafting (CAD) was developed for 2-D drawings. As time passed and the computer hardware and software technology advanced, CAD improved significantly including 3-D models. Also, the internet allowed online submission, review, and collaboration, rather than the old money-and-time-consuming process of sending the physical copies of the drawings by courier, especially when partners are located in different regions or countries.

Other construction-related functions were also computerized such as cost estimation and scheduling. Not only this, but they were able to communicate electronically with CAD and among themselves. This includes cost estimation, accounting, procurement, scheduling, project planning, and other functions.

Building information modeling, BIM, was a big leap forward in the design-construction process. Although its concept has existed since the 1970s, the real take-off did not happen till the past two or three decades, when computers and the internet became strong enough to carry it effectively. It helped visualize the finished project as well as the detailed (item-by-item) construction process. It allowed the construction team to optimize the process. This helped reduce the cost and duration of the project and improve the efficiency of the process. BIM also allowed for the early detection of clashes among system (structural, mechanical, electrical, etc.) and is a huge advantage that saves time, money, and headache. BIM can carry n-dimensions, having the first three basic dimensions (X, Y, Z); it can add other project information such as time as "extra dimension."

Value engineering (VE) studies became more effective with BIM and other project simulation and visualization tools. Electronic cost estimating also allowed for comparing alternative materials and systems easily. Talking about cost estimating, the power of databases, software tools, and the internet gave this process a huge boost, including easy updates for databases, flexibility in adjusting items, and other manipulations and adjustments. The process of quantity take-off also advanced rapidly. After it was completely manual till the late 1970s, technology has been adding tools and software programs to make the process faster, easier, and more accurate. Rolling pens and digitizing boards, as shown in Figures 18.3, were nice tools on the road for digitizing, and then got connected to the computers, so the shapes and numbers were automatically transferred to the software.

Technology now allows the accurate transfer of quantities from electronic design documents (CAD) to cost-estimating software without manual manipulation, minimizing the possibility of errors. Model-based estimating has also allowed for a more seamless flow of quantities from design to construction by enabling all project stakeholders to see more clearly how much material is needed for each building system which allowed the designers to optimize the design for minimum waste.

Project scheduling also took a big leap forward with advancements in computer hardware and software. The main shift was in the project breakdown structure: Before the use of personal computers, it was impractical or even infeasible to breakdown the project into hundreds or thousands of activities because the scheduler has to manually draw the critical path diagram and calculate dates. This led to combining many activities under one package, thus, stripping scheduling and project control of their

---

13 Copies were called "blueprints" because they used to be copied with old technology, using a chemical called ammonia that has bad odor, only in blue color. Later, newer technologies allowed for better copying process in black or color prints. However, the name blueprints remained in use, even when prints are not blue!

Figure 18.3 Courtesy of Construction estimators: blueprint take-offs with the "Quik-Ruler" Digitizer (constructupdate.com). Source: Courtesy of ConstructUpdate.com.

main tools for effective analysis. With the advancement of computer hardware and scheduling software programs[14], complex projects are broken down into thousands of activities, with each activity easily measured and controlled, and coded by several criteria for grouping and sorting. This allowed the detailed analysis for project control and many other functions such as crew/labor productivity, delay claim and risk analysis, and schedule acceleration.

Nowadays, there are systems that move beyond traditional scheduling programs, such as Oracle Primavera P6, to use artificial intelligence (AI) to generate multiple schedule scenarios and help the user select the optimum one, based on the user's objectives and constraints. As an example, ALICE Technologies[15] developed a program that utilizes AI to create and explore numerous possible construction schedules. ALICE users can start by importing a BIM model, or else they can enter project details directly into ALICE. Considering factors such as activities' durations, logic, resources, cost, site restrictions, and other information, the system runs thousands of possible scenarios and presents options to the user for further exploration and refinement (ALICE refers to this process as "construction optioneering.") The system works not only in creating the schedule, but also updating it as things on the ground may change. If a change order was issued, a delay occurred for whatever reason, the shipment of a critical item is delayed, a resource available amount changed, or any other update, the system can react to the new situation and re-run the scenarios, suggesting new solution to achieve the objective as efficiently as possible.

In summary, technology tools allowed project participants (owner, designer, and contractor) early enough to examine all alternatives; materials, systems, and processes, in order to optimize the project execution process, considering the project scope, constraints (time, money, etc.), and priorities.

---

14 Mainly critical path method, CPM, but also other methods such as linear scheduling, LSM.
15 The ALICE Blog | The Latest on AI in the Construction Industry (alicetechnologies.com).

Planning for Construction

Technology has helped contractors in investigating and preparing the site. Drones are used as a method of reality capture to examine and videotape the topography of the site, including hard-to-reach areas. This enables the physical world to be brought into the digital world for further analysis. Underground utilities and conditions are another important matter and always have potential for disputes. New technologies came up with tools to sense and detect underground conditions.

Ground Penetrating Radar Systems (GPRS) based in Toledo, Ohio, provides services that include utility locating, 3D laser scanning, acoustic leak detection, video pipe inspection, underground storage tank locating, concrete scanning, and drone imaging[16]. SiteMap made all of that information available in a GIS viewer and digital plan room[17].

Interestingly, GPRS that was originally in the business of concrete scanning, acquired TruePoint Laser Scanning, and it found itself with an even more expansive library of site and construction data. It is called "the Google Maps of the underworld", although SiteMap is not just for the underground space. It has the above-ground, above-grade options as well[18].

The above discussion extends to technologies used for non-destructive testing (NDT) and detection. NDT is a testing and analysis technique used by industry to evaluate the properties of a material, component, structure, or system for characteristic differences or welding defects and discontinuities without causing damage to the original part. NDT is also known as non-destructive examination (NDE), non-destructive inspection (NDI), and non-destructive evaluation (NDE).

Non-destructive testing is a group of methods using different techniques and devices such as laser, ultrasonic, magnetic field, acoustic emission, infrared thermography, radiographic, and others. Not only does it save time and money by performing the test without destruction, but in some cases, it can do what was not possible in the past.

The Construction Process

Throughout the years, many inventions and novel ideas were developed to help:

1. Speed up the work process,
2. Reduce the cost of both, the construction process and the operation and maintenance of the project during its life service,
3. Improve the quality of the project,
4. Improve the safety of workers and others,
5. Improve the aesthetics of the project,
6. Reduce any adverse impact on the environment / improve sustainability on and around the project site,
7. Improve the documentation system and reduce the potential for disputes among stakeholders,
8. Improve the project, during and after construction, in any other way.

---

16 https://www.gp-radar.com/.
17 https://www.site-map.com/.
18 https://www.enr.com/articles/56110-gprs-releases-sitemap-a-gis-for-construction-platform?oly_enc_id=8064G3090134I0M.

The technical improvement in materials includes both installed materials, whether it is totally new materials or improvement to existing materials, and materials and systems used for operations support such as concrete formwork and geogrid fabric and shoring systems.

New and improved materials have been coming to the market constantly. For example, some types of PVC/vinyl tile and siding, air-bubbled concrete blocks with high strength and light weight, permanent (Styrofoam) concrete forms that serve as both forms and insolation, self-healing concrete, and fiberglass or other non-metallic reinforcement bars. Many researchers replaced some components or added others, in concrete mix, blocks, bricks, mortar, grout, and other building materials for quality, cost, and/or other purposes, utilizing – in many cases – local and waste materials.

Concrete additives are a wide topic that continues to develop and improve. This goes for other concrete-related materials such as sand, aggregates (and their alternatives) and reinforcement bars, cables, and accessories. Concrete falsework/formwork[19] is another world that keeps coming up with new and improved products such as sliding forms and ganged/truss tables. Methods and equipment for concrete placement and finishing now have a large variety that suits different situations. Other methods were developed to hydrate concrete in hot and dry locations and heat the concrete or keep it warm in cold locations. Then, we also have new methods for stamping concrete surfaces, so they get different textures and colors, and look like natural stone and pavers, wood, tile, or other materials.

Many situations and challenges force an invention or improvement, whether it is in the design of the project itself, the surrounding conditions and climate, or other factors such as constraints of time, budget, safety, or other. For example, concrete buildings with large spans led to the introduction of waffle slabs. In waffle slabs, there was a challenge when removing the pans (domes) that stuck to the concrete. First, they tried putting oil on the pan to facilitate its separation from the concrete but later found out that oil stains concrete in an undesired way. A simple invention by a construction worker was to add a nozzle/inlet in the center of each panel, see Figure 18.4. The metal pans are removed by air pressure through the nozzle and literally blown off the concrete. Sometimes, the greatest inventions are the simplest ones.

Wood is an organic material that comes with its own characteristics, advantages, and disadvantages. Natural wood joists, beams, and columns were limited in size due to the fact it is an organic material that comes from trees with limited dimensions. Builders needed bigger and stronger sections that can carry heavier loads, span larger distances, or both. Many solutions were developed such as wood trusses, glued laminated (glulam) timber, laminated plywood sheets, medium-density fiberboard (MDF), and composite wood I-beams and joists, some sections as shown in Figure 18.5. Such creative products provided advantages that either did not exist in natural wood sections or existed at a high cost. These advantages include flexibility in size and shape including custom sizes and curved shapes, large sections for long spans, and utilizing wood particles and minimizing or even eliminating waste. Wood connections also got creative such as tongue-and-groove (finger jointing) and metal plates.

---

19 Technically, falsework and formwork are not the same. Formwork is only the part that is in direct contact with the freshly placed concrete. The falsework includes formwork as well as support elements such as shoring.

Figure 18.4 Waffle slabs using metal domes. Source: MODLAR.

Figure 18.5 Composite and glulam wood sections. Source: tamayura39/Adobe Stock; SockaGPhoto/Adobe Stock Photos.

Technology and Construction Work

Throughout the years, the technology of earthwork equipment increased significantly, including excavation, trenching, soil removal and transport, backfilling and compaction, dredging, and other works. Manufacturers introduced a variety of equipment that suits the size and type of job to be performed. Many types of equipment come with different attachments so the same equipment can do several functions with these attachments. Many of them are now equipped with sensors for safety. One of the significant advancements in earthwork was the introduction of trenchless technology, which allowed underground excavation (tunneling and micro-tunneling) without disturbing life above the ground. Today, there is a variety of these systems that allow underground work to be performed in downtown of major cities, with minor disturbances to people and traffic above the ground. Other sensors in earthwork, such as Machine Control from Trimble, utilize GPS systems through direct communications with satellites. It accurately guides the machine in the right direction and also allows for a very accurate placement of a grader's blade, for example, through the communication between the satellites and the sensors. It opens the door for full automation[20].

Talking about earthwork, brings the topic of demolition, which also benefited from new technologies including a wide variety of equipment that suits the specific type and location of the structure to be demolished. The technology of demolition by imploding the building has also developed to be a science by itself, in calculating the type, amount, and locations of explosives, as well as the sequence of detonation, to complete the mission with no or minimum impact on surrounding structures. Decontamination and removal of toxic and other harmful and hazardous materials, has also developed to scientific and professional discipline, utilizing the latest technologies to protect people and the environment.

There were many inventions that had an impact on construction, both the process and the final product. Take, for example, solar and other renewable energy sources that reduced the consumption of fossil fuels and allowed electrical operations to occur away from the grid. The invention of LED lights and the improvement of electrical motors have also helped reduce the consumption of electricity. Also, high-capacity rechargeable batteries allowed for many operations without cords and away from electrical outlets/sources, which was both safer and more convenient. Drones have been helping in several functions such as investigating the site before the contract agreement, monitoring the site for measuring work progress and security, taking pictures, and even delivering small items. Also, fixed security cameras are getting better and cheaper. They provide functions such as monitoring the site for security and production.

Many houses and other structures now have cell phone apps that control many home systems such as lighting, HVAC thermostats, and security cameras, and can even open/close or lock/unlock main and garage doors. New materials cover a variety of functions such as flooring, adhesives and insulation, countertops, and HVAC and plumbing systems. Many of these materials have advantages in functionality, durability, aesthetics, cost-savings, firefighting, and safety.

Improving the safety of human beings, workers, visitors, and people who live near the project site, as well as the property in and around the site, has also improved through the years. Perhaps this improvement is attributed to factors such as the increased role and authority of OSHA, evolving technology, and increased awareness. Safety benefits from technology of wireless devices, sensors, cameras, new materials,

---

20 Trimble Earthworks Grade Control Platform - Horizontal Steering Control.

and others, particularly as these electronic devices are getting smaller, lighter, and more powerful. There have been new functions of the construction hard hat, in addition to its basic function of protecting the head of the person. Some hard hats today are equipped with cameras, sensors, and communications devices. For example, Lantronix Inc. has introduced the Guardhat Communicator smart hardhat, which connects workers with a remote command center via sensors, cameras, and microphones[21].

Digital Twins

Digital twins are simply digital representations of physical assets, systems, and processes. The creation of digital twins requires a meticulous process of digital 3D design paired with ongoing reality capture during construction to ensure any changes in the physical space are incorporated into the highly complex 3D model (as-built modeling).

The advancement in Internet of Things (IoT), sensors, and computational power contributed to the increased interest in using digital twins in construction. There are several potential use cases of digital twins in construction. The use cases stem from the need to accelerate schedules and minimize construction defects and errors.

1. Digitization of construction document interdependency. Updating drawings and engineering documentation is inevitable during construction. As built dimensions, tolerances, and fit-up pose challenges to constructors. Accordingly, a digital twin of various interconnected documents accelerates identification of what document requires updates and provides a mean to alert all interested parties of the changes detected in the field which consequently saves on downtime and help adhering to schedule.
2. Digital twins digital scanning of modular components and critical structural members to collect as-built dimensions to detect clashes and interfaces.

Extended Reality: Virtual Reality, Augmented Reality, and Mixed Reality

As the adoption of 3D models has become more prevalent, a new way of viewing this data has emerged called Extended Reality. This is comprised of Virtual Reality (VR), Augmented Reality (AR), and Mixed Reality (MR). VR completely immerses a user into the digital environment while AR and MR allow an overlay of digital content onto the real world.

Virtual reality (VR) concept and application started several decades ago but has improved and expanded with technology convergence: the improvement of other related tools such as computers, the internet, and other gadgets. Virtual reality immerses users in a fully artificial digital environment but makes users feel real and that they are part of that artificial imaginary world. Later, two younger generations of virtual reality were developed:

- Augmented reality (AR) that overlays virtual objects on the real-world environment, and
- Mixed reality (MR) not just overlays but anchors virtual objects to the real world and allows for a semi-immersive experience as well[22,23].

---

21 https://www.lantronix.com/newsroom/press-releases/lantronix-delivers-iot-connectivity-guardhat-communicator-smart-hardhat/.
22 The Difference Between Virtual Reality, Augmented Reality And Mixed Reality (forbes.com).
23 Virtual Reality vs. Augmented Reality vs. Mixed Reality - Intel.

There are applications of virtual, augmented, and mixed reality in the field of design and construction. In fact, one major advantage of this technology is to connect the design with construction by allowing the user to imagine the design as a reality when it is not yet so. AR and MR aim to connect the virtual design and the built environment by overlaying one on top of the other. This conglomeration allows for better quality control throughout the project lifecycle through improved communication.

Artificial Intelligence (AI) and ChatGPT

AI has been around for a long time and many industries have greatly benefited from its applications. The construction industry is yet to fully utilize this technology. Various mega projects are being constructed around the world offering rich sources for lessons learned, best practices, and smart insights, throughout various project lifecycle.

Genetic Algorithms (GA) is an AI tool that can be leveraged to optimize construction schedule and structural designs given critical constraints and input. Artificial neural networks (ANN) and machine learning can leverage statistics and mathematics to develop predictive models for cost estimation and budget overruns. Specifically, AI can be used in the following areas:

1. Identify design solutions based on owner's constraints. AI can generate potential designs and evaluate their applicability based on boundary conditions.
2. Risk mitigation through developing AI models and running various scenarios considering critical parameters with major impact on construction execution and schedule. AI can identify which parameters have the most impact and can identify plausible paths for risk mitigation.
3. Improved safety through scanning photos of construction sites to identify various safety hazards and suggest solutions to maintain construction safety.

Modularization and 3-D Printing

Using pre-engineered, prefabricated (pre-manufactured), pre-assembled units is a common and relatively old practice in construction, although it is getting better and more sophisticated with the advancement of technology. These units can be 2-D or 3-D, small or large, single- or multi-work-specialty, as shown in Figure 18.6. However, the matter is not so simple when deciding whether to use such units. Using such units has its advantages and disadvantages, so the decision has to consider several factors, and it differs based on the specific situation. The most important factors are:

1. Feasibility: In some instances, it is infeasible to construct the project on-site due to physical difficulties, such as when the project is in the sea water (e.g., oil rig) or on hard or risky terrain. Feasibility also includes the option of modularization because it includes the manufacturing / assembly, transportation, handling, and installation of these units, which requires certain equipment and logistics.
2. Cost: When the job conditions financially justify the overhead cost of manufacturing these units and transporting them. This is likely the case when there is a high number of repetitions, particularly when the manufacturing plant is close enough.
3. Duration (time): Using pre-manufactured is very likely to be faster than traditional construction, especially in the case of mass production. Keep in mind that manufacturing these units usually occurs in an indoor facility, where production is faster, has higher quality, and not subject to weather conditions.

Figure 18.6 Construction using modular units. Source: Jarama / Adobe Stock.

The design of these units has to consider not only the service loads, but also the transportation and assembly / installation loads.

Modular units reached the level of entire "house package" that comes folded and can be unfolded and attached to the foundation in one day. The package, as of now, offers a small 375 SF small house[24] but well designed and equipped to allow users to occupy and utilize it in very short period of time, which can be a solution for many situations, mostly temporary. However, it is likely that such solution will offer more options in the future in terms of size, design, and internal contents (finishing and furniture).

So, with so many pieces (modular units) being manufactured and used in the construction project, can we still call it a construction project? The answer is yes, as long as there is an amount of work performed on-site. As an example, when a vehicle is said to be "manufactured" in a country, this does not necessarily mean that all its parts are manufactured in that particular country. In fact, it is likely that many parts and systems came from other plants in different countries and were assembled in the final location. The same principle applies to construction projects. We may acquire and use many manufactured components, of different sizes and shapes, and assemble them on-site to complete the project. It is still a construction project.

3-D printing is a relatively new concept that started with "printing" small objects and then expanded to as large objects as buildings. Again, the question of using this technology boils down to almost the same three questions: feasibility, cost, and time. However, feasibility here is more than what was included in the case of modularization. It also includes the ability of the 3-D printer to construct the specific design. As the technology of 3-D printing in the construction industry is still in its infancy, it is now limited to simple designs of limited sizes and experimental projects. Economic consideration also limits it to repetitive buildings. As this book is being written, a housing community of 2500 homes is planned for Wolf Ranch, in Austin, TX, with 100 of these homes to be 3-D-printed by a combined effort between the

---

24 https://www.boxabl.com/.

home builder (Lennar Homes) and Icon, a local 3-D printing company[25]. The project uses a 46.5-foot-wide "Vulcan" printer consisting of a crossbar that moves up and down between two 15.5-foot-tall towers that sit astride a foundation. Attached to the crossbar is a nozzle that shuttles from side to side, layering a proprietary concrete mixture called Lavacrete.

Drawbacks of Technology

Many people express nostalgia to the old time when things were simpler and everything was "real," including and most importantly social relationships. But away from nostalgia, emotions, and resistance to change, are there drawbacks to technology? Or, what are the drawbacks of technology? The answer differs, depending on the specific technology and area of application, and who you ask. There is no question we lost several manual and mental skills, such as mathematical intuition, handwriting and drawing skills, navigating our way to different destinations, and solving simple problems. Automobile mechanics now depend more on computerized diagnostic machines than on their experience and intuition. Physicians are almost doing the same thing. But is this good or bad for humanity?

In the past, a few decades ago, you could have a casual meeting with an architect describing the project you would like to build. That architect would be able to sketch the project for you on a plain piece of paper, or even a napkin, using a pen or pencil. He/she could likely also give you a ballpark estimate for the cost without the use of a calculator or other tools. The same professional today can give you a much better drawing and more accurate cost estimate in no time, with the help of an electronic gadget, a laptop, tablet, or even a mobile phone. This is a great advantage of technology. But the main drawbacks, in the author's opinion, are: What if you temporarily do not have access to your gadget or to the internet? And, what if you inadvertently make a mistake when using the computer so it gives you a ridiculous answer, would you be able to tell it is wrong?

This discussion raises several questions, technical and ethical. Are we going to be completely dependent on your gadgets? Are we going to be enslaved or controlled by machines? Have we thought of some "what if scenarios"?

Technology and Change Management

Many of us find our "comfort zone" and refuse to leave it even when a better alternative is introduced. We discussed this in Chapter 16, Soft Skills for Construction Project Management. It is important for the organization's leadership to be visionary and implement the most effective technology available. This implies occasional changes in the concepts and tools used by employees and workers. This is a delicate mission that can give the organization an edge. On the flip side, it may have adverse impact on the organization if the leadership fails in this mission. The biggest hurdle to advancement is not technical, but it is rather in the mentality of humans.

> The two main barriers to acquiring and applying new technologies in the construction industry are:
> 1. Fear of the unknown and attachment to "what we have been doing for years," and
> 2. Finding and hiring the competent employees who can be in control of this technology.

---

25 https://www.lennar.com/new-homes/texas/austin-central-texas/promo/auslen_3d_homes.

Is Applying the Latest Technology Always the Better Solution?

The short answer is: Not necessarily. The contractor will, almost always, have more than one option. The better choice always depends on the situation. For example, using 3-D printing and modularization may seem cool and modern, but it does not always work as the better choice. Automation is another example: Assume the contractor has to excavate a large amount of soil, but this is in a rural area where labor is abundant and inexpensive, while it takes time and high expenses to acquire large excavators. It may be better to hire laborers to excavate manually if soil conditions allow for such an option.

Also, there are many theories and concepts that evolve constantly. Many of them are suitable for certain industries or situations. The mistake that some people make is to generalize; positively or negatively. For example, the author received sharply divided opinions on using the linear scheduling method (LSM) for scheduling or earned value management (EVM), for measuring performance in construction projects. Some people described the method or tool as very valuable while others described the same method or tool as worthless. In the author's opinion, both sides may be correct, but one applied the method or tool appropriately while the other applied it in the wrong situation.

For a long time, especially in the past few years, we have been hearing about automation and that intelligent machines, robots, and systems will replace humans. AI has come a long way. ChatGPT has generated a storm of debate, technical, ethical, and professional, over that same subject, and where are these systems going. There are questions that no one can answer with full confidence, but the author and many others believe that there will always be a role for the intelligent human being, no matter how AI and other concepts progress. The human role may, however, change due to the advancement of technology, but there will never be a world with machines taking over humans. Sometimes, we underestimate the creativity, intelligence, and versatility of human beings.

Here lies the problem for us, whether we are talking about concept, method, tool, software, or any gadget. There is no magic wand, there is no one tool that works best in all situations. It is up to us to make a judgment.

Never choose a tool before analyzing the situation and comparing solutions.

A tool that worked best for one situation, may not be the best for another situation.

Our field is full with people who are trying to sell us new products and services. In most cases, the question is not how good this product or service is, but rather how good it is for this situation. However, another syndrome is equally or even more dangerous: Avoiding a tool just because I am not used to it, I am attached to what I am used to, or for fear of the unknown. Those who do not know the tools available to them cannot make an educated decision, and thus, miss many opportunities.

Avoid the emotional factor in implementing new technologies:
a. The extreme positive, when obsession with the technology pushes you to use it, even when it is not the best tool in the situation, and
b. The extreme negative, when resisting anything new and different, and resorts to "what we have been doing for years"!

Exercises

18.1 Do an internet search for new technologies in the construction industry. Pick one technology and do an in-depth coverage of it. Focus on benefits to the construction industry. (Note: the new technology may not be exclusively for construction but has an application in the construction industry.)

18.2 Do an internet search for top construction companies now and 20 years ago. You will notice that some companies dropped out the list while others climbed up. Pick one of each and do an in-depth analysis of the reasons why one dropped out and the other ascended up.

18.3 You are the CEO of a construction company. Personally, you are not technically savvy, but your IT manager is a technology enthusiast, who, luckily, knows quite a bit about construction. You want to take a balanced approach between two extremes:
a. Staying traditional: "Keep doing what we have been doing," and
b. Investing in expensive technical gadgets and systems.
   Create a plan, utilizing the expertise of your IT manager to follow a balanced approach with a focus on benefit/cost ratio and forecast for the next few years.

18.4 Conduct research on construction cost estimating over the past 50 years. Try to meet with old construction professionals to ask them how they did it at the beginning of their career. Compare the impact of new methods / tools of cost estimation to the old ones in terms of time (effort) needed, accuracy, and cost.

18.5 Do research on the use of solar power in construction operations. Was there a big impact? If yes, in what type of projects, in particular?

18.6 Do research on the use of rechargeable tools. What are the benefits of these tools in terms of cost, convenience, and safety (no dangling cords)?

18.7 Do research on the use of drones in construction operations. Was there a big impact? If yes, for what uses and in what type of projects?

18.8 Do an internet search on new construction hard hats with added safety features. Mention advantages and disadvantages.

18.9 Do an internet search on a new gadget that helped in the construction industry in one or more of these categories:
a. Improve productivity,
b. Improve safety,
c. Reduce cost,
d. Improve quality,
e. Shorten duration.

18.10 Technology of communication has come a long way in the past 50 years. Do research on this topic with chronological mention of the communication means that a typical contractor used in the 1980s, 1990s, and after.

18.11 Do an internet search on a new computer software system that helps the contractor with documentation.

18.12 Do an internet search for a new technology that helps the contractor with non-destructive testing.

18.13 Do an internet search for a new technology that helps the contractor with soil investigation.

18.14 Do an internet search on a new system that helps the contractor with automation. Do you think automation is the solution for the construction industry? (The answer is not a simple yes or no).

18.15 Meet with a few construction professionals who are familiar with the newest technologies. Ask the question: What changes do you foresee in the next 10 to 20 years?

19

Management of Remodeling, Renovation, Restoration, Expansion, and Demolition Projects

Introduction

Although the fundamentals of construction are the same, the objectives, constraints, priorities, and methods change based on the situation. When working on an existing project for remodeling, repair, or restoration, the contractor needs to worry about several issues such as the removal of old damaged or unwanted items, replacement of items, and retrofitting, while protecting surrounding area from damage. This may not be easy, especially for old buildings with items that are no longer available on the market. Historical restoration projects add further restrictions on the contractor, where the project must be restored to its original shape.

In this chapter, we will give a brief cover of such projects to just bring awareness of the challenges the contractor may face in these projects.

Types of projects to be covered in this chapter:

1. Demolition, removal
2. Decontamination, abatement, mold removal
3. Repair projects such as those damaged by fire, wind, water, earthquakes, sinkholes, and mudslides
4. Remodeling
5. Restoration, renovation
   a. Restoration of old and abandoned structures
   b. Historical renovation
   c. Road resurfacing/maintenance
6. Expansion / retrofitting

Project Management in the Construction Industry: From Concept to Completion, First Edition. Saleh Mubarak.
© 2024 John Wiley & Sons, Inc. Published 2024 by John Wiley & Sons, Inc.
Companion Website: www.wiley.com/go/nextgencpm

The General Challenges in These Projects

There are similarities between these types of projects, on one side, and traditional (new) projects, on the other side, but there are differences as well. In these types of projects:

1. The contractor is often unable to know the exact scope of work, at least until the damaged items are removed. The work scope may be more than originally expected.
2. Many of the items to be replaced may be old with no identical replacement items available on the market. Sometimes they must be custom-built, which takes extra time and costs.
3. Retrofitting may also be a challenge when new items must be attached and integrated to old members in a seamless way. There may also be a challenge in conforming to a new building codes or new technologies or upgrading the electrical wiring or plumbing systems.
4. There is sometimes physical dangers and hazards when removing old items, such as mold, fungus, bacteria, insects, rodents, and other dangerous creatures. The risk may also happen from unexpected collapse of the structure, partial or complete. The contractor must consult with a structural engineer before removing any member that may be load-bearing.

> Renovation and restoration projects often involve high degree of uncertainty because the contractor may not know the extent of the work until after starting. This is why most of these projects are contracted on the basis of cost-plus-fee.

For the above reasons, it is very difficult in most of these projects to estimate the cost or time to complete the project. It may be in the best interests of both the contractor and the owner to sign a cost-plus-fee contract, perhaps with a guaranteed maximum price (GMP), representing the contractor's best judgment. Sometimes repair projects may be done in two phases. The first phase is "surface demolition" for exploration purposes. The findings will help determine the scope of work for the second phase, which is completing the work.

Demolition

Demolition is the process of tearing down and removing buildings, partially or totally, or other structures. Demolition may be performed for a variety of reasons, such as to make way for new construction, to remove a hazardous or condemned structure, or simply to cleanup an area. There are hazards in this process that may pose danger to the demolition crew as well as the people and structures nearby the demolition project. This includes falling and flying objects, sharp edges and injury-causing objects, loud noise, dust and fumes, toxic substances, and others. This is why demolition can be a delicate type of work that requires specialized equipment and training.

Importance of Safety During Demolition Work

It is important to start by defining the scope of the work and then determine the potential hazards that require the services of a professional and/or a permit from the authorities. When there is a potential hazard, it will be best to work with a professional demolition company to ensure that the project is completed safely and efficiently. Trained professionals can better assess the situation, operate heavy machinery, and handle hazardous materials.

A well-planned demolition project must consider the safety of workers and bystanders and cause no or minimum damage to nearby property that is not included in the scope of the demolition, as well as the surrounding environment.

Demolition Methods and Common Types

There are many methods and tools for demolition. The choice amongst them depends on:

1. Type of materials and systems to be demolished.
2. The position and location of the structure or member to be demolished.
3. The size and volume of demolition project.
4. The surrounding environment and local laws and regulations.

The main considerations are safety first and then economy. The contractor must choose the most economical *safe* method. There are many methods and tools, we can combine them in these groups:

1. Simple hand tools such as sledgehammer, pickaxe, mattock, pry bar, crowbar, demolition bar, shingle remover, pliers, nail puller, snips, hammer, carpet puller, pipe cutter, and honey badger demolition fork.
2. Power tools and equipment such as jackhammer/demolition hammers, oscillating multi-tools, reciprocating and circular saws (possibly with demolition blades), and air impact hammers and drills. Those tools may be pneumatic, hydraulic, electric, or gas-powered. Pneumatic tools and equipment are powered by compressed-air generators. Hydraulic tools and equipment use similar systems as pneumatic ones except for the use of hydraulic fluid instead of air. They produce a larger amount of force and tend to be mounted as an attachment on heavy machinery. Electric and gasoline tools and equipment use motors that run on electricity (plugged or rechargeable) or gasoline to operate the tool/equipment.
3. Attachments to construction equipment such as wrecking balls attached to cranes and several different attachments to excavators and bulldozers including picks, hammers, grapples, pulverizes, thumbs, couplers, and skid steer loaders. Many attachments can do more than one function. Excavators with high-reach arms may be used for demolition at high elevations or hard-to-reach locations. Excavators with blades of different sizes and shapes are also used in demolition at ground level and moving the debris.
4. Pressurized water, using air compressors to generate pressure suitable for the type of demolition materials.
5. Blasting and imploding. Blasting is used to remove rocks mostly for tunneling, mining, roads, bridges, and other projects. The explosives are usually inserted into drilled holes at a depth that makes them most effective and also minimizes flying rocks and rubble. The contractor must be careful about flying rocks from the explosions and also about causing consequential and unintended damage to surrounding rocks that may cause safety hazards later. This is why contractors, in many of these situations, have to wrap the rocks on the side of the road with nets[1], to stabilize the rocks and prevent any loose ones from falling.

---

1 It is called rockfall netting. These nets are woven by low carbon steel wire, stainless steel wire or galfan wire with the surface of galvanized, PVC coated or galvanized plus PVC coated.

Imploding is used to demolish buildings by collapsing in a planned and contained manner. There are two ways for imploding structures:

A. Falling sideways without collapsing vertically, like a tree. This is usually done for water towers, power and communication towers, and some other buildings.
B. Falling into its own footprint by sequential collapse from top to bottom. This is usually done for residential and commercial buildings, especially high-rise.

The science of imploding buildings has been developed in the past few decades. It is performed by placing explosives in certain amounts and in certain positions in the project and then igniting them in a sequential manner starting at the top of the building. When done properly, the building collapses as planned. Even with the most accurate calculations and planning, debris may fall or fly away from the building, so the demolition contractor must take all necessary precautions to protect the surrounding area, particularly humans.

The choice among the above methods may not be the contractor's choice. It depends on the four factors mentioned earlier. For example, the demolition of a building can be relatively easy, fast, and inexpensive if the building is located in an area by itself with no major safety concerns when imploding the building. The demolition of a similar building located in a populated area may be difficult, expensive, and time-consuming.

Special Demolition

Some projects need special handling for demolition, such as nuclear power plants, oil refineries, and chemical and some industrial plants. Such projects need to be dismantled in a manner that makes their demolition safe without any unexpected side effects to people, structures, or the environment. All hazardous materials have to be removed and secured before any demolition action.

Safety Tools and Devices:

The demolition contractor has the obligation to protect people, property, and environment in and around the project. While we discuss this issue in more detail in Chapter 12, Project HSE Management, we can briefly mention some safety tools and procedures here. Personal protective equipment (PPE) includes gloves, hard hats, goggles, safety (steel toe) boots, ear plugs, dust masks, and others. In some cases, a respirator with an exhalation valve may be required. Depending on the volume and type of demolition, the contractor may need to evacuate certain areas around the project from people, animals, and valuable assets such as vehicles. Assets that cannot be moved or protected, such as buildings, must be evacuated and utmost care must be taken to cause no or minimum damage to them.

One of the side effects of demolition is dust and fumes that can move with the air to a larger circle of impact. Again, the contractor must deal with this issue with mitigating methods such as spraying water on the rubble.

Demolition Attachments | Cat | Caterpillar (https://www.cat.com/en_US/by-industry/demolition/work-tools.html)

Disposing Debris

In almost all cases, the demolition contractor is responsible for disposing of the demolition debris. This can be relatively easy, rather complicated and expensive, or a degree between this and that. It all depends on the material types, local restrictions, contract terms and client's demands, and options available. Some materials can be reused or recycled. For example, concrete chunks can be crushed into small pieces and used as aggregates for road pavements and other projects. Toxic, hazardous, radioactive, and biological materials must be dealt with according to official safety regulations.

Remodeling, Renovation, and Restoration Projects

The three terms above may sound similar. Indeed, they have similarities among each other, but they have differences as well. Remodeling transforms the purpose of an area, building, or structure from one use to another. For example, transferring a residential building to offices will require certain work that makes the place more suitable for the new use. Both restoration and renovation projects bring the place to a newer condition without changing its purpose. Renovation work usually includes updates, upgrades, and improvements while restoration simply restores a space to its original condition with replicas or original materials. Historical restoration projects usually add more restrictions that obligate the contractor to bring the building or structure to the exact original condition and look.

The one thing in common among all these types of projects is that they require some demolition to make place for new construction additions and/or replacements. It is important to define the scope of work, but it may be difficult to prepare an accurate cost or time estimate, particularly if the structure is old and the work involves the removal of old members. The contractor may not be able to determine in advance the extent of the work and the items to be replaced. Also, finding a replacement item may be a challenge, especially in historical restoration projects. This is why most of these projects are contracted based on cost-plus-fee contracts.

Another challenge that may sometimes arise is the ability of the owner to visualize the finished product at the time of the agreement with the contractor. In many cases, the finished product may not match what was imagined or envisioned by the owner or even if it matches in description, may not get the owner's satisfaction. This issue is becoming less of a problem with the rise of high-tech programs that help the designer simulate and demonstrate the finished product to the owner, perhaps with several options. The contractor is recommended to keep the communication with the owner, step by step, to make sure that work is proceeding according to the owner's intent and satisfaction.

In many situations, the contractor replaces the item with an alternate that does not match the original. This is either because a replacement of the original item may not be available on the market or because there is a better alternative available now. The owner needs to be informed of any such replacement.

The contractor must be careful in tearing down the members that need to be removed so no unnecessary damage is caused. Safety also needs to be a priority. Some old buildings may contain hazardous materials such as asbestos and lead-based paint. Also, old electrical devices and wires may need consultation with a specialist before tearing them down. The contractor may see symptoms of a problem like decay, dampness, or bad smell. It is important to diagnose the issue, identify the root cause, and treat it. Some problems like water leaks may happen only in certain events such as heavy rain, which may not allow the contractor for full investigation in a dry time.

Complying with new building code regulations is another issue the contractor may need to pay attention to. For example, installing fire and smoke alarms, sprinkler systems, and fire escapes. Improvements that do not alter the purpose of the building or structure are always encouraged but need to be communicated with the owner, preferably in the early stages of the project. This includes items such as security systems, insulating windows, solar systems for generating and storing power, and any materials or systems that can improve the efficiency of the project.

The disposal of demolished items, including any hazardous materials, has to go according to the safety regulations and terms of the contract.

When removing a column, wall, or any member that may be load-bearing, the contractor must get a structural engineer's approval before proceeding. Any uncalculated demolition can lead to a disaster.

When operating in downtown or near residential areas, the contractor may have to provide cover over the building to protect pedestrians and property from any falling objects and dust. Also, the contractor needs to provide a chute for sending the debris into containers that will be hauled away later on.

Expansion and Retrofitting

Sometimes, there is a need to expand an existing building: horizontally or vertically. Horizontal addition in this context is attached to the existing building and will be an integral part of it. The design must be approved by a certified designer and by the local authority. Vertical additions must be approved by a structural engineer who will make sure the load-bearing elements, columns, shear walls, and foundation, can take the additional loads without compromising safety.

Another issue is retrofitting, which means integrating the new construction with the existing building or structure, seamlessly in both looks and function. This can be challenging, depending on the parts to be integrated and the availability of compatible parts in the market. All systems, such as electrical wiring and HVAC must be evaluated to assure sufficiency after the addition. In some cases, the old system is replaced with a newer and higher-capacity system that can serve the expanded building or structure.

Construction in Disaster Areas

After the event of a natural disaster or other destructive events such as wars, the authorities usually start with rescue operations and providing safety and basic necessities to affected people. Thus, it will also be important to restore transportation routes and communications means. Clearing up the roads serves two important objectives: Remove any hazards such as falling powerlines, trees, and debris and allow the movement of vehicles so rescue and restoration operations can proceed. In some cases, draining flood water from roads may be necessary for the accessibility of traffic, especially for emergency vehicles. Restoration and re-construction will likely require demolition of damaged buildings and structures before any building or structure is repaired, restored, or rebuilt.

During this phase and despite the focus on the rescue, cleanup, and restoration, it is necessary to collect and record data that may help organize the current operations and mitigate such events in the future.

After the initial cleanup, it is time for damage assessment and restoration. The project objective will be restoring the area including its infrastructure, buildings, services, and perhaps more. The damage may be caused by a flood, earthquake, tornado, hurricane, explosion (under whatever cause), fire, sinkhole, or other causes. Work usually starts with assessing the situation and estimating the scope of work. This may be a challenging task since much of the damage may be buried and invisible, at least partially. In case the area impacted by the disaster is relatively small such as in cases of sinkholes and some cases of fires and tornadoes, the estimate can be based on expert observation and inspection of the impacted properties. It may be performed by the owners of the property affected, contractors, local authorities, insurance companies, or others.

In cases of widespread damage, such as earthquakes and hurricanes, authorities usually run an initial estimate of the cost of repairing and restoring the damaged area by using approximate methods. The easiest and quickest method divides affected areas into blocks or grid squares after taking aerial pictures using a helicopter or drones. The damage assessment is given several grades from none to catastrophic (first column in Table 19.1), with a rough estimate to restore per grade of damage per block or grid square (second column in Table 19.1). The product of these two numbers represents the cost estimate for all blocks belonging to the same degree of damage (third column in Table 19.1), which will give us the grand total when added up. Of course, the accuracy of the estimate may be somewhat improved by making the

Table 19.1 Cost estimate of an area hit with a natural disaster.

Damage	Cost/block $	No. of blocks	Total cost
None	5,000[a]	5	25,000
Light	500,000	7	3,500,000
Moderate	2,000,000	10	20,000,000
Severe	5,000,000	8	40,000,000
Catastrophic	8,000,000	9	72,000,000
	Grand Total	39	$135,525,000

[a] We assumed a minimum cost of $5,000 for cleanup, despite the no damage classification.

squares smaller, which results in a larger number of squares. The assessment and classification of squares can be automated using artificial intelligence (AI) programs.

The cost must include all items such as buildings, roads, sidewalks, public transportation lines and stations, power lines and poles, communication towers, parks and vegetation, and infrastructure. It must also include cleaning up the area from all debris.

Again, there will likely be hazards during the assessment and inspection phase to the impacted areas such as live power lines that fell to the ground, unexploded bombs (in wars), unstable structures, sharp objects (as a result of items broken), and contamination.

Although disasters have plenty of negative consequences, including casualties and property losses, they may also be an opportunity for starting all over again with better and updated replacement. Many areas were built a long time ago, and they may not comply with current building and environmental codes and regulations. Also, those buildings and structures may not be built to accommodate new technical systems. So, the disaster may provide an opportunity to start again with new infrastructure and buildings that are better in several aspects.

There are other cases where a catastrophe or calamity happens somewhere else and people (refugees and others) flee to a safe area. The authorities in this safe area may be faced with the unexpected and unplanned influx of people who need shelter and necessities of life. This may be a short- or long-term issue. In many instances, it is assumed to be short-term but turns out to be longer than expected. Providing immediate housing for thousands of people is a huge challenge. Depending on the number of people, the resources available, and the authorities' policies and plans, the housing provided may be tents, modular units, or traditional units built in an accelerated way. It is also possible that tents, trailers, or other temporary units, are used for a transitional period till the permanent units are built. The solution must take into account many factors, economic and humanitarian, short- and long-term, but construction professionals are intimately involved in these solutions.

Accelerating methods will likely be utilized in such cases. But even when housing comes from manufactured units (modules), there is a need for infrastructure construction, to provide utilities and roads in the area, and likely foundation for the mobile housing units. Also, manufactured housing units usually take some time to be built, transported, and set up, especially when it is demanded in high numbers. This can be a classic program that includes several projects, perhaps in multi-phases.

Specialty Construction:

Most contractors specialize in the type of construction they do: residential, commercial, heavy, or other. There are some types of construction that require specialization and expertise such as oil refineries, dams, water desalination plants, power plants, stadiums, amusement parks, and others. Although construction projects have a lot in common in the way they are managed, every project requires a certain technical knowledge and expertise. Those who are involved in managing the project must be qualified with project management concepts and processes as well as the specific technical aspects of the project.

Exercises

19.1 Demolition is a crucial part of the construction industry but there are many specialties within the demolition work. Do a search to find some of these specialties.

19.2 Methods and costs of the demolition of a structure may vary significantly based on location. Explain this statement.

19.3 Mention some of the challenges the contractor faces in estimating the cost and time of renovation and restoration projects.

19.4 What are the reasons for the contractor to prefer cost-plus-fee contracts in many renovation projects?

19.5 You are a demolition contractor. You contracted to demolish a small building in a crowded residential neighborhood. Create a plan outlining the steps, methods, and safety precautions you will take.

19.6 Conduct an interview with a renovation and remodeling contractor. Ask about the challenges such as:
   a. Uncertainty about the scope of work,
   b. Need to replace old items that are no longer in the market,
   c. Estimating completion time,
   d. Dangers to workers, and
   e. Compliance with historical renovation rules and laws.

   Also, ask the question: What are some of the lessons learned? What would you do differently if you could do this project (or these projects) again?

19.7 Do a search on demolition accidents and mishaps. Analyze the causes and make a list of these causes. If you have access to a demolition contractor, ask them about any previous accidents and what lessons learned from those accidents.

19.8   Interview a renovation or remodeling contractor who did a project in a crowded neighborhood. Ask about the precautions taken to protect the public from falling debris, dust, noise, and other issues resulting from the construction work.

19.9   Do a search, internet or other, for a variety of specialty construction. Make a list of them and find out the names of local and regional specialty contractors.

19.10   Conduct an interview with a specialty contractor and discuss a unique project they did. Ask about the challenges and lessons learned. What are the advantages and disadvantages of being a specialty contractor?

19.11   Conduct an interview with a FEMA (Federal Emergency Management Agency) official. Find out about FEMA's response to a natural disaster that forced a large number of people to flee their homes. How did FEMA provide temporary housing for them? Focus on solutions that require construction. (Note: The United Nations High Commissioner for Refugees (UNHCR) deals with this issue at the international level. It would be great if you could do this interview with one of their officials.)

19.12   Search for a contractor who does abatement (removing carcinogens and other harmful materials from old buildings, such as asbestos and lead-based paint). Discuss the challenges the contractor faces and the safety precautions they must take.

19.13   In your interview, in any of the previous questions, ask the contractor on materials disposal. How would the contractor dispose debris, including hazardous materials, in a professional, legal, and efficient way?

20

Real Estate Development

Introduction

Real estate (or property) development is a business process involving activities aiming to turn this property into a new use, as defined by the developer. In this chapter, we will focus on land development, a type of real estate development, which is the process of transforming land into construction-ready-to-build site(s) for a specific project type. This land can be originally either raw or previously used for a different purpose.

The process includes changes and improvements that make the land suitable for new use. It also makes sure this land integrates well with its neighborhood and environment. A baren or agricultural land may be turned into residential compound. An old warehouse may be turned into a shopping center. An old apartment building may be turned into a modern office building and so on.

Although real estate development technically is considered a business process and not construction, it intimately relates to construction and the construction industry. It provides the ready-to-build property that will later become the construction site. Also, real estate development often involves some types of construction work such as the required infrastructure (utilities, stormwater drainage, roads, common facilities, etc.), demolition, and more.

The process involves the approval of the authorities including all relevant departments. In many places, this also requires the approval of the majority of the neighbors of that land through public meetings, particularly when the purpose or use of the property changes.

Why Real Estate Development?

Real estate development contributes to the progress of society. It helps local economies grow by improving the status quo and creating jobs and opportunities. It helps add value, not only to the land or development itself, but likely also to the surrounding areas. It may also respond to a shortage in supply which helps balance the demand and stabilize the market prices.

Project Management in the Construction Industry: From Concept to Completion, First Edition. Saleh Mubarak.
© 2024 John Wiley & Sons, Inc. Published 2024 by John Wiley & Sons, Inc.
Companion Website: www.wiley.com/go/nextgencpm

The Business Side of Real Estate Development

Developers have to go through a certain process for developing land or other real estate. The process has legal, technical, and financial sides and it often involves risks before the new development is completed. One of the major risks in this process is the official approval of the rezoning. For this reason, many developers add a condition in the land sale contract that this sale is contingent upon the rezoning approval. The contract usually gives time (e.g., 3 months) for completing the rezoning process. If the rezoning is approved, the sale becomes final, and the deal closes, and the developer can go ahead with the development. If the rezoning is not approved, the contract becomes null and void, and the investor gets back the deposit but loses the expenses of the feasibility study and rezoning application and effort. Of course, this type of contract is likely to increase the prices somewhat compared with a fast-closing non-conditional contract for the risk and uncertainty the owner is taking during the wait time.

The developer must provide infrastructure, making the land suitable and ready to use/build for future users. This includes but is not limited to:

- Utilities (electricity, potable water, sewage, and internet cable)
- Roads, sidewalks, and landscape
- Stormwater drainage system
- In some, but not all, cases, the developer must also build common amenities such as playgrounds, parking, clubhouse, etc.

This also requires the local authorities to upgrade the infrastructure (outside the development) and public services to meet the new demand (roads, schools, hospitals, emergency services, libraries, parks, etc.) Local authorities usually charge the developer an impact fee to cover the cost of these upgrades. An impact fee is typically a one-time payment imposed by a local government on a property developer. Impact fees are determined based on the size of the new development, and the cost of implementing and integrating it with the existing area. The developer will either pay these fees as lump sum and add them to the total cost or finance a loan (bond) that will later be transferred to the new individual owners, based on the value of their parcels.

Who is Involved in Real Estate Development?

The development process requires orchestrating the services and skills of several professionals. Some of them can be combined, subject to legal restrictions:

1. Land planners to address the project design. This may include architects, civil, traffic, and environmental engineers.
2. Market consultants to determine demand for the project's economics.
3. Lawyers to handle the legal process and regulatory approvals.
4. Environmental consultants and soil experts to analyze the project's environmental impact on the site.
5. Land surveyors and title companies to provide legal description of the property and title search and insurance.

6. Lenders/investors to provide needed financing.
7. Contractors, general and specialty, to perform the site development and construction.
8. Specialty consultants.

The developer may act as a financier and also as a general contractor. After the development process is completed, the developer can either sell the land as buildable lots (e.g., home sites) or act as a general contractor and build and sell the project as units. Some corporations, particularly large ones, may develop their own properties so they retain ownership of the real estate after the development process.

Phases of Real Estate Development

1. Preliminary Research and Concept Phase: The developer starts with the project conception using the "comprehensive land map" which is set and approved by the local municipality. The developer will think of the best use for the project land within the restrictions. Timing and economic conditions, current and forecasted, must be taken into account. The developer will perform market analysis and preliminary feasibility study and make preliminary financial analysis.
2. Contract Phase: Negotiations start between the developer and the land/property owner. The developer (land buyer) will send a letter of intent (LOI) to the landowner with the main deal points. If they both agree on the terms, they sign a contract. The contract closing hinges on the conditions contained in it. The developer will start making the development team.
3. Due Diligence Phase: The developer will start with the property analysis, current and proposed zoning, and title review and title commitment. Next, several services will be performed such as land survey, environmental and geotechnical study, and traffic/utilities/roads study. Also, the developer starts the financial analysis, project financing, and market research.
4. Entitlement – Zoning Phase: The zoning will depend on several factors including the comprehensive land map and the impact of the growth and the integration of the new development. The zoning, or practically the rezoning, process will also involve meetings that allow the public to speak out, in favor of or against the development, and negotiate with the zoning committee. The zoning committee will make its decision, which can be approval as applied, denial, or approval but not as applied. As an example, the developer applied for a residential compound with four single-family houses per acre, but the committee only allowed three single-family houses per acre. There will also be imposed conditions such as the upgrade to the existing infrastructure and pay government fees (called impact or mobility fees) for the upgrades to infrastructure and services the government will do as a result of this new development.
5. Design and Plan Approval phase: Once the developer gets the entitlements for the property, the engineering phase starts. The engineer needs to deal with the different regulatory agencies and their demands and restrictions such as those regarding wetlands, endangered species, and protected trees. Construction plans are prepared for approval and permitting. During this phase, there will be many challenges that require negotiations with different parties. After the engineering is completed, the contractor selection starts.
6. Construction Phase: Now that the developer has secured the zoning and permits, the land purchase and acquisition must be finalized. After that, the pre-construction services start including site and

infrastructure work and coordination with local utilities. Inspections follow the work. The developer is now wrapping the project: They prepare to submit the "as-built" plans to the authorities, go through the final punch list, and final acceptance. Once the final plat and Certificate of Occupancy are issued, the work is completed.

Development Cost Includes:

Investors look at land development as an investment opportunity that carries some risks. The investment involves upfront costs.

1. The purchase of the land or property.
2. Surveying the land and soil inspection.
3. The feasibility study before the purchase of the land. This includes legal, environmental, and economical aspects.
4. Government fees such as application fees, permitting fees, rezoning fees, (dividing into parcels), and impact fees.
5. Legal fees.
6. Civil and environmental engineering fees (design and review) including floodplain studies.
7. Environmental remediation including wetlands delineation.
8. Clearing the land from unwanted vegetation, rocks, and old structures.
9. Infrastructure construction costs, including stormwater drainage system, utilities, roads, and more.
10. Inspections.

Stormwater Drainage

Most legal regulations have certain requirements regarding stormwater drainage whenever someone changes the condition or the nature of the land surface. The main reason is because natural land is porous, allowing a portion of the rainfall to infiltrate the soil and go to the aquifer. The amount of the infiltration depends on various factors such as the amount of precipitation, the duration of the rainfall event, the permeability of the soil, and the landscape of the site. However, when part of the land is covered by a non-porous or low-permeability material, such as concrete, pavers, or asphalt, the water over this covered land drains to the closest permeable land that can drain the water down to the aquifer. This action will make the flow of water on this permeable land higher than its natural flow-down (infiltration) rate. A drainage system needs to be built to prevent flooding and accumulation of water. Such system is to be designed by civil engineers considering all factors involved. The system must drain the water, with certain slope, to covered and uncovered waterways to water collection bodies; either existing lakes or rivers, or ponds excavated specifically for this purpose.

Excavating a pond, from the developer's perspective, has potential advantages in addition to being a fulfillment of a regulatory requirement. It creates water bodies, call them ponds or lakes[1], that can be

---

1 There is a difference between a lake and a pond. The lake is bigger in size and deeper, and usually natural, which makes it permanent, despite the water level fluctuation. Ponds are smaller and shallower. They are usually created for the purpose of storm water drainage. So, they may be dry in parts of the year, depending on the amount of precipitation and other factors.

attractive to buyers, especially with landscape added. Lots overlooking a pond or a lake can be sold for a higher price as "waterfront lots." In addition, the excavated soil is likely to be used as a fill material, especially when certain parts of the site need to be raised. If the excavated soil is sand, it can be used for multi-purposes or even sold, since sand is an attractive and demanded material.

Exercises

20.1 Define real estate development.

20.2 Do you consider real estate development as a type of construction? Explain your answer.

20.3 Why does the developer provide infrastructure or infrastructure improvements to the area he/she is developing?

20.4 Why do the local governments charge the developer impact fees?

20.5 What parties are usually involved in the process of real estate development?

20.6 Mention the phases of real estate development.

20.7 Mention the types of development costs.

20.8 Excavating ponds can be an obligation for the developer, but it can also be a benefit. Explain this point.

20.9 Why is stormwater drainage a concern the developer needs to address and resolve, while it was not a concern before the development?

20.10 Conduct an interview with a real estate developer. Discuss issues like:
    a. The process of real estate development.
    b. The upfront cost of real estate development.
    c. The duration of real estate development and phases during this process.
    d. Risks the developer faces.
    e. Advantages and disadvantages of combining the developer's and contractor's roles in one entity.
    f. Reflections and lessons learned after the completion of development.

20.11 Would the developer change his/her plans if the economy unexpectedly changes during the process (for better or worse)? How?

21

Construction Management from Owner/Client Perspective

Construction Project Management from Owner's Perspective

Introduction

Both the contractor and the owner have a stake in executing the project successfully, but each party has its own objective, constraints, and priorities. The contractor is in the business of building projects to make profit and enhance their organization's financial status and reputation, especially in the long-term outlook. There are usually other considerations such as public relations, public image, environmental cleanliness, and welfare for own employees. Also, the contractor may be juggling the resources and constraints of the project with those of other projects under their organization, current or upcoming.

On the other hand, the owner's focus is on completing the project within the major constraints (budget, schedule, and quality) and other constraints, with the important notion that the priorities of these constraints vary from one situation to another. The owner is also concerned with the successful operation of the project after completion and during its service life.

There is a stigma of an adversarial relationship between the owner and the contractor, which is not necessarily true. Many projects are completed with both sides satisfied, which leads – in many cases – to continuous relationship between the two parties; a big advantage that will bring benefits to both. Such a good relationship comes as a result of the good and professional practice of each party in their role and being honest and fair.

Project Cost to an Owner

Before we discuss the project cost to an owner, we need to define the owner's project because, technically, it differs from the contractor's and the designer's projects. The contractor's project, depending on the contract terms and method of delivery, is basically to construct the project as described in the design drawings and according to the terms of the contract. The owner's project is more than that. It starts from the project's conception and feasibility studies and continues after the physical completion of the project. In fact, we can divide the owner's project into three phases:

1. Planning and feasibility studies. This is the "pre-project" stage when the owner is still formulating their project concept against the constraints and the results of the feasibility study.

Project Management in the Construction Industry: From Concept to Completion, First Edition. Saleh Mubarak.
© 2024 John Wiley & Sons, Inc. Published 2024 by John Wiley & Sons, Inc.
Companion Website: www.wiley.com/go/nextgencpm

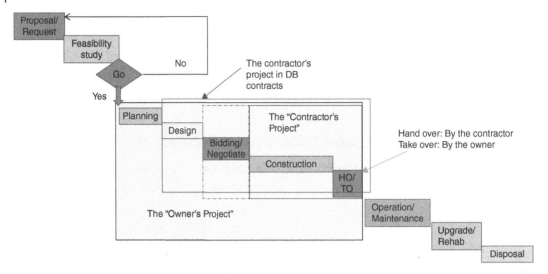

Figure 21.1 Lifecycle of the project.

2. Design and construction. This is the main stage where the project will be designed and built. Depending on the contract terms and method of delivery, the design and construction may be combined in one contract. This stage ends with the physical completion of the project and fiscal close out of the contract. The project should then be ready for occupancy and operation.
3. Operation and maintenance. This stage includes periodical and occasional upgrades and renovation. When the project life ends, it must be disposed of, in an appropriate way.

So technically, the owner's project starts from the point of approval (Go) till the takeover from the contractor, as shown in Figure 21.1. However, the owner's role differs during the three phases. During the design and construction phases, the owner delegates the responsibility to the designer to create the design documents, and to the general contractor to build the project, although will still being intimately involved in both stages. After the completion of the project and the takeover by the owner, it goes to the hands of the owner for operations and maintenance. The owner must expect these items as part of the project cost:

1. Lost opportunity cost: This represents the amount of profit that would have been earned if the money was invested differently.
2. Land, right-of-way (access), easements. In addition to the cost of the land, the owner may have additional related expenses such as right-of-way or easement through a neighboring adjacent land.
3. Financing (cost of borrowing money). The owner may pay financing costs for the project in more than one way. There are owner's purchases and services including land and construction usually require financing. The traditional way of contractor spending and owner reimbursement usually results in additional financing costs to the contractor, which is passed to the owner.
4. Design, including architectural, civil, structural, mechanical, electrical, and other design services. These fees range usually between 5% and 12%, depending on several factors[1]. It is also very common,

---

1 In non-Western countries, this percentage may be significantly less.

especially in residential construction, to reuse the same design and save on design fees. Owners can choose and buy complete home designs (complete set of drawings and specifications) from among hundreds of designs from internet sites and other sources[2], for as little as a few hundred dollars. This will save thousands of dollars but will not allow the owner to make major modifications unless approved by a design professional, which will likely reduce or wipe out the savings.

5. Permits and licenses fees. Such fees may be initially paid by the owner, the design consultant, or the contractor; but eventually, they come out of the owner's pocket. Developers usually pay impact fees to local governments as compensation for the impact of adding this new development on the existing infrastructure and services the government provides.

6. Construction. This includes all expenses, direct and indirect, associated with the construction process.

7. Project/construction management services. This is yet another service the owner may need to ensure that the project is being executed according to the contractual agreement and that the owner's interest is protected. The term project management services is wider and more comprehensive than construction management services as it starts earlier—perhaps as early as the project is initiated. Construction management services on behalf of the owner may be performed by the owner's team, by the designer firm, or by a third party construction management consultant. In all cases, it is an additional expense.

8. Furniture and equipment. This cost has to be considered by the owner or someone else (possibly the tenant for residential and commercial projects). Depending on the style, quality, level, and other factors, this can be a significant expense.In some commercial projects, a tenant offers a long-term lease from the owner, on condition that the owner finishes the space in a way compatible with the tenant's business.

9. Start-up. In many industrial projects, the owner requires the contractor to do the start-up, which means the contractor delivers the project to the owner in a working condition. It might include training personnel also.

10. Warranties. Most building codes and local laws require builders to warrant their projects for a certain period of time, although the components of the project such as the HVAC system, usually have their own warranties with different terms and lengths. Some builders offer a longer or more comprehensive warranty as an attraction or confidence booster to the clients. The owner may also require an extended warranty, but of course, the longer this period is, the more expensive it will be even though this cost might not be conspicuous to the owner.

11. Operating and maintenance. This expense comes after the completion of the project and during its occupancy, which may be a major expense to the owner. It includes operating and maintenance not only of the structure itself, but also of the equipment and main systems affiliated with the building.

12. Rehabilitation / upgrading. In industries that change rapidly with technology, this can be a major expense. For example, if an office building has personal computers, they will need an upgrade at an average of three years. In many cases, the infrastructure of the computers or other equipment may also need upgrading/updating.

13. Disposal. Disposal can be a major headache and expense, especially if the project contains chemicals, or toxic, hazardous, or radioactive materials. This might also include demolition in a crowded downtown area, which is another major expense and liability. Hauling and dumping fees may also be substantial.

---

2 For example, https://www.houseplans.com/.

The total cost that includes these 13 items constitutes the life-cycle cost for a project, which may be necessary to estimate before approving the project in order to calculate the benefit/cost ratio[3]. Professionals and experts sometimes conduct value engineering (VE) studies to optimize the project's total life-cycle cost. This may result in slightly increasing the upfront cost in exchange for bigger future savings and/or better performance. Also, BIM started to be utilized to simulate the lifecycle (design, construction, and operation) of the facility to optimize cost or other criteria.

Planning and the Decision-Making Process

Planning is an extremely important phase[4]. It must not be taken lightly by the owner. During the planning, design, and construction phases, the owner must make many decisions, minor and major. The decision-making process must be defined and controlled, especially when this is done in a situation with several people contributing to or influencing the decision. The owner must resist the emotional instinct and, instead, focus on reconciling the needs with the budget. In most situations, reversing or changing the decision later may be costly and disruptive. The decision, especially when major, must be discussed with experts who can explain the pros and cons of each available option.

When the owner thinks of adding an option to the project, he/she may do a benefit/cost or sensitivity analysis. This is basically to answer the question: Is it worth it to spend the money on this option? The answer must take the lifecycle of the project into account. For example, a newlywed couple who are building a house and plans to have children in the near future. Also, a couple in their 60s may think of "after retirement." An owner who is building the project for investment must think from the buyer's perspective. When the owner has an item in the "wish list" such as a swimming pool but cannot afford it at the time the project is approved, may want to include the infrastructure for that item. Thus, it will be easier and less expensive to add it later.

Scope creep can happen to the project as a result of poor planning, poor decision-making, or both. It has a negative impact on both cost and schedule. It may also strain relationships between the owner and the contractor. In fact, the number and amount of change orders, which are directly related to scope changes/scope creep, is one of the indicators of project success. Successful projects should have no or few/minor change orders.

The Owner's Organization – PMO's

Owners who conduct frequent construction projects may have their own teams for design / design review, preconstruction services, and construction supervision. These teams may not be qualified or equipped to perform detailed design or construction operations, but they help the owner in several ways:

- Feasibility studies,
- Schematic design,

---

3 Sometimes, investors do a "pro forma" which is an analysis of the project's expected cost, initial and on-going, and revenue to help decide on its approval. The pro forma uses certain projections and assumptions.
4 Was explained in Chapter 4.

- Review detailed design,
- pre-construction services,
- Contractor procurement, bidding negotiation, and award processes,
- Construction supervision / management,
- Managing construction-related issues such as change orders, progress payments, and claim resolution,
- Closeout and start-up processes, and
- Archiving.

This in-house project management unit can expand and shrink depending on the need. The Project Management Institute (PMI) calls it Project Management Office (PMO), and defines it as "a management structure that standardizes the project-related governance processes and facilitates the sharing of resources, methodologies, tools, and techniques. The responsibilities of a PMO can range from providing project management support functions to actually being responsible for the direct management of one or more projects[5]."

In general, and regardless of what this unit is called, it needs to look after the owner's interest during the design, bidding, and construction phases.

Public Projects

Public projects are those created and managed by government agencies, and traditionally financed with public (taxpayers) money. There are many types of public projects that a public agency may build, such as

- Roads, bridges, and other transportation/traffic projects
- Infrastructure projects such as underground utilities
- Parks, libraries, and educational institutions
- Water projects (potable, sewage, and stormwater)
- Solid waste treatment plants
- Power plants
- Commercial buildings and office spaces
- industrial facilities

Some of these projects are classified as capital improvement projects (CIP), which are projects that provide improvements in one form or another to the assets in that area under the jurisdiction of the public agency. These projects come with the same variety mentioned above. They can be new construction, improvement, upgrades, expansion, or renovation, but they do not include the operation and maintenance of these projects. They are usually part of the capital improvement program, which is funded by the capital improvement budget.

Funding public projects is a complex topic. Public agencies get funding from numerous sources such as taxes and fees people in that district pay, grants by private and public agencies, partnerships with private individuals or entities (Public Private Partnership, PPP), and other sources.

Planning for public agencies is necessary but it is a delicate and challenging matter because many of the projects planned for the future are funded by money that is not always guaranteed, at least in amount.

---

5 The Project Management Body of Knowledge by the Project Management Institute, 7th edition, 2021

Property and other taxes and fees fluctuate, so the amounts of money planned to be received, may not come as expected. In addition, many of the planned projects are to be built on land that has not been acquired yet. The government set a budget for acquiring the land, but prices may spike unexpectedly, which puts the government plans in jeopardy. Also, grants often come with a variety of conditions. In some cases, they are not paid till the final completion of the project and after the granting party assures the compliance of the project with its conditions. This not only carries risk for the public agency performing the project, but also puts the burden of financing the project, which may come from different sources, till its final completion.

Most government agencies have set budgets for different programs and a list of proposed projects along with their cost. Almost always, the cost of the proposed projects exceeds the set budget, so the agency needs to prioritize planned projects based on the available budget. This prioritization is usually based on multiple criteria, such as the cost and urgency of the project and the consequences of postponing the project. Public pressure and politics may also play a role. To explain the prioritization point, we may have a need for a new road as the existing one has deteriorated. We can do the new construction at a cost of $7.5 million and life expectancy of 15 years or resurface the same road at a cost of $1.5 million and life expectancy of 3 years. Some projects must be done without delay, some can be postponed, and others may be canceled. However, often a delay for a project results in an increase in its cost.

There could be several reasons for the increase in actual cost over the estimate, especially because the estimate is usually performed several years before the start of the project. Inflation and escalation play a role as well as the market conditions, which influence the bidding competition.

Public agencies manage their projects, after completion, under their operation and maintenance units. In many cases, the agency may not do proper maintenance to the project or facility, because of low budget, neglect, or other. This results in higher costs for maintenance, repair, or replacement later. There must also be a system for asset management that covers all assets owned and/or managed by the agency with continuous updates. The different departments must communicate and integrate among themselves as new projects are created, existing projects are operated and managed, some projects need repair/renovation/rehabilitation/upgrade, and other projects need disposal.

Most public agencies have engineering and construction departments to study the project in the planning phase, before and after approval, manage the bidding/award process, and monitor its execution. This is similar to the concept of the PMO. It is also possible that such departments, engineering, and construction, directly perform small jobs without contracting with an external party.

Choosing the Contract Type and Delivery Method

Part of the owner's planning is to choose the contract type and delivery method. The choice depends on several factors, mainly how much risk the owner is willing to take, which, in turn, has to do with the owner's expertise and readiness in supervising and managing the construction work. This is the most important step, because many events, responsibilities, and liabilities depend on it. Also, it is very difficult and costly, and sometimes impossible, to change it later on. No owner should take on a role, which naturally comes with responsibility, liability, and risk, if not prepared, qualified, and dedicated.

Not only the contract and delivery method but perhaps other options such as fast-tracking the project (overlapping construction with design) must be studied well to see if it is suitable for the situation.

Every arrangement comes with pros and cons and costs, so it is necessary for the owner to study the situation well and start the project on the right foot.

Requiring and Approving a CPM Schedule

Owners usually require several documents from bidding or negotiating contractors. In fact, these documents come at two major stages for bidders: One set of requirements from all bidders, and then another set required only from the selected contractor. These requirements usually include a schedule, likely using the critical path method (CPM) and a specific software program. The purpose of the schedule is to show their work plan and their ability to complete the project on time. This is standard in almost all public agencies. It is common to see the owner requires a CPM schedule from all bidders and then have the winning bidder resubmit the schedule, perhaps with adjustments and tweaking.

The owner has the authority and responsibility to review and approve (or reject) this schedule. Approving the schedule shifts some liability to the owner. Some contractors submit schedules with errors, exaggerations, and inaccuracies. Perhaps this is because the contractor rushed before the bid deadline and did not put enough effort when creating it under the premise that "We will correct the schedule later as we execute the project." This situation puts the burden on both the owner and the contractor. The owner must have the schedule reviewed by a professional and must not accept or approve a schedule with errors. Things may later go smoothly on the ground, but when an event happens that creates a dispute between the two parties, these errors will come back to haunt both sides: The contractor for submitting a schedule with errors and inaccuracies, and the owner for approving the schedule with the errors. I like to say, "Everything you submit or approve, may and will be used against you!"

So, does the owner need to require a CPM schedule from the contractor? The answer is yes, but if the owner is not an expert in scheduling, he/she must hire a professional to examine the schedule. Dropping the schedule requirement from the contract documents is not a good idea because the owner will not have a tool to measure the contractor's performance against. The submitted schedule, once approved, becomes the baseline schedule, which will be the yardstick to measure performance.

The Owner's Involvement During the Design and Construction Processes

The owner must always be involved and engaged during both the design and construction phases, but the important questions are: How? And how much? The owner's involvement during the design phase must be more active and intimate because it is much easier to make changes there than later during the construction phase.

The owner's involvement during the construction phase is different from that during the design phase. In the design phase, the owner is dictating what the project should look like, within the legal, professional, and financial constraints. In the construction phase, everyone knows what the project should look like, so the owner's role is basically in monitoring the contractor constructing the project per the design and other contract documents.

As we learned in Chapter 3, there are several types of contracts and delivery methods. We will revisit them in the context of the owner's role in each situation. In general, the more risk owners take, the more obligation they have to be present on site: monitoring work and getting involved in making decisions about matters within their responsibility.

> Both owners and contractors have an interest in the successful completion of the project. However, each party has its own objectives, constraints, and priorities.

There are several things the owner may have to keep an eye on:

1. The quality of work and its conformity to the specifications.
2. Work production and expenses in the situation of cost-plus-fee contracts.
3. Work performed, particularly quantities in unit cost and cost-plus-fee contracts.
4. Pace of work as measured by both percentage complete and hitting set milestones on schedule.
5. Unknown/suspicious visitors or activities on site.
6. Unsafe or illegal practices by any of the working parties.

In lump-sum contracts, the owner may focus on the quality of work since the project's price tag is already set, except for change orders. Unit-price contracts may require more involvement in monitoring the quantities since the final payment will be based on actual (measured) quantities, not those estimated in the contract. Cost-plus-fees contracts require even more involvement from the owner since nothing is known in advance and the stakes are higher. Of course, as mentioned in Chapter 3, even these types of contracts may have clauses that shift back some of the risk to the other party.

The owner must balance their involvement between the two extreme approaches:

A. The passive/absent owner: Not getting involved enough or at all. When absent or unmindful from site, some contractors and/or subcontractors may take advantage of the situation. Remember that most of the construction work will be covered later, so you cannot check on many items after the concrete, drywall, or brick veneer is placed[6]. Even if the owner trusts the contractor and subcontractors, this trust must never reach a "blind trust" level, as shown in Figure 21.2. Permanent presence or occasional (random but frequent) visits to the site is a must, sending a subliminal message to the workforce: I am here, I am watching!
B. The micromanager/annoying owner: Getting too involved to the point of micromanaging and annoying the general contractor and subs, especially when the owner gives work instructions to the crews or subcontractors. The owner must not interfere with the work process unless there is a worthy issue like safety or clear violation of the law. Otherwise, the owner not only will annoy the contractor and impede work, but also may pick liability.

In general, there are two ways of having someone do work for you: Materials and methods or end results. In the first approach, the owner provides the materials and methods to achieve results. The workers just follow the boss's (owner) instructions. In this case, if results are not satisfactory with the work

---

6 Of course, you can do so but at a high cost!

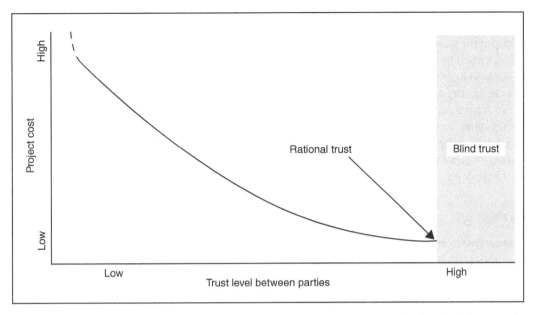

Figure 21.2 Conceptual cost–trust curve. Courtesy of CII, reports RS24-1 and RT-024, CII - Topic-Summary-Details (construction-institute.org).

performed, the burden may be on the owner. The other approach, end results, is asking the professional for a well-defined product or service in exchange for a set price. The professional is totally responsible for knowing how to get satisfactory results. The owner does not interfere in the process but makes sure the final product or service is done according to the agreement. The owner, in almost all construction work, is strongly recommended to use the "end results" approach.

> The owner must always have a presence on the project site. The form and extent of this presence depend on the project's contract type and delivery method.

The owner must also abide by the site rules, mandated by the law as well as the general contractor's instructions. This includes, but is not limited to, wearing safety gear (usually and depending on the situation: hard hat, safety boots, and safety glasses. In some cases, a bright vest, face mask, or other items may be required) and following on-site traffic rules. The owner must be careful when getting in or close to hazardous operations such as table-saw, welding, or handling chemicals / hazardous materials. In many cases, the owner must inform the general contractor if they intend to bring visitors to the site.

Sometimes, the owner acts as a general contractor (pure agency project management delivery method). This situation requires the highest involvement from the owner and lays responsibility and liability on them. This delivery method is recommended only when the owner has a professional construction management team that can coordinate and supervise subcontractors' work.

Owning and Managing Float

Ownership of the schedule float has been a contentious issue between owners and contractors. Generally, the float is owned by the party doing the work, which is the general contractor. Some owners insist on adding a clause to the contract, giving them ownership of the float. The motive is usually avoiding contractor's change order claims for delay to non-critical activities: It did not delay the completion of the project, so why should I compensate you? However, this practice by the owner annoys contractors because the delay, even to non-critical activities, disrupts their work plan and may cause financial losses resulting from having to rearrange crews and other resources. The author strongly recommends leaving the float to the general contractor since taking ownership of the float by the owner is a type of interference in the work process that may make the owner liable.

Weekly Meetings

One of the owner's important responsibilities is to conduct an onsite weekly meeting. The meeting serves several purposes such as:

1. Updating everyone, especially the owner, on work progress. This is one of the ways the owner shows presence and concern about the project.
2. Discuss any issues or problems and try to find solutions before they escalate.
3. The general contractor or a subcontractor may need to consult with the owner on an issue that needs the owner's approval. The meeting will give a face-to-face discussion between them that settles the issue quickly and effectively.

The owner must have the meeting minutes written and sent to all parties who attended the meeting for review. Usually, but not always, the approval of the meeting minutes occurs in the next meeting, but parties can also send comments by email. The meeting minutes will be part of the project documentation record. It is okay if the owner audio/video records the meeting only if such action is announced to and approved by the attendees.

The meeting must include at a minimum the general contractor and the owner[7]. The general contractor usually invites major subcontractors and those subs involved in work relevant to that meeting agenda. For example, the doors and windows and other finishing subs may not need to be present in the early meetings of the project when the structural skeleton is being erected, unless an issue that relates to their specialty is to be discussed. Also, if there is a third-party project management consultant, they must attend too. Occasionally, the general contractor or owner invites a guest, an attorney, vendor, or other to the meeting. But it is highly advisable to inform the other main party of such an invitation.

Owners who have no construction background must either delegate this role to a representative who is professional or accompany such professional to the meeting to make sure any technical matter is explained by a member of their own team.

The owner must not fall to the "everything is going fine, so there is no need to meet" syndrome. The meeting must take place regularly and without skipping. If there is a holiday or no work for any reason that day, the meeting can take place the day after. It is better to have the meeting in person but in certain circumstances, an online meeting by video is okay.

---

7 Every time we mention the owner of client, it can be them personally or their representative.

The owner is highly recommended to do a walk through the project site, accompanied by the general contractor, before or after the meeting and investigate any issues that need explanation. Such a walk-through, even when brief, adds another sign to the presence of the owner in the project.

The owner or general contractor may call for an emergency or urgent meeting if an issue arises and cannot wait for the next regular meeting.

Financing and Cash Flow – Progress Payments

The owner must plan their own cash flow and be prepared to make progress payments to the contractor on time. These payments may vary, by a certain margin, from the planned cash flow diagram depending on actual quantities (actual performance), change orders, and other factors. The owner also has the obligation to review the progress payment request accurately and on time. Even when there is a dispute, the owner must be fair in not holding more than the disputed amount. Making these payments to the contractor on time allows the contractor to continue the construction operations, paying their own expenses and subcontractors. Failing to do so may have negative consequences for all parties and the project performance.

In some situations, the owner may demand schedule acceleration from the contractor. If this acceleration is requested for a matter other than contractor's lack of proper progress, then it may have an additional cost to the owner for this acceleration[8]. This has an impact on progress payments for two factors: The total cost of the project will likely increase because of the acceleration effort. Also, the amount of work per period will be more. The owner must be aware of this point before demanding acceleration.

The owner is recommended to add a contingency allowance that is not included in the project's budget and the contractor has nothing to do with it. This is different from the contractor's contingency, which is usually included in the project's price tag. The owner's contingency allowance covers unexpected expenses such as change orders resulting from unforeseen conditions or owner's new requirements.

Communications and Documentation

The communication system and protocol are expected to be mentioned in the contract. It is in the best interest of all parties to agree on a robust and effective communication system and protocol. The general contractor as well as subcontractors may have requests for information or clarification during the construction process. It may be more efficient to allow direct communication with the technical party that is responsible for this matter, provided that the owner and general contractor are copied into the correspondence.

When the owner has a question or request from a subcontractor, it is more proper to go through the general contractor, unless there is a direct contract between the owner and that subcontractor.

Documentation is another very important task for the owner. The owner must keep records of all documents and correspondence in a well-organized and searchable manner. This includes but is not limited to:

1. All contract documents including baseline schedule and baseline budget that the owner approved: electronic and printed copies.
2. All updates of the schedule, periodic and other.
3. Change orders: including administrative and technical correspondence.

---

8 Acceleration was discussed in Chapter 7, Time Management.

4. Owner's project manager's daily log.
5. Submittals along with their records.
6. Record of any transmittals such as requests for information (RFI), requests for clarification (RFC), test lab results, warranties, certificates, etc.
7. Correspondence with the general contractor, architect/ engineer, subcontractors, or other contracting parties.
8. Meeting minutes.
9. All safety records, including citations and OSHA reports.
10. Any pictures or video clips from the site, especially when related to an incident.

Give it to the Professionals!

One common problem with some owners is trying to save money by cutting corners to avoid paying professionals' fees. In most cases, this leads to issues that cost a lot more than the fees supposedly saved, in addition to possible delay and headache. This may happen when the owner decides to play the role of the general contractor and hire the subcontractors directly. This can be normal and appropriate when the owner has a dedicated qualified team or person to do the job, mainly supervising work and coordinating among subs. However, this can be a disaster when the owner does not have such a team or person.

There are other examples such as when the owner has to decide in a technical matter that he/she has no knowledge of it. Always, consulting a professional is recommended especially when the stakes are high. Take, for example, the choice of a security system the owner is thinking of installing in a new house. Brand names, options, types of service, and costs differ widely for security systems. Another example is choosing appropriate flooring type among a huge variety of types in the market. The owner may prefer a certain type of hardwood or marble for an office building but the contractor or consultant would recommend another type that is more suitable for that purpose and/or more cost efficient. This is why it is important to sit with a neutral expert, not just a salesperson, and make an educated decision. Sometimes, an experienced owner may be able to extract information from different sources, dissect and analyze them, and then make a rational decision that fits their needs and budget.

In some cases, people have false self-confidence or overconfidence, perhaps boosted by self-pride, social status, macho image, or other. They may claim knowledge or expertise in a certain area when indeed this is false or exaggerated. This may lead to wrong or even disastrous decisions. In Chapter 16, we spoke about the "professional democracy" concept, which is the ability of a person, employee, worker, or other, to give a truthful opinion to a superior even if this opinion goes against the superior's opinion or wishes, without any fear or retribution or repercussions. It also includes the superior's willingness to provide the freedom atmosphere to the other side, listen to their opinion respectfully, and accept this opinion if convinced.

One of the worst things we may do is to ask someone else for their opinion only for the purpose of supporting our opinion. It is very common for a CEO or a large company, or another person in high position, to listen to the opinion of an employee, several tiers below him/her, when this employee knows more about the subject matter than the CEO.

General Recommendations for the Owner:

1. Planning is extremely important. Resist the temptation to rush the start of the execution of the project, until you do the proper planning.
2. Know your options before deciding. Consult with the designer, the contractor, or the technical party who is related to this matter or an expert in it.
3. What worked in one situation may not be the best tool or solution for another one. Always look for the best solution for *this* situation.
4. Try to resolve claims before they become disputes, i.e., in a proactive way. First, you fulfill your obligations in the most professional manner. Second, make sure to have a mechanism for resolving disputes in the contract. The Dispute Review Board, DRB, is an excellent way to deal with claims. You can also add arbitration as a secondary method. Third, make sure you have a good documentation system for use in case of a dispute.
5. Have a balanced presence on-site. Do not be absent but do not micromanage the contractor.
6. Trust is important in business relationships, but it must not be blind. The owner must check regularly or occasionally on performance, financial items, and others.
7. Avoid scope creep and numerous change orders as much as possible.
8. Do not be picky on every little issue. Often, there are some variations between the design and reality. Sometimes it is okay to accept certain variations if they do not adversely affect the appearance or the functionality of the project.
9. Insist on doing regular schedule updates even when things go as planned. Insist also on doing them using the formal procedures mentioned in the contract.
10. When you do not have the expertise in a technical matter, consult with a professional. Do not try to save money by acting as a professional when you are not one, especially in major matters.
11. Do not cut corners. Always do things properly and professionally. Look at the consequences, especially in the long run. In most cases, you get what you pay for and sometimes you get even more than that.
12. Finishing both on schedule and on budget is important, but you need to make your top priority clear to yourself and the general contractor. In some cases, you may find the schedule slipping at no fault on the contractor's side. Accelerating the schedule to finish on time may cost extra money, so you need to decide if the extra cost is justified.
13. While construction is underway, the site belongs to the general contractor. The owner needs to acknowledge this point.
14. Documentation is very important. Remember that winning a dispute or case, in arbitration or litigation, may depend on documents you are able to present.
15. Remember that your objective of the project is different from the designer's or contractor's objectives. They are providing services to you for a fee, but you are the ultimate owner of the project. After it is completed, you will have the project in your hands, and they will be out.
16. Do not make the contractor your adversary. A good relationship with the contractor is a benefit to both parties.

Exercises

21.1 The owner has every right to protect their investment (the project being built) and interest, but it is also best for them to be in a good relationship with the contractor and minimize conflicts and confrontations. The owner and contractors must look at each other as partners, not adversaries. Do you agree with this statement? Explain your answer.

21.2 Are the "owner's project" and the "contractor's project" the same? Explain your answer.

21.3 What are the three phases of the owner's project?

21.4 What are the possible project costs to the owner?

21.5 Choosing the type of contract and delivery method is one of the most important decisions the owner makes, why?

21.6 Risk is always a thorny issue that owners like to transfer to the contractor. How can you help in allocating project risk between the owner and the contractor? Is there one rule in such allocation or does it depend on the owner's situation?

21.7 Is there something as "shifting the risk at no cost"? Explain your answer.

21.8 Planning is of utmost importance to the owner. Can you convince an owner, who likes to rush to execution, to do proper planning first?

21.9 What makes public projects particularly important from a project control perspective?

21.10 The owner's involvement in the project must be balanced. Elaborate on this point.

21.11 An owner wants to build a house. Instead of hiring a general contractor, he/she believes he/she can save 10% of the project cost by hiring subcontractors directly. The owner is not familiar with the construction process. Make a few arguments to convince the owner to hire a GC.

21.12 When the owner is monitoring the work on-site, there are several things he/she must keep an eye on, such as what?

21.13 An owner is pushing the contractor to finish on time despite some delays because of unexpected extreme weather conditions. The owner also has a tight budget. Discuss time–cost trade-off with this owner, informing him that cost and time must be prioritized, with one of them – not both – as a top priority.

21.14 A particular owner has done several projects with the same general contractor. They built a strong trust between them. How can the owner save money with such a trustworthy contractor? Is there a limit to this trust? Explain.

21.15 How important is it for the owner to have a good documentation system for the construction process? What are some of the documents that you expect to see stored in this system? Do you have some tips for such a system?

21.16 You are the general contractor's project manager for a custom home project. The contractor signed a lump-sum contract with the owner, but the owner continues to make changes. The problem is that sometimes the owner's wife calls you and demands certain changes, which her husband may later disapprove. This matter is causing a headache for you and racking up the bill for the owner needlessly. You decided to have a meeting with the owners (husband and wife) to resolve this issue. Outline your plan and argument.

21.17 Should an owner require that the contractor, during the bidding or negotiation process, submit a CPM schedule? What are the advantages and precautions of such a requirement?

21.18 What is scope creep? Talk to a project owner about it and its negative consequences on the project and its cost.

21.19 You are the owner of an office building project. The contractor sent you a memo informing you of a new product in the market that can replace a product specified in your project (has not been installed yet.) There is a minor increase in the cost, but the contractor is talking about great benefits. Since this is a technical subject that is outside your knowledge, what would you do?

21.20 You are the financial manager in a company that is planning to sign a contract with a general contractor to build a new headquarters. You are afraid that the company's financial situation may not be able to support such a large project. You developed a rough cash flow diagram for the project, and you are using it to convince the upper management to either scale down the project or wait till the financial situation gets better. Discuss the importance of the cash flow diagram for an owner through this example.

21.21 When does a PMO become a good idea for an owner's organization? Why?

21.22 "Give it to the professional" is great advice. Mention a practical example when you are trying to convince a friend who wants to sign a major construction contract. He is making many decisions on matters he does not have much knowledge in them. How would you convince such an owner?

21.23 Are weekly meetings with the contractor important for the owner? Will you change your answer if things are going fine? Discuss your answers.

21.24 The owner attends all weekly meetings, but she feels overwhelmed when some technical issues are being discussed as she is not knowledgeable on these issues but does not want to admit or show it. Give this owner some advice on what she should do.

21.25 The general contractor is complaining about a pattern of delay in your progress payments (you are the owner). You do not take his complaint seriously. You believe a few days of delay in payment should not matter with such a large construction company. Suddenly, you received a memo from the contractor informing you that there will be interest charges and penalties with any further delay in payments. Furthermore, there will be severe actions for extended delay, as per the contract. Do you think the contractor is right about this issue? What should you do?

Index

Note: Page numbers followed by "n" denote footnotes.